Synthetic Biology and Chemistry

Synthetic Biology and Chemistry

Dr. Ajay Kumar

ARCLER
PRESS

www.arclerpress.com

Synthetic Biology and Chemistry

Dr. Ajay Kumar

Arcler Press
4164 Lakeshore Road
Burlington, ON L7L 1A4
Canada
www.arclerpress.com
Email: orders@arclereducation.com

ISBN: 978-1-77956-949-3 (Hardcover)

ABOUT THE AUTHOR

Dr. Ajay Kumar, Associate Professor at the School of Pharmaceutical Sciences, Chhatrapati Shahu Ji Maharaj University (accredited with NAAC A++, categorized in UGC grade 1 and has QS ranking), Kanpur, is a distinguished academician in School of Pharmaceutical Sciences (ranked NIRF in 2024). Dr. Ajay Kumar is M.Pharm. in Pharmaceutical Chemistry, and completed Ph.D.(Pharmaceutical Sciences) from Sam Higginbottom Institute of Agriculture, Technology and Sciences in 2016. His area of research spans Pharmaceutical and Medicinal chemistry, focusing on the design, synthesis, and evaluation of novel antimicrobial, antitubercular, anticancer, and anti-inflammatory compounds with potential therapeutic applications. He has guided more than 15 M.Pharm. and currently supervising 05 Ph.D. research scholars and published approximately 27 papers in Scopus indexed impactful journals. He is a life member of APTI. Apart from sound academician, he has quite experience in university administrative work.

TABLE OF CONTENTS

LIST OF FIGURES

PREFACE

The union of biology and chemistry is one of the most groundbreaking developments in modern science. Biology provides the grammar of life, where DNA, proteins, and cells form the template of living things, and chemistry provides the tools with which to manipulate matter at the atomic and molecular levels. Together, these two sciences are no longer studied in isolation but are now combined to offer the possibility not just of studying life but of engineering and building new forms of it. This union offers the prospect of medical breakthroughs, agricultural advances, environmental cure, and energy. It offers one chance to rethink how nature functions so that human necessity can be met and world issues solved.

Synthetic biology, along with chemistry, propels science from observation to design. Not just about how to replicate what nature already does, but programming living systems and even designing entirely new biological functions. With the ability to code for DNA as easily as programming code, researchers are building biological systems that perform precise functions, from the production of drugs and biofuels to detecting toxins or repairing damaged tissue. Chemistry provides the components and tools of analysis needed for this progress, so that molecular-level design can be done with predictability and precision. This union is transforming how scientists perceive biology: not just as a descriptive science, but as a discipline of engineering with applications.

Synthetic Biology and Chemistry has been authored to introduce students, scientists, and professionals to an organized understanding of how chemistry and biology interact in this new discipline. It articulates fundamental principles and advanced ideas, relating them to applications that have an impact on society and industry. The book attempts to define important topics in a clear and useful manner so that readers gain a theoretical understanding of the equipment that makes synthetic biology possible. From molecular basics to engineered cells and proteins, the book focuses on the function that chemistry plays in facilitating the synthesis, assembly, and optimization of synthetic biological systems.

Chapter 1 provides an introduction to synthetic biology and chemistry. It presents the list of arguments for why synthetic biology and chemistry were consolidated, the manner in which synthetic biology evolved as a distinct discipline of study, and how it has become more relevant in research and industry. This chapter lays the background by setting the field's scope and focusing on real applications, such as engineered bacteria for pharmaceutical production or chemical approaches to stabilizing synthetic gene circuits.

Chapter 2 explains the fundamental principles of molecular biology that form the basis of synthetic biology. The central dogma of molecular biology, the structure and function of DNA and RNA, structure and function of proteins, and the mechanisms of gene expression are all introduced to readers. By explaining the processes, chapter 2 establishes the biological foundation which synthetic biology is trying to manipulate.

Chapter 3 describes the essential chemistry concepts that are at the heart of understanding synthetic biology. It explores chemical bonding, reaction kinetics, thermodynamics, and catalysis as the foundation for how biological molecules interact and behave. The chapter emphasizes the importance of chemical understanding in designing and predicting the behavior of engineered biological systems.

Chapter 4 covers genetic engineering tools, from the classical techniques like restriction enzymes and plasmid cloning to the most recent advancements like CRISPR-Cas9. The tools are the core of the synthetic biology application, enabling researchers to modify genomes, control gene expression, and build synthetic circuits. The chapter highlights the role of chemistry in altering enzymes and in reagent design that improves accuracy and efficacy.

Chapter 5 focuses on DNA synthesis and sequencing. It informs us about how chemical synthesis is so refined now that one can synthesize long DNA pieces with precision and how next-generation sequencing platforms provide the feedback necessary to qualify these constructs. This chapter emphasizes the synergy between chemical methodology and biological engineering, demonstrating how the two complement each other.

Chapter 6 addresses synthetic gene circuits, which are man-made arrays of DNA sequences that are designed to perform specific functions in cells. It outlines the design principles, such as logic gates and feedback loops, and how these can be used to control cellular functions. Examples include circuits that detect environmental signals, produce therapeutic molecules, or regulate metabolic pathways.

Chapter 7 deals with metabolic engineering, where entire biochemical pathways are remade or reengineered to produce useful chemicals. The chapter shows how enzymes are optimized, pathways are balanced, and host organisms are reengineered to provide maximum levels of pharmaceuticals, biofuels, or industrial chemicals. Chemistry is found to be an essential partner, providing insights into enzyme mechanisms and reaction efficiencies.

Chapter 8 discusses synthetic cells and bare-bones genomes. It considers efforts to construct lean cells that contain only the genes necessary for survival, as well as strategies to create completely artificial cells from molecular components. These efforts illustrate the ultimate goal of synthetic biology: designing life at its most fundamental level with chemical principles to stabilize and control the resulting systems.

Chapter 9 places a strong focus on protein engineering, wherein protein structure and function are manipulated to create molecules of new or improved properties. Directed evolution and rational design methods are addressed, with case studies for engineered enzymes, antibodies, and structural proteins. Chemistry's impact on the understanding of protein folding, stability, and activity is a recurring theme.

Chapter 10 concludes by examining chemical biology in synthetic systems. This chapter shows the ways in which chemical tools, such as small-molecule probes, non-natural amino acids, and chemical switches, are integrated into synthetic biology initiatives. This chapter also examines how chemical insights push the boundary of what synthetic biology can accomplish, enabling scientists to engineer systems that exceed natural biology.

This is an appropriate book for students, academics, researchers, and professionals interested in the intersection of biology and chemistry in synthetic biology. Undergraduate and graduate students will learn a lot about the fundamentals of the subject. Biotechnology, bioengineering, and chemistry researchers will be assisted by the systematic explanation and discussion that guides them toward applications. Research experts and practitioners in the pharmaceutical, agricultural, and energy industries may cite it to identify where synthetic biology can advance in their specific industries. Through the integration of theoretical principles and practical approaches, the book is designed to be a handbook to support learning, instruction, and research at various levels.

CHAPTER 1

INTRODUCTION TO SYNTHETIC BIOLOGY AND CHEMISTRY

Contents

INTRODUCTION

Chemistry-synthetic biology convergence is a new and growing field with broad potentialities for revolutionary innovation in various disciplines like medicine, agriculture, and the environment. Synthetic biology combines elements of biological sciences and engineering in designing and constructing new biological components, devices, and systems. By employing a rational design approach, it is possible for scientists to manipulate genetic circuits and metabolic pathways to produce biomolecules that were previously inaccessible. The dynamic nature of the field requires that one be aware of the chemical processes involved that control biological interactions and metabolism and use them as the basis for innovative applications. With research unfolding within the realm of science, the study of complex carbohydrates is an area where synthetic biology can play a crucial role in the development of future therapeutics, necessitating interdisciplinary research and collaboration within this rapidly advancing area.

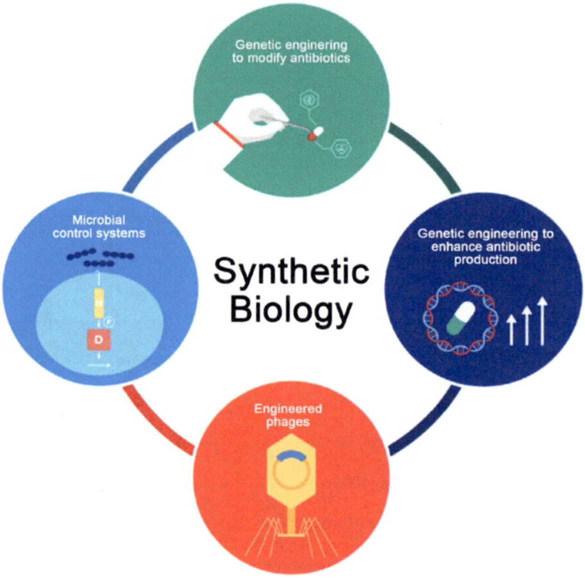

Figure 1.1. Recent works have demonstrated the impact of synthetic biology and the tools and techniques that the field has developed to allow genetically modifying antibiotics, enhancing antibiotic production, developing engineered phages, and designing microbial control systems.

Source: https://www.frontiersin.org/files/Articles/869206/fbioe-10-869206-HTML-r1/image_m/fbioe-10-869206-g001.jpg.

Chemical synthesis consistently presents challenges because biological molecules, particularly those carbohydrates that enable crucial biological interactions, are complex as well as diverse. The traditional methods of carbohydrate synthesis are slow in keeping up with the demands of their synthesis in large quantities of complex molecules. Recent advances in the technologies of chemical and enzymatic synthesis procedures have offered sufficient alternatives to enable the overcoming of these constraints. For instance, the methods of delivering functionally modified carbohydrates by stereoselective conjugation to proteins and lipids are fundamental to the pharmaceutical industries. The ability to functionalize linkers at specific positions in carbohydrates expands their promise as drugs. By allowing glycoconjugation to vaccine carriers, such strategies open doors to new carbohydrate therapeutics that may address significant health challenges, thereby illustrating the exciting potential for synthetic biology to serve chemistry (Gervay-Hague et al., 2018). Apart from the progress in carbohydrate synthesis, the use of engineered biocatalysts has transformed the production of polysaccharides and their derivatives, which are absolutely vital to our lives. The discoveary of a chemoenzymatic method with engineered *Bacillus megaterium* cytochrome P450 demethylases represents a clear demonstration of how synthetic biology can overcome the long-standing challenges of producing intricately complex biopolymers. The high regioselectivity of the process enables efficient access to a range of useful intermediates, which can then be converted into a variety of substituted polysaccharides and monosaccharides. This new synthesis method depicts the key function provided by synthetic biology in the production of compounds pertinent to industrial, medical, and biochemical research. Moreover, the preparation of custom carbohydrate structures not only advances our understanding of biochemical processes but also creates new possibilities in the realm of health and disease management, emphasizing the need for bringing chemical principles to bear on biological knowledge in this rapidly evolving field (Arnold et al., 2009) (Figure 1.1).

Definition of Synthetic Biology

The definition of synthetic biology is multifaceted, integrating various disciplines such as genetics, engineering, and computer science in a bid to create new biological systems or modify existing ones. At its core, synthetic biology seeks to deconstruct and redesign biological components in order to attempt to achieve defined functionalities, bridging the gap between natural biological organisms and designed constructs. This emerging discipline has drawn significant attention due to the enormous variety of possible applications, ranging from environmental sustainability to human health. Still, attempting to subscribe to an accepted, generally applicable definition of what synthetic biology is remains elusive. Scholars argue that operational definitions can more helpfully serve the scientific community than definitive definitions by providing structures for allowing space for flexibility and responsiveness in various research contexts, particularly in light of synthetic biology's ongoing progress (Bich and Green, 2016). The manufacture of biological parts tends to raise philosophical questions about the essence of life itself. Since scientists try to engineer living things, they are confronted with the complexities of what defines a living system, and the issue about synthetic biology's position in science and ethics becomes further complicated. Varied definitions bring to mind pragmatic approaches which integrate operational parameters for life, considering how the definitions impact research practice and popular conception of synthetic biology. By reflecting on life and technology at the crossroads, scholars advocate for definitions that go beyond categorization, arguing that operational definitions find utility in constructing theoretical models that are capable of producing observable operations. This must be taken into account while navigating the ethical issues of synthetic biology along with pushing the boundaries of scientific inquiry (Bich and Green, 2016). Furthermore, as synthetic biology continues to advancement, it becomes more and more seen as an impetus for transformation in the chemical manufacturing sector. An integration of sustainability into chemical engineering practice calls for a re-definition and re-prioritization of existing definitions and approaches. Compressing time scales and the need for solutions to be sustainable prompts chemists and engineers to adopt new ways of thinking which reflect the new paradigm that underpins the industry. This movement encourages the embracing of transparency reporting, driven by the Triple Bottom Line approach to account for the environmental, social, and economic impacts of synthetic biology applications. Definitions that evolve to meet the needs of sustainability, they serve as a foundation for educating the next generation of engineers to be competent to meet the challenges posed by synthetic biology and its implications for the wider field of chemical manufacture (Figure 1.2).

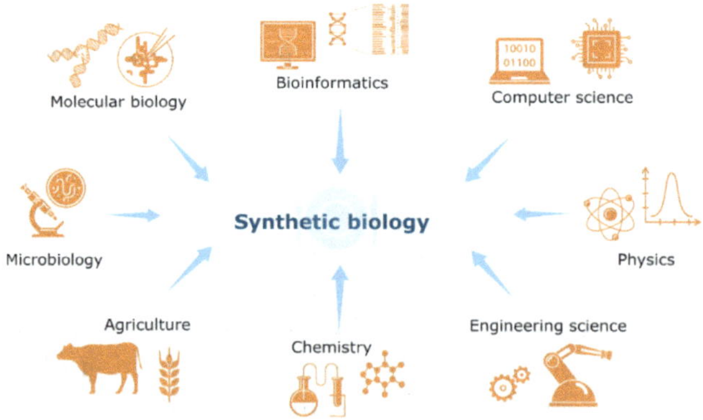

Figure 1.2. Synthetic biology.

Source: https://www.theaiops.com/wp-content/uploads/2024/08/image.png.

Overview of Chemistry's Role in Synthetic Biology

The synergy between chemistry and synthetic biology has become a fertile ground for discovery and innovation that facilitates the development of new biomolecular systems and new synthetic pathways. One of the fronts for this synergy is in the synthesis of glycoconjugates and complex carbohydrates, which are of core relevance to biological interactions.

Through advances in chemical and enzymatic synthesis methods, researchers have improved access to these key biomolecules, enabling their broader application in therapeutics. Stereoselective conjugation of functionalized linkers with proteins and lipids is increasingly viewed as pivotal in drug discovery. These developments are unfolding the pathway for carbohydrate-based drugs, showcasing the confluence of chemistry principles and biomedical applications, signifying the potential of chemistry in addressing biomedical challenges and transforming therapeutic potential (Gervay-Hague et al., 2018). Furthermore, the use of chemistry in synthetic biology extends to the science of data and information management in the face of increasing research result complexity. With ever-growing big data development, attention to the efficient sharing of data has burst into the scientific arena.

Organizations such as the Royal Society and the European Commission have advocated for open access to research results, promoting knowledge with the most impact when shared. However, certain areas, such as chemistry, have been slow to adopt these principles, at times avoiding to curate their data for broader dissemination. The integration of cheminformatics, an area that overlaps between computer science and chemistry, seeks to advance the curation and sharing of chemical data, thereby facilitating collaborative research. The relevance of chemical information to the purposes of communication and functional use cannot be overstated, as it lies at the heart of progress in synthetic biology and its associated methodology (Bird and Frey, 2013). Finally, the emergent relationship between chemistry and synthetic biology means that future progress will increasingly rely on interdisciplinary collaboration.

As synthetic biologists design new biological systems, chemistry provides the paradigms and tools needed for designing and constructing these systems. The complementarity of synthetic biology with open science principles also highlights the importance of cooperation among researchers from a broad spectrum of disciplines. By embracing these principles, the scientific community is able to share not only successful outcomes but also failures and mishaps, thus accelerating the process of learning and stimulating innovation. For this purpose, chemical information needs to be preserved, curated, and made accessible to enable research operations. Consequently, the destiny of synthetic biology is inextricably intertwined with advances in chemistry, underscoring the need for an integrated, multidisciplinary strategy across related disciplines (Figure 1.3).

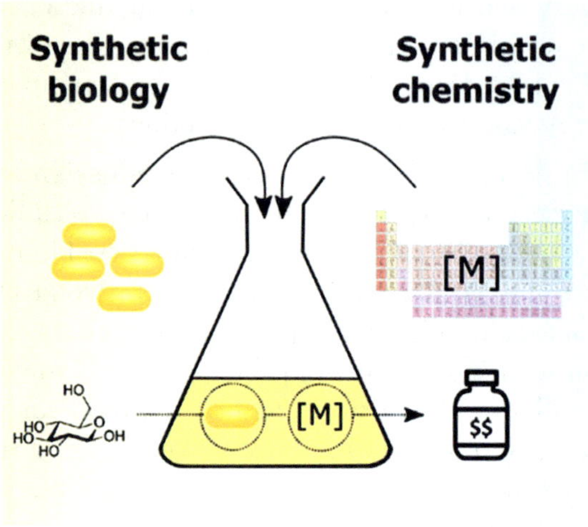

Figure 1.3. Synthetic biology and chemistry.

Source: https://chemistry-europe.onlinelibrary.wiley.com/cms/asset/5a62c934-d960-44e6-85dd-ac6ee4b8a9f1/cbic202000418-toc-0001-m.jpg.

Importance of the Intersection Between Synthetic Biology and Chemistry

The intersection of chemistry and synthetic biology is core in driving our understanding and manipulation of biological systems. Synthetic biology, as a collaborative approach, enables scientists to design and create new biological pathways, molecules, and organisms to satisfy specific needs such as drug production and environmental sustainability. Utilizing chemical concepts, synthetic biology can enhance the production of biochemicals and drugs to be more efficient and less expensive. Furthermore, the integration of chemistry maximizes the potential to create new biomaterials with novel properties and numerous applications in medicine and materials science. This convergence of these disciplines, when coupled, not only gives rise to innovative solutions for complex biological problems but also inspires the creation of new scientific strategies, paving the way for fresh advances in medical as well as environmental studies. Furthermore, the potential of the union of synthetic biology and chemistry is in diagnostics and therapeutics. For instance, the production of synthetic biomolecules with the capacity to selectively detect disease markers is an example that portrays the role of this union in revolutionizing diagnostics.

Using chemically synthesized building blocks, researchers are developing more sensitive and accurate diagnostic tools, enabling earlier disease detection. Moreover, the formulation of living systems to produce therapeutic molecules straight answers drug efficacy and delivery issues. Such an innovation is highly influenced by an understanding of molecular interactions as well as principles of organic and inorganic chemistry, which are very important during the process of optimizing drug-like properties. The result is a combined strategy that not only enhances therapy techniques but also leads to personalized medicine, where treatments can be tailored for individual patient needs. The importance of this intersection is also highlighted by its ability to address global challenges like climate change and resource depletion. Chemical innovation-driven synthetic biology has the potential to produce sustainable biotechnological substitutes, such as biofuels and renewable feedback stock-based biodegradable materials. The process of interdisciplinary integration leads to the production of green alternatives that offset the ill effects of traditional petrochemical technology. Additionally, studies of chemical pathways in synthetic biology can be used to extend our understanding of natural biosystems so that designed organisms can be created to aid bioremediation efforts. This capability addresses the significance of chemistry in ensuring the greatest potential of synthetic biology, the necessity for ongoing research and collaboration between such fields to create sustainable and useful solutions for our world's most critical challenges.

1.1. HISTORICAL BACKGROUND

Historically, the origins of synthetic biology and chemistry date back to the early understanding of biological systems, as researchers sought to master nature to satisfy human needs. The lineage of the two disciplines is associated with the emergence of molecular biology in the mid-20th century, during which the structural elucidation of DNA became a possibility of manipulating gene materials. This initial breakthrough spawned a paradigm shift; researchers soon turned to explore the interfaces of chemistry, technology, and biology. Realization that biological systems can be engineered along comparable lines to traditional chemical synthesis was the genesis of the modern synthesis of biomolecules. With the passage of time, research collaboration among scientists from diverse fields has been shown to enhance research output to support the case that interdisciplinary collaboration, especially in synthetic biology and chemistry, drives innovation and invention (Abramo et al., 2018). With steady advancement, synthetic biology became a mature discipline with a systematic approach in the design and assembly of new biological parts and biological systems. The employment of engineering concepts in biology introduced the development of synthetic circuits, which work in biological organisms as electronic circuits work in machinery. The systems are proof of synthetic biology's ability to excel beyond traditional chemistry in delivering functionalities that do not occur naturally. A hallmark of such an evolution is mounting reliance on hybrid strategies that combine natural and synthetic processes to create novel solutions. Of specific interest to such an evolution is the emergence of the term biomimicry, since practitioners sought nature's evolved designs as solutions to difficult real-world challenges (Battisti and Persiani, 2018).

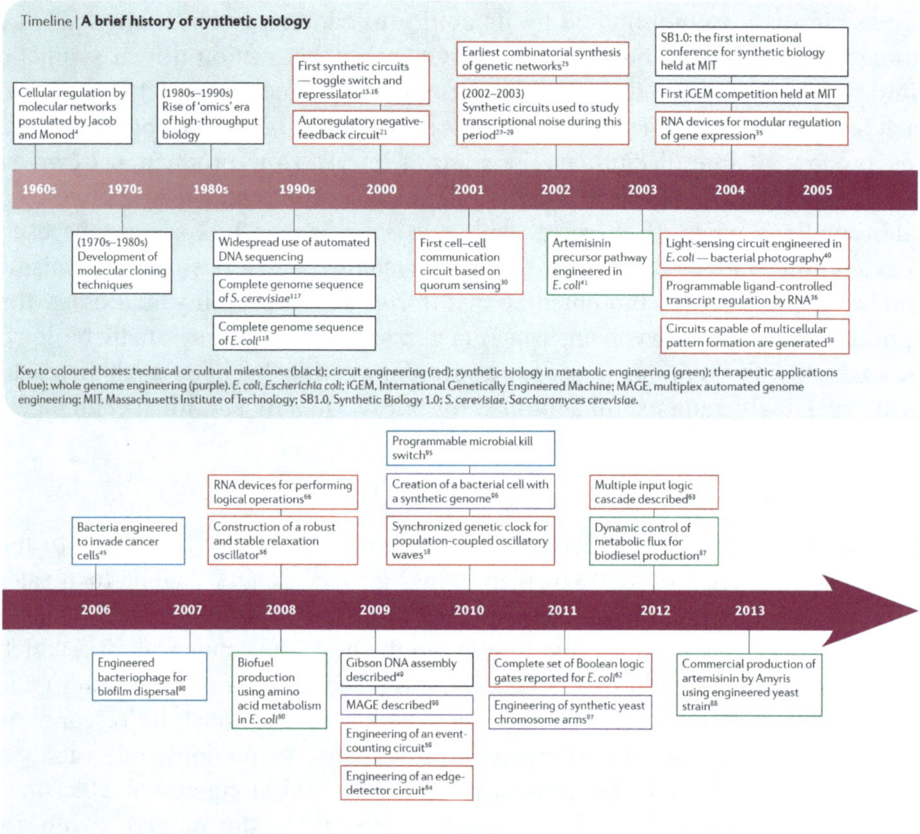

Figure 1.4. A brief history of synthetic biology.

Source: https://media.springernature.com/full/springer-static/image/art%3A10.1038%2Fnrmicro3239/MediaObjects/41579_2014_Article_BFnrmicro3239_Fig1_HTML.jpg.

Such innovative integration of biological insight and engineering design is a substantial leap forward in the evolutionary trajectory of synthetic biology.

Secondly, the advent of engineered membranes is a turning point where synthetic biology meets chemistry, an expansionist wave towards discovering sustainable solutions in nature-inspired technologies. Across the centuries, membranes have had all types of functions ranging from filtration to structural components, and today they are a higher form of synthesis between biological understanding and man-made material. The flexibility of membrane technology illustrates synthetic biology's ability to reform established chemical practice by incorporating ecological awareness and multifunctionality. Such developments have pertinent implications in areas such as environmental sustainability and materials science. The history behind such hybrid technologies is proof of the evolution of scientific thought, where different fields converge, propelling innovation in synthetic biology and chemistry towards new applications and solutions to complicated problems facing the world (Battisti and Persiani, 2018) (Figure 1.4).

1.1.1. Early Developments in Synthetic Biology

The early innovations of synthetic biology mark an extraordinary convergence of a number of distinct sciences, representing a new threshold of comprehension of biological systems. The earliest pioneering efforts date back to the start of the 20th century, when scientists began to articulate what could be done with engineering biological components. The evolution of fundamental concepts in genetics, with the progress in biochemistry, gave the stimulus for synthetic biology to establish itself as a new field distinct from standard biology. By tapping into the toolkit of molecular biology, scientists began an unprecedented approach to designing and assembling DNA sequences, and therefore to governing the building blocks of life. This shift toward engineering biology was also aided by technical advancements in DNA sequencing and synthesis, enabling greater experimentation and exploration of genetic structures and paving the way for future advances in synthetic biology. With the advancement of synthetic biology, scientists no longer focus solely on investigating biological systems but also sought to manipulate and engineer new organisms with novel capabilities. This desire had its roots in the juncture of physical sciences and biology, showing how ideas from chemistry could be applied to biological engineering. Stephane Leduc's work in the early 20th century is especially striking; he suggested that biology could be understood with physical forces acting on living material, an idea that would find later resonance within synthetic biology as it discussed self-organization and emergent properties of biological systems. Back in the 1950s and 1960s, it became possible to synthesize simple biological building blocks, such as proteins and ribosomes, thereby demonstrating that biological processes could indeed be engineered and manipulated for a particular purpose, expanding the range of chemical applications in biology considerably. The discovery of synthetic biology propelled science into the field of the ethics and implications of biological engineering. While the capability to redesign life became more viable, controversy surrounding the ethical aspects of such work became more dominant. Groups and institutions began to consider the potential impact of synthesized organisms on the environment and public health, and affected the regulatory climate in which synthetic biology operated. In addition, the interaction among scientists, ethicists, and policymakers became increasingly important for addressing these issues, fostering a conversation that sought to balance innovation and responsible stewardship of biotechnology. Thus, the early achievements in synthetic biology not only opened the way for revolutionary uses but also put the demand on gaining a thorough comprehension of the ethical issues that accompany the manipulation of life itself. It fostered an integral approach towards the field that is just as true today (Figure 1.5).

Figure 1.5. Stéphane Leduc.

*Source:*https://upload.wikimedia.org/wikipedia/commons/4/49/St%C3%A9phane_Leduc.jpg.

1.1.2. Key Milestones in Chemical Synthesis

The field of chemical synthesis has witnessed revolutionary advances that have progressed synthetic chemistry and biology. One breakthrough is the creation of membrane-based artificial cells, which leverage lipid structures to engineer cell-like spaces where biological processes are enabled. Synthetic cells have been fascinating due to their biocompatibility and ability to encapsulate biological machinery, hence playing a very critical role in research and potential therapeutic applications. One issue in this area is the demand for controlled, high-throughput production of artificial cells. Advances in microfluidic technology in recent times have presented innovative solutions to address the issues at hand and make precise dimensions and biomolecular composition of artificial cells possible. This development not only testifies to the innovativeness inherent in synthetic biology but also indicates a crucial turning point between chemical synthesis and biological engineering (Elani, 2016). Another principal breakthrough in chemical synthesis has emerged through an investigation into bioinspired nucleation of silica, which underlines the position of organic additives for control of silica synthesis. This accomplishment is particularly noteworthy when translated to the creation of materials that mimic natural processes found in biological systems. It was found that the synthesis of silica under nearly-neutral pH, aqueous conditions, and room temperature can impart the material with desirable properties. These breakthroughs are important as they open up new technologies across a variety of applications, including sensors, coatings, and drug delivery systems. By applying the principles observed in biogenic silica formation, scientists are not only synthesizing new materials but also addressing global issues of sustainable development and biotechnology. This exploration of bioinspired processes also reflects a broader trend in chemical synthesis that unites natural processes and synthetic processes (Patwardhan, 2011). The constantly evolving nature of chemical synthesis is evident in the convergence of chemistry and synthetic biology aimed at creating more functional and complex systems. The co-option of synthetic routes with biological templates has yielded spectacular innovations, particularly in drug development. Advancements in high-throughput chemical synthesis libraries combined with advances in computational chemistry allow rapid identification of valid therapeutic leads. This interdisciplinarity synergy not only accelerates the process of discovery but also enhances the precision with which synthetic compounds can be designed to have designated biological activity. With synthetic biology and chemistry becoming virtually indistinguishable in many ways, the future of a world of highly advanced synthetic methodologies and bioengineered systems is increasingly within reach, positioning scientists to address some of the most pressing issues in health and environmental responsibility.

1.1.3. Evolution of Techniques in Both Fields

Technique advancement in synthetic biology and chemistry has highly enhanced our understanding and control of biological systems. One of the important techniques in advancement is protein engineering, which has evolved into a highly significant tool with an enormous range of applications, from medicine to industrial applications. Scientists are able to create proteins with desired functionalities using techniques such as rational design and non-rational design. Site-directed mutagenesis, the conventional technique, is a knowledge-based technique reliant on an understanding of protein folding principles and requires a multidisciplinary set of skills spanning biocomputing and advanced data analysis from large-scale sequencing projects. Directed evolution or non-rational mutagenesis allows direct introduction of random mutations, with the subsequent creation of enormous libraries of mutant proteins. This approach not only enhances the variety of variants produced but also facilitates the selection of proteins of interest, even when the exact structural changes remain to be determined (Khan and Akhtar, 2011). As synthetic biology techniques continue to evolve, researchers are now working on the creation of artificial cells, which is one of the greatest points of convergence of chemistry and biology. Applying the bottom-up approach, researchers attempt to rebuild living organisms using simple building blocks. These cell mimics, especially those encapsulated in lipid membranes, are highly biocompatible and can accommodate advanced biological machinery. A significant challenge in this area is the controlled and high-throughput fabrication of membrane-based artificial cells. It is extremely important that the structures possess well-defined parameters, such as size and biomolecular composition, for them to operate properly. New microfluidic technologies have recently provided new solutions to address such challenges and have made systematic assembly and characterization of artificial cells with pre-defined, specified characteristics possible (Elani, 2016). Another revolutionary trend is the use of computational tools in synthetic biology and chemistry, focusing on advancing techniques in these two sciences. Computational aids enable researchers to model biological processes and forecast protein structures from provided data, significantly facilitating the rational design of synthetic biological systems. The union of artificial intelligence with traditional approaches has enabled the identification and production of new compounds at a record pace. Such convergence of disciplines not only accelerates the pace of research but also fosters a deeper understanding of the complexity of biological processes. As the evolution of techniques in synthetic biology and chemistry progresses, interdisciplinary collaboration will remain essential, paving the way for groundbreaking innovations that could redefine our approach to biological design and engineering, ultimately changing the landscape of healthcare and biotechnology.

1.2. FUNDAMENTAL CONCEPTS

The fundamental key concepts that underpin synthetic biology establish the foundation for how this interdisciplinary science brings together chemistry and the biological sciences. Essentially, synthetic biology seeks to engineer and construct novel biological components, devices, and systems with specified functions. Synthetic biology relies significantly on chemical principles, particularly those relating to molecular interactions and self-assembly processes. The theory of autopoiesis, where life is characterized as a self-organizing and self-replicating system, forms the backbone of designing synthetic organisms. The Luisis monograph indicates the correlation, demonstrating that the principles of abiogenesis can be applied in synthetic biology to design life-like structures that mimic natural processes, ultimately bridging the gap between chemical origins and constructed life forms (Gradoff, 2017). Through this, the analysis of these fundamental principles serves to give a better understanding of the nature of complexity in life and how it can be synthetically replicated. The second most important aspect within this discussion is the ongoing controversy over what constitutes life and what its requirements are.

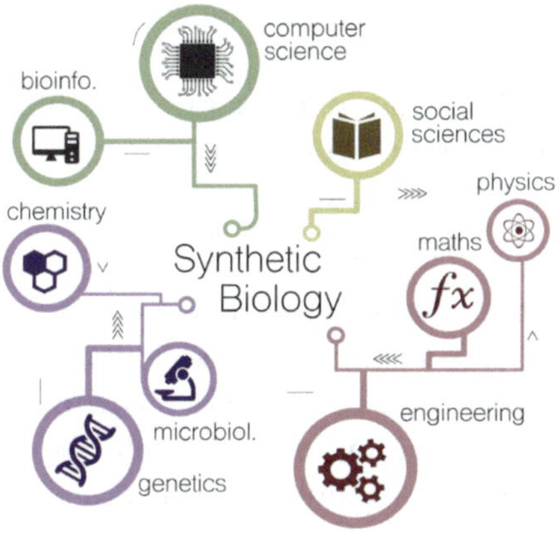

Figure 1.6. Synthetic biology is an interdisciplinary research field that involves many different areas of knowledge.

Source: https://www.researchgate.net/publication/345446068/figure/fig1/AS:11431281179425593@1691186189251/Synthetic-biology-is-an-interdisciplinary-research-field-that-involves-many-different.png.

The lack of total consensus on what constitutes life has significant philosophical and practical concerns within synthetic biology. Scholars emphasize the importance of operational definitions that address the diverse scientific environments in which synthetic biology operates, e.g., origins-of-life research and astrobiology. By thinking of life less as a classificatory term but as a conceptual tool influencing research agendas and approaches, scientists will be better able to cope with the complexities of biogenesis and synthetic life design. The life study under this approach emphasizes the useful utility of definitions that can facilitate experimental design and enable innovative strategies in understanding and manipulating living systems (Bich and Green, 2016). This is why answering these fundamental definitions will be important in further developing the discipline. Lastly, the overlap between ethics and foundational concepts in synthetic biology cannot be overstated. This capacity to engineer living organisms raises core ethical issues that are transdisciplinary in scope, overlapping social, environmental, and philosophical concerns. As scientists design new synthetic systems, they must address questions of biosecurity, ecological impact, and ethics of establishing forms of life that could compete with natural organisms. The complexity inherent in biological systems, as highlighted by Luisi, emphasizes the unpredictable nature of synthetic entities and their repercussions for life on Earth. As synthetic biology advances, continuous ethical discourse is necessary to guide responsible innovation while ensuring scientific progress aligns with societal values. This cross-disciplinary knowledge of fundamental principles in synthetic biology not only enhances the sciences' credibility but also provokes a responsible exploration of its vast potential and related obligation (Figure 1.6).

1.2.1. Genetic Engineering Principles

The concepts of genetic engineering establish a cornerstone on which synthetic biology is based, enabling the manipulation of biological systems in ways that were unimaginable before. They are founded on practices such as modularity and standardization, which enable the design and integration of novel genetic components. Modularity allows researchers to break down complex biological systems into simple components that can be assembled in unique configurations, and standardization allows for these components to be readily integrated and duplicated. This systematic approach not only allows for efficiency in experimental design but also speeds the assembly of innovative biological systems that can address urgent societal issues. By application of these principles, scientists aim to deploy the ability of living organisms for purposes ranging from medicine to earth sustainability in order to construct a more robust future as conceived in Eissa (2025). A crucial aspect of genetic engineering is that it can be applied to manipulate DNA, hence it is central to the construction of synthetic biological systems. Techniques such as polymerase chain reaction (PCR) and CRISPR-Cas9 allow for the precise modification of DNA, thereby allowing one to engineer recombinant DNA capable of expressing desired traits. The manipulation is not only limited to the alteration of organisms but is also applicable to metabolic pathway design that maximizes the yield of crucial products such as drugs or fuels. The integration of computational models enhances predictability for these kinds of systems, which allows for the modeling of metabolic networks and gene interactions. With this knowledge of how to design these biocircuits, scientists can develop targeted drugs and novel biotechnological products that will lead to revolutionary advances in medicine and industry, as noted by der Leopoldina et al. (2010). In addition, the interdisciplinary nature of synthetic biology emphasizes interdisciplinary collaboration among science, computer science, chemistry, and biology. This convergence not only facilitates innovative research but also enhances the capacity to solve complex global challenges, including food security and climate change. The application of artificial intelligence (AI) and machine learning (ML) also demonstrates the possibility of revolutionary paradigm breakthroughs from the convergence of numerous different disciplines, enabling scientists to analyze large data sets, predict protein structure, and optimize experimental protocols. The integration of such methods not only shortens the research time but also makes synthetic biology advances economically advantageous as well as socially equitable. Generally, the pillars of genetic engineering are fundamental components of synthetic biology and provide the tools to construct effective solutions for a sustainable future, emphasized in Eissa (2025) (Figure 1.7).

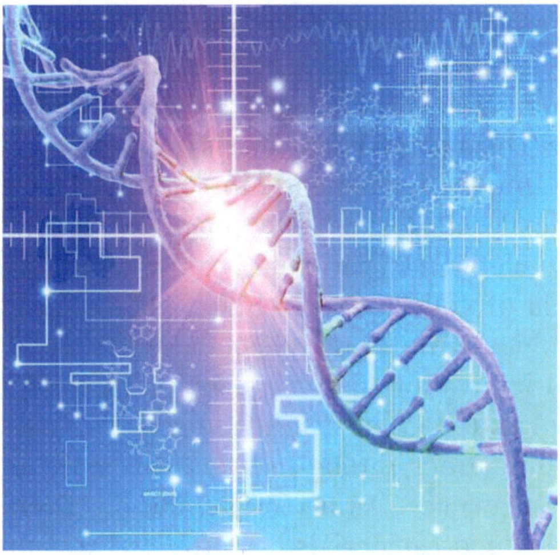

Figure 1.7. Principles of genetic engineering.

Source: https://cce.iisc.ac.in/wp-content/uploads/2023/12/genetics-1.jpg.

1.2.2. Chemical Synthesis Methods

Techniques of chemical synthesis are the basis for the advancement in synthetic biology and chemistry and allow the design and construction of complex molecules. A great number of applications for these methods include the synthesis of oligodeoxynucleotides (ODNs), which are required in many biotechnological processes, e.g., gene synthesis and biopharmaceutical production. The traditional methods of purification, however, have encountered broad limitations, particularly in the achievement of long ODNs with precise properties. The limitations of purification combined with mistakes have a tendency to undermine the utility of conventional methods. Therefore, the evolution of chemical synthesis techniques, particularly those that are followed by non-chromatographic purification, has been a breakthrough. With these constraints overcome, researchers can now efficiently synthesize long ODNs, thereby making it feasible to develop more sophisticated genetic constructs critical for solving complex biological questions and advancing innovation in this field (Yin, 2024). The history of methods of chemical synthesis also heavily overlaps with the efforts to construct artificial cells, a prime frontier in synthetic biology. Control over the assembly of complex proteins and lipids, the scaffolding of such artificial systems, is required to be capable of synthesizing cell mimics from minor constituent building blocks.

Microfluidic techniques have been at the forefront of this endeavor, providing high-throughput potential for the formation of lipid membranes around biological machinery. These membranes need to be tightly controlled in size, composition, and spatial organization to replicate the activity of living cells. Hence, the chemical synthesis technique involved in polymer and lipid synthesis remains fundamental in determining parameters necessary for functional artificial cell design. The convergence of such methodologies portends thrilling developments, particularly as researchers aim to develop more sophisticated models of cell behavior (Elani, 2016). Challenges in chemical synthesis methodologies in the future reflect the active development of the field, requiring continued innovation and optimization. A good example is the use of controlled pore glass (CPG) as a solid support for ODN synthesis that has proven to possess limitations through steric hindrance, resulting in mistakes within synthesized sequences. As a remedy, the discovery of novel surface-functionalized solid support materials has provided ways of developing these barriers so that direct synthesis of very long ODNs is possible. Furthermore, although the catching-by-polymerization (CBP) strategy has great promise, economic factors involved in the materials needed for polymerizable tagging phosphoramidites (PTPs) have hindered further development. Overcoming such economic limitations by using scalable synthetic methods allows researchers to leverage chromatography-free approaches, thereby accelerating the efficiency and accessibility of chemical synthesis in synthetic biology and chemistry more broadly (Yin et al., 2024) (Figure 1.8).

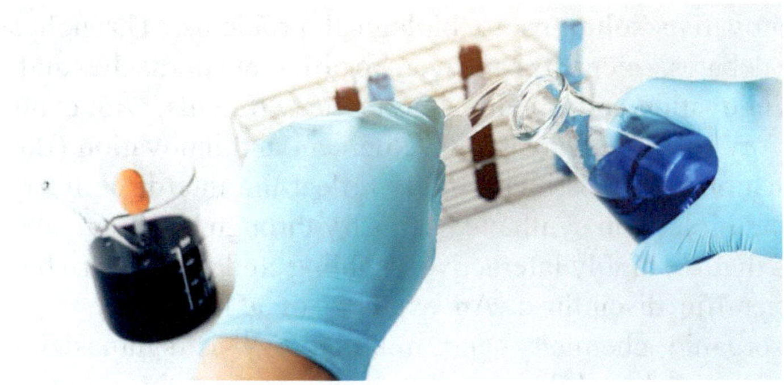

Figure 1.8. Chemical synthesis methods

Source: https://capitalresin.com/wp-content/uploads/2023/12/CapitalResinCorporation-262165-About-Chemical-Synthesis-blogbanner1-1000x500.jpg.

1.2.3. Interdisciplinary Nature of Synthetic Biology and Chemistry

The interdisciplinary nature of synthetic biology and chemistry becomes increasingly significant in enhancing our understanding of complex biological systems. Synthetic biology unites engineering strategies with biological information, ultimately leading to new approaches to drug discovery and disease treatment. For instance, studies of the mechanism of penicillin's interaction with the target protein have indicated the significance of interdisciplinary cooperation in research into antibiotic resistance, particularly with such strains as methicillin-resistant *Staphylococcus aureus*. All such endeavors to enhance antibiotic efficacy require that an integrative approach be adopted, which combines organic chemistry's methodological toolbox with microbiology's experimental verification. Such interdisciplinary interaction not only accelerates the discovery of new compounds but also contributes to the educational system, developing an environment where students with varying scientific backgrounds interact on substantive issues, working on problems in society (Rehberger, 2015). Secondly, the case of synthetic biology as an integrated system illustrates the necessity to overcome boundaries across various scientific disciplines. An analysis of a large database has demonstrated that consistency in this field stems not from a single overarching theory, but from cooperation between separate research communities pulling on shared tools and ideas. This is typical of trading zones, where sub-communities create interfaces so that different disciplines can work together. For instance, the application of DNA microchips or network theories illustrates how synthetic biology and chemistry come together to generate innovative solutions to biological problems. Through such interdisciplinary debates, scientists can leverage different approaches and know-how to advance innovation and development in the two fields, with communication playing the key role in advancing development and innovation (Bastian et al., 2012). Educational models increasingly reflect the interdisciplinary nature of chemistry and biology in synthetic chemistry through pedagogical cooperative experiments that are highly interactive, enabling students to learn by combining different scientific disciplines. An example of a laboratory experiment for combining organic chemistry and microbiology is synthesizing different penicillin compounds while simultaneously testing their potency against bacterial strains. This experiential approach facilitates experiential learning where the students share their knowledge with each other, thereby increasing their expertise in chemical synthesis as well as biological screening. Such interdisciplinary exercises not only impart practical knowledge to the students but also insight into how scientific knowledge is interconnected. This combined model of learning is essential to prepare the next generation of scientists who will be in a position to tackle and contribute to complex problems in synthetic chemistry and biology (Rehberger, 2015).

1.3. APPLICATIONS IN MEDICINE

The intersection of chemistry and synthetic biology has revolutionized medical applications, particularly through the development of new therapeutic strategies. An emerging development is the use of engineered microbes in the production of medicines and biologics with effectiveness. Through the regulation of gene pathways in yeast or bacteria, scientists are able to manufacture insulin or human growth hormone more efficiently than traditional methods. The tools of synthetic biology not only maximize yield but also allow drug characteristics to be designed, thus enabling the development of drugs that suit patients' individual needs. Moreover, the ability to design organisms to biosynthesize complex molecules is a revolutionary landmark towards green pharmaceutical production, reducing the reliance on artificial organic chemistry that is founded on toxic solvents and hazardous chemicals. This union of synthetic biology and chemistry thus unlocks the doors to a new era of medicines, founded on efficiency, individualization, and safety. Nanotechnology also demonstrates the promise of synthetic biology in medicine, particularly in drug delivery systems. New advances have designed nanoparticles to deliver therapeutic drugs directly to target cells, minimizing side effects and enhancing the efficacy of treatment. For instance, mesoporous silica nanoparticles or liposomes can encapsulate drugs for controlled release and targeted action, particularly in cancer treatment. By adjusting the surface properties of such nanoparticles, researchers improve their biocompatibility with cell membranes and cellular uptake. In addition, nanomedicine is poised to transform diagnostics, too, where nanoparticles can serve as imaging contrast agents or biosensors to identify biomarkers of disease. The interdisciplinary nature of the field, blending nanotechnology with chemistry and synthetic biology, is leading to more effective medical treatments and fewer side effects of conventional medicines (Anurogo et al., 2017). Another potentially groundbreaking application of synthetic biology in the medical field is the engineering of biological material, such as plant virus capsids. These capsids present highly ordered nanostructures and can serve as biotemplates for the synthesis of materials. Their unique properties, including high precision and monodispersity, make them ideal candidates for creating nanoscale architectures essential to the development of new therapeutic agents. Importantly, since plant virus capsids cannot replicate in mammalian cells, they pose minimal safety concerns for medical applications. It is this characteristic that makes them suitable for use in a number of biomedical applications, e.g., drug delivery systems or vaccine design, where tight control over drug delivery is required. Utilization of these biologically derived materials validates the potential of combining biological design with synthetic strategies and opens up prospects for innovations in targeted therapies, therefore enhancing treatment efficacy and safety within the healthcare industry (Aljabali et al., 2015) (Figure 1.9).

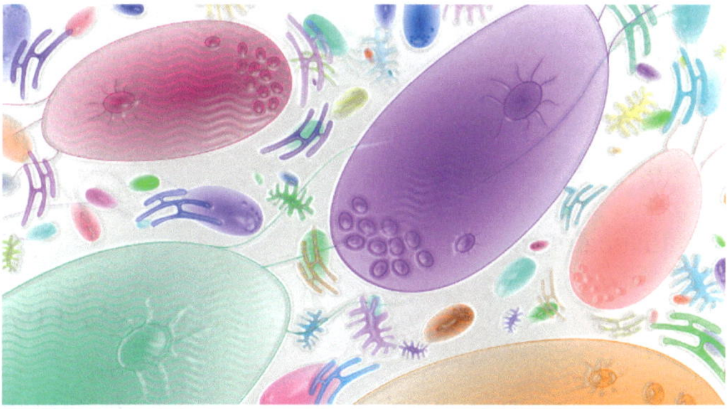

Figure 1.9. The promise of synthetic biology in medicine and healthcare.

Source: https://hudsonlabautomation.com/wp-content/uploads/2023/10/Hudson-bio2-1.png.

1.3.1. Development of Synthetic Drugs

Synthetic drug design is a principal intersection of synthetic biology and chemistry that revolutionizes the landscape of pharmaceutical research in its very foundation. In the past, drug discovery has emphasized the discovery of naturally occurring chemicals, which has conventionally led to slow timelines and limited reach. Through synthetic biology, researchers can now precision-craft new compounds. In conjunction with technologies such as CRISPR and pathway engineering, scientists can design and engineer drugs to target specific biological pathways or interactions, thus being more effective with fewer side effects. The new paradigm not only accelerates the discovery process but also allows modulation of intricate metabolic networks, enabling the creation of therapeutics specifically tailored to individual genetic backgrounds. Among the interesting avenues of de novo drug design is the application of computational techniques to enhance predictive capabilities in drug development. Of special importance is the incorporation of regularization models like the Generalized Elastic Net (GELnet).

Figure 1.10. Synthetic drugs.

Source: https://footprintstorecovery.com/wp-content/uploads/2019/08/synthetic-drugs-manufacture.jpg.

It incorporates information from gene pathways in feature selection and leads to solutions for groups of correlated genes. Thus, GELnet is not only predictive but also reveals genetic determinants of drug response, such as in cases predicting the response of breast cancer cell lines to specific compounds. Knowing such genetic determinants, researchers can engineer synthetic drugs that effectively bypass the intricacy of personalized tumor profiles, making cancer treatment more personalized (Baertsch et al., 2016). In addition, synthetic drug use can extend beyond individual treatment models to address broader public health concerns. Targeted gene knockout as a methodology of repairing metabolic function in deficient cells is a revolutionary leap forward in synthetic biology. These technologies are innovations from traditional gene therapy approaches in that they suggest the deletion of individual genes as a method of correcting pathological states rather than the replacement of faulty genes. This observation points to a future direction for educated drug design, especially in creating antagonistic drug combinations that avoid mechanisms of resistance, particularly in the treatment of cancer, antibiotics, and metabolic illnesses. While scientists explore these compensatory interactions, they open avenues for therapeutic breakthroughs that might reshape treatment paradigms in contemporary medicine (Motter et al., 2010) (Figure 1.10).

1.3.2. Gene Therapy Advancements

The last decade or so has seen significant advancement in gene therapy, a breakthrough in the field of synthetic biology and chemistry. By applying various strategies, scientists have developed methods to manipulate genetic material successfully and cure a range of genetic disorders. Enhanced knowledge of gene expression and function, coupled with technological improvements in delivery systems such as viral vectors, has enabled targeted therapy to become a reality, allowing precision medicine to be specific to the individual. Such advances hold promise not only for treating genetic disease but also for combating diseases such as cancer, where targeted gene therapy can be used in conjunction with traditional therapies. As in recent publications, the crossroads of synthetic biology with such advances have unveiled novel therapeutic avenues, showcasing the potential to breach enduring bottlenecks in treatment efficacy and selectivity, thereby revolutionizing patient outcomes heretofore deemed impossible (Bivona et al., 2019). Additionally, the discovery of CRISPR-Cas9 technology has revolutionized the field of gene therapy, providing unprecedented ease in genome editing. This flexible technology makes efficient alterations at specific locations in the genome possible, providing a flexible approach to the correction of hereditary and acquired mutations that are the basis for most diseases. The ease and low cost of CRISPR have opened gene therapy research to more individuals, and it is being used in many different fields, including regenerative medicine and cancer therapy. The therapeutic potential of CRISPR has also been demonstrated in efforts to overcome drug resistance in cancer therapies. As stated, using engineered bacteria to target anticancer drugs, researchers are finding new solutions to fight the rampant problem of drug resistance that has bedeviled traditional chemotherapy approaches (Bivona et al., 2019). The achievement underscores the transformative power of synthetic biology in therapeutic innovation, consolidating the ongoing momentum for medical innovation. Another gene therapy frontier on the rise is regenerative dentistry, which applies synthetic biology principles to treat oral disease. The field has experienced dramatic advancements in employing stem cells and gene delivery methods for dental and oral tissue repair. Researchers have explored various biomaterials, growth factors, and scaffolds to enhance tissue engineering outcomes, with an interdisciplinary approach that unites molecular biology and materials science. While traditional dentistry has depended on replacement and restoration of dental tissues, regenerative approaches offer a more holistic approach with the potential for longer-term outcomes. The latest innovations indicate that gene therapy, in combination with such materials, may bring about partial regeneration of complicated dental structures, albeit with difficulties before full tooth engineering can be realized. These advancements are the epitome of synergy between therapeutic use and synthetic biology, paving the way for the future advancement of the field (Figure 1.11).

Figure 1.11. Gene therapy advancements.

Source: https://assets.technologynetworks.com/production/dynamic/images/content/361066/recent-advances-in-gene-therapy-361066-960x540.jpg?cb=11749712.

1.3.3. Role of Synthetic Biology in Vaccine Production

The use of synthetic biology in vaccine development has revolutionized the field by enabling the design and application of novel vaccine platforms. These systems employ engineered biological elements, such as nucleic acids and proteins, which can be crafted with exquisite precision to trigger specific immune reactions. The most spectacular innovation in vaccine technology has been the use of recombinant DNA technology for the construction of subunit vaccines or viral vector vaccines. These vaccines contain no more than the necessary proteins or genes of the disease-causing agent and, therefore, reduce the potential of adverse effects compared to whole-virus vaccines. This application of synthetic biology is not only able to speed up vaccine development but can also improve the precision of vaccine targeting, ultimately maximizing efficacy and safety in immunization strategies. Furthermore, the development of sophisticated carbohydrates as vaccine adjuvants has turned into a profitable area of synthetic biology to bypass the constraints of vaccine effectiveness. It has been shown that carbohydrate forms can mediate crucial biological interactions underpinning vaccine immunogenicity and can enhance immune responses when used in vaccines. Improved chemical and enzymatic glycan synthesis enables research with glycan-based systems, and it is becoming increasingly feasible to design glycoconjugate vaccines. By enhancing the conjugation of glycoconjugates to lipids and proteins, researchers are able to enhance the immunogenicity of vaccine constructs, which is important in guaranteeing effective immune recognition. This integration of synthetic biology and carbohydrate chemistry underscores the potential for developing more effective vaccine platforms that utilize the specificities of glycoconjugates in stimulating immune responses and public health intervention (Gervay-Hague et al., 2018). Furthermore, synthetic biology also takes advantage of the ability of phage display technology to generate peptide libraries, a requirement to identify mimotopes—a new vaccine design strategy. Mimotopes are utilized as artificial mimics that can replicate the binding properties of natural antigen epitopes, which is a new method of eliciting strong immunogenic responses. For instance, the identification of dengue virus mimotopes from phage-displayed peptide libraries demonstrates the effectiveness of this strategy in producing vaccines targeting a given disease. The peptides can be manipulated to induce antibody secretion that reacts with virus-protein-related proteins and form a foundation for an effective vaccine candidate. By combining solar synthetic biology with these state-of-the-art techniques, researchers are poised to move more rapidly in discovering and developing multivalent vaccines, continuing to advance immunization science in the battle against infectious disease (Acosta et al., 2013).

1.4. ENVIRONMENTAL IMPACT

When synthetic biology is paired with chemistry, it holds vast potential to address environmental issues. By genetically modifying microorganisms to produce biofuels or bioplastics, scientists expect to end the use of fossil fuels and reduce plastic waste in the environment. These biotechnologies can provide green production processes that not only make resources more efficient but also restrict the production of greenhouse gases. Particularly in the region of Florida, where marine biotechnology is becoming increasingly significant, there is potential for the interface of academia and industry to become pivotal. Based on reports, the trend towards marine biotechnology continues to grow, and developing a network of research and education partnerships can effectively facilitate innovation in novel solutions to mitigate footprints in the environment and ensure sustainability in aquatic systems (Seaman and Whitehouse, 2001). However, the application of synthetic biology tools must be performed with care, particularly with regard to environmental security. Release of the engineered organisms to ecosystems must undergo rigorous assessment to avoid unwanted consequences, such as compromising local biodiversity. Current findings regarding the ecotoxicological effects of nanoparticles in water bodies indicate the difficulties and potential risks involved. For instance, research on gold nanoparticles has shown differential uptake in amphibians, offering a demonstration of how engineered materials can exert complex ecotoxicological effects based on the environmental context and species interaction (Andresen et al., 2017). This calls for integrated risk assessments to ensure that synthetic biology applications make a positive contribution to environmental health without, in doing so, raising new ecological issues. Also, the long-term environmental impact of synthetic biology operations must consider not only the immediate ecological impact but also on soil and water quality. As synthetic biology continues to advance, the potential use of bioengineered microorganisms in agriculture is being increasingly scrutinized, particularly with respect to their influence on soil microbiota and aquatic ecosystems. Agroecologically sustainable practices must prioritize the conservation of natural ecosystems while achieving high agricultural yields. The potential of synthetic biology is the creation of more robust ecosystems, but the harmony between innovation and preservation of ecology must continue to be a point of discussion. By creating an educated discussion between scientists, policymakers, and the public, the safe application of synthetic biology technologies can be steered towards a green and sustainable future.

1.4.1. Bioremediation Techniques

Bioremediation techniques are an important overlap between synthetic biology and chemistry, where biological processes are used to combat environmental contaminants. Bioremediation techniques primarily employ microorganisms or vegetation to degrade, transform, or mobilize toxic compounds from polluted environments. The effectiveness of bioremediation relies on microbial community diversity and their metabolic landscape, which can be enhanced with the use of genetic engineering. Through the illumination of these organisms' genetic pathways, researchers are able to create strains that are specially designed to degrade particular pollutants, like heavy metals or petroleum hydrocarbons. Not only does this highlight the promise of innovation in synthetic biology, but it also demonstrates a cost-efficient model for environmental rehabilitation, aligning with sustainable development priorities for ecological restoration and public health. Besides, bioremediation has two primary methods: *in situ* and *ex situ* remediation. *In situ* means treating the contaminated site itself, at times reducing the logistics of hassle of moving contaminated water or soil. Techniques such as bioventing, which injects air into soil to support microbial degradation, are examples of this technique. Conversely, *ex situ* methods involve removing contaminated material for treatment, which allows controlled environments and potential synergism with other cleanup technologies. Both methodologies can use naturally occurring bacteria or genetically engineered organisms to stimulate higher degradation rates, demonstrating the diverse applications of synthetic biology in environmental chemistry.

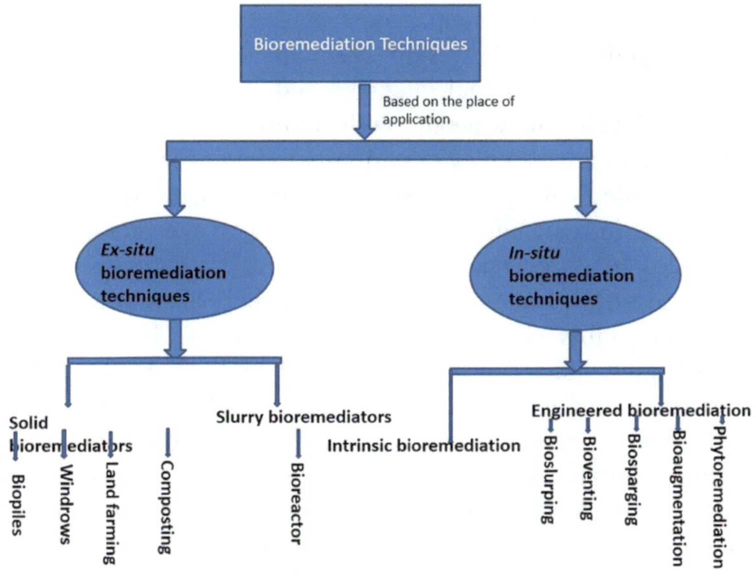

Figure 1.12. Bioremediation techniques.

Source: https://www.frontiersin.org/files/Articles/937186/fsoil-02-937186-HTML/image_m/fsoil-02-937186-g001.jpg.

The ongoing pursuit of developing these techniques, including the use of engineered microbial consortia that are synergistically active, testifies to the innovative character of the discipline. Supplementing the traditional bioremediation techniques, the addition of new biotechnological approaches has opened up new frontiers to the discipline. For instance, application of microbial fuel cells in bioremediation is both a solution to pollution as well as a source of energy, exemplifying a double benefit of sustainability and waste management. Studies on extremophiles, which inhabit extreme environments, have also led to the discovery of unusual metabolic pathways with a potential to degrade complex compounds. As explained in marine biotechnology reports, industry-academia interaction programs can facilitate better bioremediation processes greatly, with novel treatments tailored to respective pollutants in specific ecosystems (Seaman and Whitehouse, 2001). Ultimately, these advances in bioremediation processes significantly contribute to the overall narrative of how synthetic chemistry and biology can complement each other towards environmental sustainability and rehabilitation (Biswas, 2015) (Figure 1.12).

1.4.2. Sustainable Chemical Processes

The call for sustainable chemical processes has increased further with the emergence of synthetic biology and chemistry. As traditional methods of production tend to be fossil fuel-intensive and generate significant waste, it is now time to develop new means with minimal environmental footprints. Synthetic biology provides a new opportunity to engineer chemical production by employing biological systems to create biofuels, bioplastics, and other valuable products in a sustainable manner. Through the metabolic processes employed by designed microorganisms, researchers can create manufacturing systems that are not only highly efficient but also eco-friendly. This revolutionary capability demands a reformulation of chemical engineering paradigms, merging industrial application and biological knowledge, which is critical to solve climate change and resource depletion (Hall et al., 2012). Besides, the interfacing of synthetic biology with conventional industrial biotechnology facilitates the design of sustainable chemical processes through a cycle of systematic design engineering. In contrast to traditional methods that are inclined towards a focus on short-term economic worth, synthetic biology encourages a holistic view of the entire production process. This extends to energy input, raw material sourcing, and product disposal at the end of life, supporting a circular model economy. Process engineers, with their expertise in scaling and optimizing chemical reactions, are uniquely positioned to lead this new field. They are able to bridge the gap between synthetic biology's theoretical framework and real-world implementation, ultimately producing products that are not only cost-effective to manufacture but also socially feasible and environmentally friendly (Boyen et al., 2025). When planning curricula for future process engineers, it is necessary to incorporate principles of synthetic biology along with traditional chemical processes. Instructors must surmount the unique challenges of teaching these cross-disciplinary principles, equipping students with the knowledge and skills they need to innovate in sustainable chemical production. Institutions of learning must also collaborate closely with industry participants to remain attuned to emerging trends and technological advances. This cooperative approach not only enhances the learning process but also prepares graduates to address critical issues in sustainable development. Sustaining sustainability in the chemical engineering curriculum will equip a new generation of professionals to prioritize ecological integrity as they strive for the development of the next generation of productive, cleaner chemical processes (Hall and Howe, 2012).

1.4.3. Synthetic Biology in Biofuels

The overlap of biofuels and synthetic biology is an innovative solution for meeting the world's energy requirements. Synthetic biology's capability to engineer microorganisms to produce fuel more efficiently can make sustainable fuels possible as an alternative to fossil fuels. The capacity to manipulate the metabolic pathways in organisms like yeast and bacteria allows scientists to produce more of the biofuels ethanol and biodiesel, thus making the process cost-effective. In addition, the synthetic biology engineering design cycle permits iterative improvement of these organisms to optimize their operation for the production of energy. This activity is quite contrary to industrial biotechnology, highlighting the uniqueness of value added through synthetic biology. Overcoming great upscaling challenges of such processes requires collaboration between synthetic biologists and process engineers, emphasizing the applied value of their endeavor in biofuels and the potential for having a significant impact on energy sustainability (Hall and Howe, 2012). Furthermore, the ethics of synthetic biofuels must also be subject to rigorous critical analysis. With the revolution in biofuel production brought about by synthetic biology, significant questions about environmental impacts, safety, and public acceptance are raised to the forefront. Practitioners are not only interested in creating effective biofuel-making microbes but also in understanding the broader implications of what they do. Where scientific inquiry intersects ethics is particularly pointed in synthetic biology, where reconstructing biological pathways involves foresight into ecological consequences. This is a philosophical inquiry encompassing the social, ethical, and epistemological facets of synthetic biofuels production, urging stakeholders to adopt responsible practices and guidelines. It illustrates that microbial system design towards fuel production must be carried out alongside a commitment to ethical evaluation, such that innovations do not come at a price unacceptable to society and the environment (Clark et al., 2018). In addition, with the advance of synthetic biofuel development, education and training in synthetic biology have become necessary for future leaders in the field. Institutions of higher education and research institutions have begun to include synthetic biology in curricula, confirming the necessity of interdisciplinarity in education. Students trained in engineering and biology will have a better chance of dealing with the complexities of biofuel production.

This educational change is essential not only for scientific progress but also to develop a workforce capable of addressing current energy concerns in the real world through innovative means. As this discussion regarding synthetic biology in biofuels persists, it is more critical than ever to emphasize the importance of an educated populace to participate in both the science and ethical discourse surrounding these advances. By training future scientists and engineers in synthetic biology, we can offer sustainable energy solutions with the imagination and concern they deserve (Hall and Howe, 2012).

1.5. ETHICAL CONSIDERATIONS

There are ethical concerns in synthetic biology and chemistry, as they are the major motivations for their development and use. Though future uses of synthetic biology have revolutionary promise in medicine, agriculture, and industry, they also pose significant ethical concerns. The power to engineer biological systems challenges the morality of altering life forms and ecosystems. Fears about biosecurity, biodiversity loss, and the likelihood of unforeseen consequences demand rigorous ethical scrutiny. When synthetic biologists face the challenge of designing and engineering biological systems, embracing an ethical framework that prioritizes responsibility and safety is imperative to guaranteeing progress does not come at the expense of public trust or ecological integrity (Hall and Howe, 2012). Further, the engineering principles behind synthetic biology necessarily lead to concerns about equity and access. As the technologies change, the potential to cause enormous inequalities between individuals who can utilize these innovations and those who cannot becomes a pressing concern.

Monopolization by large biotech firms could deter smaller entities from innovating in the field, thereby decreasing diversity in research and applications. An ethical position demands a firm position on inclusiveness, advocating policy that enhances equitable access to synthetic biology innovation. Taking into account who will benefit from these technologies is crucial in order to avoid circumstances that exacerbate existing inequalities within society and between nations, consequently leading to a more just and balanced application of synthetic biology solutions (Freemont et al., 2014). Finally, educating future generations about synthetic biology raises its own ethical challenges. With the designing of new curricula that address the complexities of engineering biological systems, educators need to go out of their way to incorporate ethical dimensions into scientific training. Forming a robust appreciation of the ethical dimensions present in synthetic biology guarantees a sense of responsibility on the part of future scientists. Incorporating ethical discussion at the undergraduate level and in research will empower future synthetic biologists to respond more effectively to the moral landscape of their discipline. This shift in learning, directed toward an ethical orientation as well as technical skills, can not only drive innovation but also ensure that synthetic biology develops so as to respect societal values as well as ecological subsystems (Hall and Howe, 2012).

1.5.1. Biosecurity Concerns

Synthetic biology is a milestone in scientific innovation, but it necessarily comes with pressing biosecurity concerns calling for sober reflection. As scientists employ synthetic biology techniques to design biological systems for beneficial purposes, always at stake is misuse. The dual-use feature of technology being employable constructively and destructively implies that regulatory frameworks and rigorous control must be present to prevent risks. Current biosafety and biosecurity laws have serious loopholes that would facilitate inadvertent as well as intentional release of engineered organisms or their components. Addressing these risks involves policymakers going back to existing agreements and potentially reshaping treaties to encompass better the new threats posed by synthetic biology, as has been brought out in recent controversies surrounding regulatory regimes for these new technologies (Rabitz, 2014). The effects of unregulated access to synthetic biological technologies extend beyond mere laboratory mishaps; they encompass the risks of malicious applications that undermine public security and international security. Non-state actors, including terrorists, can exploit advances in synthetic biology to their nefarious purposes, creating additional complexity for the governance problem. While there is general international consensus that access needs to be denied to technologies that simplify nuclear weapons production, there is a similar consensus missing on biotechnology (Harris et al., 2016).

This gap illustrates the requirement for sweeping regulatory measures, especially designed to address synthetic biology's unique risks and attributes, so that its benefits to society are not the outcome of increased exposure to biological hazards. The dynamic nature of synthetic biology also makes it more difficult for traditional institutions of governance, as they fail to keep up with the accelerating process of technological progression. The complexity of synthetic organisms and their uncalculating behavior underscores the necessity of adaptive regulative structures that can keep up with new tendencies. Policy-makers must engage in visionary dialogue with scientists, ethicists, and industry stakeholders in developing a multi-layered regime aimed at promoting biosecurity. Steps that may be implemented include developing interpretive statements under existing treaties or new treaties that span a broader category of dual-use technologies (Harris et al., 2016; Rabitz, 2014). Ultimately, the harmonization of global efforts to control and monitor synthetic biology will be pivotal for maintaining the balance between security and innovation, whereby scientific progress benefits do not unwittingly create new threats to public health and safety.

1.5.2. Ethical Implications of Genetic Modification

The ethical implications of genetic manipulation are a dominant topic of discussion across the entirety of synthetic biology and chemistry, particularly as advances in these disciplines continue to unfold. The manipulation of genetic components raises deeply serious questions about the ethics of altering life forms on behalf of humanity. These moral issues typically stem from biodiversity, ecological harmony, and long-term impact on the settings in which engineered organisms reside. Further, the lessons of history illustrate that public opinion plays a significant role in ethical debates; for instance, the majority of non-experts view GMOs as posing inherent dangers, even though there is scientific consensus to the contrary. Thus, building an inclusive public debate regarding the facts and myths of genetic modification is critical to making informed decisions in this revolutionary scientific area (Acevedo-Rocha et al., 2016). Additionally, the ethical context of genetic modification is also complicated by socio-political considerations, most notably concerns of accessibility and equality. As synthetic biology evolves, access to advanced genetic technologies may be limited to particular demographic groups or nations, expanding the potential for new genetic elitism.

This potential adds to existing inequalities in medicine and farm production, as those with the means to pay might disproportionately benefit from improvements in genetic modification. Further, the terms employed to confront patenting and owning genetically modified life pose fundamental questions of who owns these innovations and whether ownership will be expanded to life. As the stakeholders attempt to respond to such challenging questions, ethical frameworks must shift so that benefits in genetic modification are distributed equitably (Clark et al., 2018). The interactive dynamic between scientific innovation and ethical reflection will eventually determine the fate of genetic modification. By incorporating interdisciplinarity, synthetic biologists must engage with ethicists, policymakers, and the wider community to unravel the ethical dimensions of their work. The issue is how to react to not only the technological aspects of genetic engineering but also to the moral frameworks that guide its applications. For instance, in the development of new biotechnologies, there is a need to consider different values and perceptions of genetically modified life entities, as these play a major role in public acceptance and the regulatory framework. As the previous discourses have shown through their outcomes, promoting constructive debate on the ethical aspects of genetic modification can facilitate greater understanding and ensure responsible practice in synthetic biology, culminating in developments that are beneficial to society as a whole (Acevedo-Rocha et al., 2016).

1.5.3. Public Perception and Societal Impact

Public perception of synthetic biology is heavily impacted by the overlap between scientific progress and societal narratives. Popular media increasingly depict synthetic biology in highly skewed representations that often mirror the realities of synthetic biology sporadically. For instance, couching synthetic biology in dramaturgical language—often its capacity to create life or create GMOs—is likely to cause immense misapprehension among the layman. Research has shown that lay individuals attending an interdisciplinary summer school overwhelmingly identified synthetic biology with its most renowned personalities and underscored the primacy of prominent persons over an organized grasp of the field. This reliance on coverage by the media can lead to unrealistic expectations and unsupported fears regarding synthetic biology, making public discussion becomes challenging and informed decision-making in policy and regulation is hindered (Acevedo-Rocha et al., 2016). The outcomes of public opinion extend far beyond personal misinformation; they also significantly impact policymaking and governance concerning synthetic biology. For effective governance, policymakers must understand the scientific principles and ethical considerations underlying advances in this field. Still, it is found from the research that synthetic biologists overestimate the scientific literacy of the public, which might further amplify communication gaps. Within such an environment where public views are directed by ethical reasoning, misinterpretation of words and scientific capabilities are bound to make constructive debate difficult. Hence, the development of better communication models between scientists and society is essential to bridging understanding gaps and ensuring that ethics are suitably placed in science and technology policy (Chaturvedi et al., 2015).

Second, the social implications of synthetic biology are necessarily embedded in how ethical concerns are resolved with multiple cultural backgrounds. Comparative studies report various public perceptions of synthetic biology dominating in different regions such as Europe, India, and China, influenced by distinct historical, ethical, and cultural science and technology narratives. Such differences in perceptions demand context-tailored responses to public engagement, highlighting the importance of inclusive discourse respecting local values and beliefs. As the global context of synthetic biology continues to shift, developing templates for ethical regulation—without ignoring variation in local attitude—will be essential. These will help enable responsible scientific progress that delivers on society's expectations, ultimately forming a more harmonious relationship between synthetic biology and the public (Acevedo-Rocha et al., 2016; Chaturvedi et al., 2015) (Figure 1.13).

Figure 1.13. Navigating public perception in synthetic biology.

Source: https://images.unsplash.com/photo-1661267589862-15cf53bb97ab?crop=entropy&cs=tinysrgb&fit=max&fm=jpg&ixid=M3w3MDgyMDR8MHwxfHN-lYXJjaHwxfHxzeW50aGV0aWMlMjBiaW9sb2d5JTJDJTIwcHVibGljJTIwc-GVyY2VwdGlvbiUyQyUyMGNvbW11bmljYXRpb24lMjBzdHJhdGVnaWVzfGVufDB8MHx8fDE3NTEzMjE0Mjh8MA&ixlib=rb-4.1.0&q=80&w=500&?utm_source=numberanalytics&utm_medium=referral.

1.6. FUTURE DIRECTIONS

Perhaps the most exciting of the future directions as synthetic chemistry and biology continue to improve is the union of artificial intelligence (AI) with biological purposes. This synergy has the potential to transform not only how we approach biological engineering but also how we design intelligent systems to assist with experimental design and data analysis. With the aid of AI algorithms that replicate biological activity, scientists can better forecast and identify the most appropriate paths toward utilizing synthetic biology. Further, research into human intelligence-machine interaction, as suggested with AI studies, can reveal more efficient approaches, laying bare hidden biological paradigms that can redefine our concept of life itself. Maneuvering the complexities inherent in this integration while maintaining ethical issues will be critical to attaining innovation without accountability in the burgeoning world of synthetic biology (Ahuja et al., 2013). Another potential area for future research is enhancing the techniques used in genetic profiling and biobanking. As forensic applications become increasingly dependent on genetics, it will be increasingly essential that they be enhanced in their discriminatory capacity.

Advances in areas such as proteomics and epigenetics hold the potential to take current sophistication in forensic science to the next level and allow for more subtle analysis of biological samples. By focusing on understanding biological factors, scientists can devise novel methods that not only bypass limitations in existing DNA profiling techniques but also yield further insights into the origin and contexts of biological samples. This is particularly crucial in promoting precision in a variety of applications, from criminal justice and environmental tracking, thus fostering public trust and confidence in the science behind this (Eilerts, 2016). In addition, synthetic chemistry and biology are predicted to collaborate to create novel materials and systems capable of meeting major global needs. Tomorrow's technological innovations could make it possible for the creation of microorganisms that produce clean biofuels or bioplastics, reducing the carbon footprint of humanity and alleviating ecological problems. These innovations require an interdisciplinary approach, bringing material science, biology, and chemistry knowledge together to create systems that both function and are ecologically sound. As this field continues to grow, the focus on equally important social and ethical implications will be used to guide the researchers and policymakers to make sure that the technologies developed play a positive part in society and do not pose hazards of synthetic biology. Attention on responsible innovation will be important to unlock the fullest potential of synthetic biology and chemistry to build a sustainable future.

1.6.1. Innovations in Synthetic Biology

The rapid evolution in synthetic biology is driving deep integration between chemical engineering and biological science, bringing innovations with revolutionary impact across sectors. Central to this revolution is the integration of molecular biological approaches with principles of engineering to create biological systems capable of solving complex problems. Such convergence not only maximizes the performance of existing organisms but also enables one to create entirely new biological devices, circuits, and pathways for specific functions. Various engineering approaches, including top-down and bottom-up strategies, enable researchers to control and reorganize biological elements rationally. Synthetic biology, therefore, is not just confined to its traditional applications; it is opening up new frontiers into pharmaceuticals, renewable energy, and so on, thereby solidifying its place in contemporary scientific research and industrial development (Afzal et al., 2023). Moreover, the charm of synthetic biology is that it has the ability to provide practical solutions to topical global problems. Technologies in this domain are continually evolving to improve chemical manufacturing processes, reduce environmental impacts, and increase agricultural yields. Microbes, for instance, may be engineered to create biofuels or bioplastics as green alternatives to conventional materials. Synthetic genome technologies are one of the best examples of how researchers can engineer and optimize organisms to synthesize useful chemicals or perform intended functions to promote circular economy systems. Moreover, the advent of protocell synthetic biology paves new ways to design minimalist life forms that are programmed to perform specific processes, which can revolutionize fields like drug delivery and environmental cleanup (Afzal et al., 2023). Therefore, not only does the innovative production of synthetic biology advance science, but it also offers utilitarian approaches for applying to real-world scenarios. The geographical concentration of top companies in the synthetic biology industry also captures the interaction between national institutions and innovative abilities. Historical examination demonstrates that nations, for example, Germany, have used their scientific progress to control industries, for instance, the manufacture of synthetic dyes in the late 19th century. This captures the coevolutionary processes by which industry and national scientific environments interactively support one another. Firms gain advantages from close relationships with universities, experiencing an atmosphere abounding with collaborative research and commercialization. This leads to ecosystems that allow successful innovative firms to flourish under respective national policies and infrastructure supporting scientific research and development. Consequently, encouraging these institutional collaborations could be key to speeding up the competitive edge of nations in synthetic biology, as well as ensuring its high growth and global contributions in the future.

1.6.2. Potential Breakthroughs in Chemical Synthesis

In the field of synthetic biology, the potential for breakthroughs in chemical synthesis has never been more evident. The ability to design microbes to form complex molecules has the potential to transform the chemistry of how chemicals are made. By leveraging the inherent properties of specific yeast or bacteria, researchers can develop new methods to produce essential compounds, ranging from pharmaceuticals to fuels. This biodegradable method not only minimizes the environmental cost associated with traditional chemical synthesis processes but also addresses the pressing demand for sustainable solutions in the wake of increasing environmental demands. In view of the escalating public health concern of antibiotic resistance, the demand for innovative chemical synthesis processes that can offer potent new antibiotics is imperative. Thus, synthetic biology innovations can bring the catalyst that will shatter the deadlock in drug discovery and usher in a new age of medicinal chemistry (Coukell et al., 2016). Apart from pharmaceuticals, innovation in chemical synthesis based on synthetic biology is also likely to significantly increase agricultural productivity and sustainability. Plant breeding for the production of biopharmaceuticals or biodegradable plastics is a rapidly developing field of research. These technologies have the potential to reshape the agricultural sector by integrating chemical synthesis capability directly into crop systems, allowing for on-demand synthesis of valuable chemicals. Such advances have the potential to substitute some synthetic fertilizer and pesticide applications with more sustainable practices. Besides, as the world transitions towards greater environmental accountability, the integration of synthetic biology in the synthesis of crop chemicals is making efforts to tackle the double challenge of climate change and food security more effectively. Developments here could not only make crops more resistant but also reduce farming's environmental footprint by boosting efficiency and reducing waste (N/A, 2015). Moreover, the advent of novel biotechnological techniques, such as CRISPR and other gene-editing tools, significantly boosts the prospects for state-of-the-art achievements in chemical synthesis. These tools allow for targeted manipulation of metabolic pathways of organisms to enhance their ability to generate target chemicals economically and more efficiently. Scientists can optimize routes in microbes to maximize yield, reduce by-products, and reduce the overall energy input into chemical synthesis processes. This biotechnological potential can lead to significant advances in the production of high-value chemicals and materials used by various industries. Combined with machine learning and artificial intelligence, such biotechnological applications have the potential to accelerate new route and synthesis discovery, ultimately tapping into the full potential of synthetic biology for innovative chemical solutions to societal needs (Coukell et al., 2016; N/A, 2015) (Figure 1.14).

Figure 1.14. Potential breakthroughs in chemical synthesis

Source: https://www.deskera.com/blog/content/images/2023/02/Assembly-To-Order-3--1.png.

1.6.3. Collaborative Research Opportunities

Opportunities for interdisciplinary research in synthetic biology and chemistry are now prime innovation and problem-solving avenues. As the complexity of biological systems and chemical reactions becomes better understood, multidisciplinary solutions are essential in advancing scientific knowledge. Colleges, universities, and private industry now have a better chance than ever before to collaborate with each other in a manner that leverages diverse expertise (Murmann, 2025). Such collaborations allow for collective solutions to significant challenges, such as antibiotic resistance, the severity of which has increased over recent decades, since the plateauing of new antibiotic research (Coukell et al., 2016) has been well demonstrated. By resource, expertise, and technology pooling, researchers can pilot new concepts that integrate biology with chemistry, leading to breakthroughs that may not have been achieved in isolated paradigms. In addition, the combination of synthetic biology and chemistry possesses promising potential for furthering solutions to environmental issues and sustainability. Collective action has much to offer towards initiatives in ocean preservation, where innovative biotechnological solutions can fight against pollution and overfishing impacts. As identified in the 10 Grand Challenges for Oceans Conservation, the research projects must incorporate diverse perspectives and expertise to develop integrated approaches, i.e., engineering ecological resilience on the coast and recovering from marine litter (N/A, 2015). The use of collaborative research not only helps address such vital issues but also helps understand the interconnections among ecosystems, human activities, and artificial substances, thereby shaping future sustainable practices. In addition to addressing individual issues, collaborative research opportunities provide a more resourceful learning process and promote capacity building for researchers. With coordination between professionals from various fields, individuals have first-hand experience in the integration of knowledge between biological and chemical sciences. This integration enables researchers to adopt a creative approach, which is essential for innovation in synthetic biology and chemistry. Collaboration also facilitates the sharing of resources, such as special instruments or knowledge, which can be vital in making experiments succeed. With the increasing convergence of the sciences today, their capacity to enhance creativity and advance the capabilities of research teams cannot be exaggerated, ultimately leading the way in the field of synthetic biology and chemistry and changing the world in revolutionary fashion.

SUMMARY

In summarizing the success and impact of synthetic biology and chemistry, it must be observed that the transformative capacity of these technologies on science and society is enormous. The ability to engineer biological systems allows one to design breakthrough applications, from biofuels to drugs. They are not mere ideas; they are concrete solutions to some of the most pressing global problems, including pandemics and climate change. By applying the principles of synthetic biology, researchers can engineer organisms with specific functions, thereby pushing biotechnology to new horizons. This summary highlights the importance of continued investment and study in synthetic biology because of its potential to transform our solution approach in the biological sciences. Additionally, the challenges involved in synthetic biology, particularly in the design and construction of artificial cells, cannot be overemphasized. As noted, transcending limitations such as controlled manufacture of membrane-based cells is essential in advancing the field. Techniques such as microfluidic generation have shown extremely promising potential in addressing these issues, enabling the control of size and biomolecular composition within artificial cells with precision (Elan, 2016). Creating these technologies is a significant achievement in synthetic biology, bridging the gap between theory and practice. The ability to mimic biological processes in a laboratory setting has the potential to drive innovation and provide information on cellular processes that have not been possible to acquire. Finally, the use of computational techniques in synthetic biology, in particular, in retrosynthetic analysis, is a groundbreaking technique in the field. As demonstrated, machine learning-based advanced models can predict retrosynthetic pathways with the same precision as traditional expert systems, while overcoming their inherent limitations (Gomes et al., 2017). This intermingling of the biological and computational disciplines not only advances the pace of research but also streamlines drug discovery and the synthesis of chemicals. Together, the future of chemistry and synthetic biology depends on collective forces between interdisciplinary approaches, pointing out that the combined power of biology and technology is to blame for unleashing unprecedented innovation for mankind.

During an inquiry into synthetic biology and chemistry, one of the major observations pointed out is that there must be a definition of life in scientific processes. Despite countless attempts at marking boundaries of clear definitions of life, there has not been scientific concurrence. This is a dilemma that has caused some researchers to doubt the utility of these definitions overall. However, another way out would be to re-evaluate our image of definitions as working tools rather than strict classificatory tools. By retaining in sight the instrumental applications of these definitions, scientists can utilize them to construct theoretical models for which experimentally accessible work is possible and that facilitate important inquiry across domains such as synthetic biology and astrobiology. This refined perspective acknowledges the crucial role definitions of life must have in guiding research programs and evaluating scientific outcomes and therefore stresses their value over mere taxonomy (Bich and Green, 2016). More information on synthetic biology reveals a growing interdependence of chemical principles and biological functions. The inclusion of chemistry in the synthetic biology paradigm lends credence to innovation in biological design, with applications that extend to medicine, agriculture, and the environment. While scientists engineer genetic codes and biochemical pathways, they leverage the intersection of chemistry and biology to design new organisms and compounds that were not hitherto attainable. These advances not only enhance our understanding of simple biological processes but also assist in the formulation of solutions for current global issues. The growing amount of research in this field necessitates regular interplay among chemists, biologists, and bioengineers to ensure that developing technologies are coupled with ethical and societal standards of the complexity involved in messing around with life itself (Andrade et al., 2018). Lastly, institutions' funding and support indicate the interdisciplinary nature of synthetic biology and chemistry studies. The efforts of institutions such as the NIH and NSF through numerous grants and programs indicate the combined effort toward creating this rapidly emerging branch of science. Through these funding institutions, interdisciplinary work becomes feasible, and scientists can bridge fields to pursue innovative avenues to merge synthetic biology and chemistry. This not only promotes scientific discovery but also expands the possibilities for functional applications that benefit society as a whole. While scientists plow through the complexities of creating new biological systems, the assistance from government and private institutions is the critical cornerstone that makes ongoing exploration and ethical consideration possible. These donations, through their support for various researchers and projects, attest to the significance of interdisciplinary work in addressing the multifaceted issues facing synthetic biology (Andrade et al., 2018).

REFERENCES

1. Abramo, G., D'Angelo, C. A., & Di Costa, F. (2018). The effect of multidisciplinary collaborations on research diversification. http://arxiv.org/abs/1810.12626 (accessed on 3rd October 2025).
2. Acevedo-Rocha, C. G., Chizzolini, F., Rothschild, L. J., Verseux, C., et al. (2016). Misconceptions of synthetic biology: Lessons from an interdisciplinary summer school. https://core.ac.uk/download/pdf/76424512.pdf (accessed on 3rd October 2025).
3. Acosta, A., Aguilar, A., Alvarez, M., Amin, M., et al. (2013). Immunogenicity of NS4B dengue 3 virus mimotope presented to the immune system as multiple antigen peptide system. https://core.ac.uk/download/19785886.pdf (accessed on 3rd October 2025).
4. Afzal, A., Bin Sajid, H., Raheem, M. A., Awan, M. H., Tabassum, N., & Fatima, S. (2023). An overview on synthetic biology: Its classification, engineering approaches, and applications of synthetic biology. https://core.ac.uk/download/586552785.pdf (accessed on 3rd October 2025).
5. Ahuja, P. (2013). Man and machine: Questions of risk, trust and accountability in today's AI technology. http://arxiv.org/abs/1307.7127 (accessed on 3rd October 2025).
6. Aljabali, A. A. A., & Evans, D. (2015). Bionanomaterials from plant viruses. https://core.ac.uk/download/151161135.pdf (accessed on 3rd October 2025).
7. Allen, D., Allenby, B., Aragon-Correa, J. A., Armstrong, R., Batterham, R., Benoit, L., et al. (2010). Sustainability of the chemical manufacturing industry–Towards a new paradigm? https://core.ac.uk/download/339962.pdf (accessed on 3rd October 2025).
8. Amrollahi, P., Seifi, A., Shah, B., Tayebi, L., et al. (2016). Recent advancements in regenerative dentistry: A review. https://core.ac.uk/download/213080485.pdf (accessed on 3rd October 2025).
9. Andrade, R. B., Bougie, D., Boxer, M. B., Brimacombe, K., et al. (2018). Canvass: A crowd-sourced, natural-product screening library for exploring biological space. https://open.bu.edu/bitstream/2144/37033/1/Canvass%3a%20A%20Crowd-Sourced%2c%20Natural-Product%20Screening%20Library%20for%20Exploring%20Biological%20Space.pdf (accessed on 3rd October 2025).
10. Andresen, K., Bury, T. B., Carfagno, G. L. F., Fong, C., et al. (2017). Differential uptake of gold nanoparticles by 2 species of tadpole, the wood frog (*Lithobates sylvaticus*) and the bullfrog (*Lithobates catesbeianus*).

https://cupola.gettysburg.edu/cgi/viewcontent.cgi?article=1063&context=biofac (accessed on 3rd October 2025).

11. Anurogo, D. (Dito), Ikrar, T. (Taruna), & Parikesit, A. A. (Arli). (2017). Bionanomedicine: A "panacea" in medicine? https://media.neliti.com/media/publications/179998-EN-none.pdf (accessed on 3rd October 2025).
12. Arnold, F. H., Bastian, S., Bennett, C. S., Chen, M., et al. (2009). Chemoenzymatic elaboration of monosaccharides using engineered cytochrome $P450_{BM3}$ demethylases. https://core.ac.uk/download/4883076.pdf (accessed on 3rd October 2025).
13. Baertsch, R., Carlin, D. E., Paull, E. O., Sokolov, A., et al. (2016). Pathway-based genomics prediction using generalized elastic net. https://core.ac.uk/download/323064028.pdf (accessed on 3rd October 2025).
14. Bastian, M., Börner, K., Fox–Keller, E., Galison, P., Kessler, M. M., Leydesdorff, L., Nicolis, G., et al. (2012). Complex systems science: Dreams of universality, reality of interdisciplinarity. https://core.ac.uk/download/49887865.pdf (accessed on 3rd October 2025).
15. Battisti, A., & Persiani, S. (2018). Frontiers of adaptive design, synthetic biology and growing skins for ephemeral hybrid structures. https://core.ac.uk/download/188829284.pdf (accessed on 3rd October 2025).
16. Bich, L., & Green, S. (2016). Is defining life pointless? Operational definitions at the frontiers of biology. https://core.ac.uk/download/131211791.pdf (accessed on 3rd October 2025).
17. Bird, C., & Frey, J. G. (2013). Chemical information matters: An e-research perspective on information and data sharing in the chemical sciences.
18. Biswas, K. (2015). Biological agents of bioremediation: A concise review. https://core.ac.uk/download/131208038.pdf (accessed on 3rd October 2025).
19. Bivona, T. G., Chang, S., Hernández, A. C., Keasling, J., et al. (2019). Overcoming the challenges of cancer drug resistance through bacterial-mediated therapy. https://core.ac.uk/download/287622369.pdf (accessed on 3rd October 2025).
20. Boyen, C., Børresen, T., Dobson, A., Höfle, M., et al. (2025). *Marine Biotechnology: A New Vision and Strategy for Europe*. https://core.ac.uk/download/pdf/29241634.pdf (accessed on 3rd October 2025).
21. Chaturvedi, S., Ladikas, M., Stemerding, D., Zhao, Y., et al. (2015). *Science and Technology Governance and Ethics–A Global Perspective from Europe, India and China*. https://core.ac.uk/download/42136809.pdf (accessed on 3rd October 2025).

22. Cheng, C., Collini, E., Engel, G. S., Gilmore, J., Ishizaki, A., Kocsis, L., et al. (2012). Molecular response in one-photon absorption via natural thermal light vs pulsed laser excitation. http://arxiv.org/abs/1109.0026 (accessed on 3rd October 2025).
23. Clark, A., Clark, A., Clark, A., Minelli, A., Pickering, A., Calcott, B., Goodwin, B., et al. (2018). Grounding knowledge and normative valuation in agent-based action and scientific commitment. https://core.ac.uk/download/186331467.pdf (accessed on 3rd October 2025).
24. Coukell, A., Shore, C., Jungman, E., Talkington, K., Kothari, P., & Urahn, S. K. (2016). A scientific roadmap for antibiotic discovery: A sustained and robust pipeline of new antibacterial drugs and therapies is critical to preserve public health. https://core.ac.uk/download/75785719.pdf (accessed on 3rd October 2025).
25. Eilerts, D. F. (2016). Beyond DNA: Epigenetics and proteomics in forensic science. https://core.ac.uk/download/70427288.pdf (accessed on 3rd October 2025).
26. Eissa, M. E. (2025). *Dawn of Synthetic Biology Engineering Life at the Microscopic Scale: A Review*. https://core.ac.uk/download/649350795.pdf (accessed on 3rd October 2025).
27. Elani, Y. (2016). Construction of membrane-bound artificial cells using microfluidics: A new frontier in bottom-up synthetic biology. https://core.ac.uk/download/77011265.pdf (accessed on 3rd October 2025).
28. Freemont, P., Kelwick, R., MacDonald, J. T., Webb, A. J., et al. (2014). Developments in the tools and methodologies of synthetic biology. https://core.ac.uk/download/76995845.pdf (accessed on 3rd October 2025).
29. Gervay-Hague, J., Hsieh, H. W., & Park, S. S. (2018). Anomeric O-functionalization of carbohydrates for chemical conjugation to vaccine constructs. https://core.ac.uk/download/323071581.pdf (accessed on 3rd October 2025).
30. Gomes, J., Ho, S., Kawthekar, P., Liu, Y., et al. (2017). Retrosynthetic reaction prediction using neural sequence-to-sequence models. http://arxiv.org/abs/1706.01643 (accessed on 3rd October 2025).
31. Gradoff, O. (2017). Autopoiesis concepts for chemical origins of life and synthetic biology [Stenogram of the popular lecture on the foreign bibliographic seminar]. https://core.ac.uk/download/186331807.pdf (accessed on 3rd October 2025).
32. Hall, G. M., & Howe, J. (2012). The impact of synthetic biology in chemical engineering–Educational issues. https://core.ac.uk/download/6112945.pdf (accessed on 3rd October 2025).

33. Harris, E. D., Lin, H., & Acton, J. M. (2016). *Governance of Dual-Use Technologies: Theory and Practice*. https://core.ac.uk/download/75785647.pdf (accessed on 3rd October 2025).

34. Khan, I., & Akhtar, M. W. (2011). Different approaches for protein engineering in industrial biotechnology. https://core.ac.uk/download/pdf/289804.pdf (accessed on 3rd October 2025).

35. Motter, A. E., Baba, T., Blagosklonny, M., Bollenbach, T., Bourgeois, T., Burgard, A., Burgard, A., et al. (2010). Improved network performance via antagonism: From synthetic rescues to multi-drug combinations. https://core.ac.uk/download/pdf/8391385.pdf (accessed on 3rd October 2025).

36. Murmann, J. P. (2025). The coevolution of industries and national institutions: Theory and evidence. https://core.ac.uk/download/pdf/6496841.pdf (accessed on 3rd October 2025).

37. N/A. (2010). *Realizing European Potential in Synthetic Biology: Scientific Opportunities and Good Governance*. https://core.ac.uk/download/629036665.pdf (accessed on 3rd October 2025).

38. N/A. (2015). Launching the grand challenges for ocean conservation. https://core.ac.uk/download/84911929.pdf (accessed on 3rd October 2025).

39. Patwardhan, S. (2011). Biomimetic and bioinspired silica: Recent developments and applications. https://core.ac.uk/download/9032722.pdf (accessed on 3rd October 2025).

40. Rabitz, F. (2014). Regulatory gaps in the global governance of synthetic biology. *IES Policy Brief*, Issue 2014/11, December 2014. https://core.ac.uk/download/76803338.pdf (accessed on 3rd October 2025).

41. Rehberger, K. R. (2015). Interdisciplinary research-based learning in organic chemistry and microbiology laboratories: Synthesis and biological testing of novel penicillin derivatives. https://core.ac.uk/download/143957564.pdf (accessed on 3rd October 2025).

42. Seaman, W., & Whitehouse, J. H. (2001). Florida marine biotechnology: Research, development and training capabilities to advance science and commerce. https://core.ac.uk/download/11016354.pdf (accessed on 3rd October 2025).

43. Yin, Y. (2024). Synthesis of long oligodeoxynucleotides. https://core.ac.uk/download/651410122.pdf (accessed on 3rd October 2025).

CHAPTER 2

FUNDAMENTAL CONCEPTS IN MOLECULAR BIOLOGY

Contents

INTRODUCTION TO MOLECULAR BIOLOGY

Molecular biology is a core science that gives the foundation necessary for the understanding of complex molecular interactions that drive biological processes. By understanding the molecular mechanisms of DNA, RNA, and proteins, researchers can unravel the complexity of gene expression and regulation. Molecular biology plays a central role in describing the transmission and expression of genetic information in living organisms. New technologies, such as next-generation sequencing and CRISPR gene editing, have made it possible to explore the molecular intricacies of life to an unprecedented extent. The promise of these technologies is more than mere biological insight; they allow for the manipulation of genetic material and hold potential therapeutic avenues for genetic disease as well as for understanding developmental processes in various organisms (Baianu et al., 2010). Hence, molecular biology not only improves knowledge of biological systems but also identifies scientific innovation with therapeutic use in medicine and biotechnology. In addition, the interaction among biomolecules is better understood as dynamic and multidimensional networks rather than single events. This perspective encourages understanding of molecular biology within the framework of interactomics, which involves the study of cellular interactions that regulate cell signal transduction pathways and functional responses.

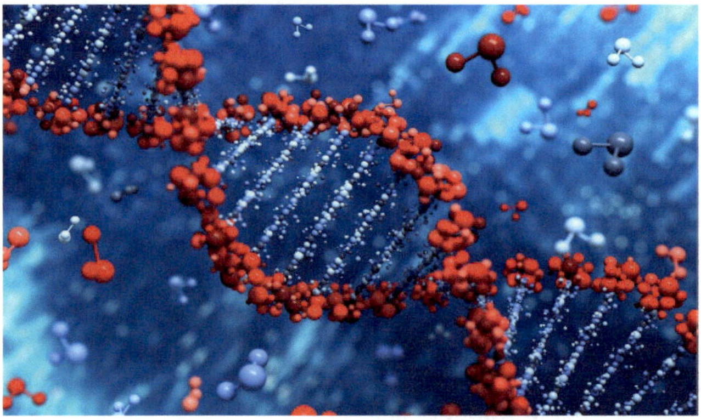

Figure 2.1. Understanding molecular biology

Source: https://praxilabs.com/en/blog/wp-content/uploads/2023/05/what-is-molecular-biology.jpg.

For instance, sophisticated quantum models of automata and their modeling of cellular behavior allow scientists to visualize how noncancerous cells differentiate and grow, forming the basis for a better understanding of diseases such as cancer. Because cancer cells undergo intense genetic mutation, the development of their interactomics profoundly changes their operation (Baianu et al., 2010). Such understanding is crucial not only in the context of understanding cancer development but also for the design of targeted therapies that can keep pace with these dynamically evolving genomic environments. Last, but certainly not least, the convergence of molecular biology with advanced materials science is transforming our methodology for understanding biology and for treatment. Materials based on semiconductors have become key elements for bridging the gap between biological systems and permitting real-time monitoring and control of cellular processes. These developments enable scientists to explore new signaling pathways and interactions in living organisms. By introducing these materials into experimental studies, scientists can better understand molecular dynamics across multiple scales. This integration of molecular biology and materials science reflects the potential for new solutions to basic biological issues, reinforcing the idea that developments within one field can further understanding within another to a significant extent. Thus, the molecular biology science not only forms the backbone of biological sciences but also encourages inter-disciplinary research that has the potential to provide revolutionary insight into life itself (Figure 2.1).

Definition and Scope of Molecular Biology

Molecular biology is a science that studies the chemical composition and processes of living organisms, with particular focus laid upon the interaction among cellular systems. This field of biology is mostly the investigation of how nucleotide sequence in DNA relates to protein structure and function, a key principle for understanding biological function. Molecular biology is not just genetic material but also encompasses a number of biomolecules, like RNA and proteins, and their roles in cellular processes such as replication, transcription, and translation. The intricate dance of molecular interactions lies at the heart of processes like cell signaling and gene regulation. As the technology has progressed, more and more computational biology and bioinformatics methods and concepts have been integrated into it, which are now absolutely essential in handling large biological data sets and interpreting molecular mechanisms at an unprecedented level (Ekmekci et al., 2016). The field of molecular biology also intersects with other sciences, like genetics, biochemistry, and cell biology, to offer a multidisciplinary strategy that more accurately elucidates life at the molecular level. The advances in genomics have demonstrated, for instance, how molecular biology can elucidate significant data on disease mechanisms, patterns of heredity, and evolutionary patterns. Synthetic biology, a new discipline, applies molecular biology methods to design new biological entities or reengineer old ones in order to realize beneficial functions. Such an endeavor poses philosophical and ethical issues of defining life itself, and hence scholars have been led to challenge and reconsider how such definitions influence science. The versatility of these definitions attests to the overall significance of molecular biology, taking its application beyond academic study into the sphere of daily life, e.g., medicine and ecology (Bich and Green, 2016). As molecular biology progresses, so too does its reach extend to computational approaches for studying high-scale datasets resulting from modern biological inquiry. The increasing reliance on computational biology makes the role of programming skills among scientists even more important. For example, Python is now an essential tool in carrying out complex analyses ranging from structural bioinformatics to genomics, which allows researchers to interpret and visualize molecular interactions effectively. The integration of computational approaches not only reinforces the analysis of data but also fosters interdisciplinary research, further expanding the field of molecular biology. Therefore, careers in this field in the present era need a combination of computational and biological abilities, focusing on the significance of cross-training at advanced education levels (Ekmekci et al., 2016) (Figure 2.2).

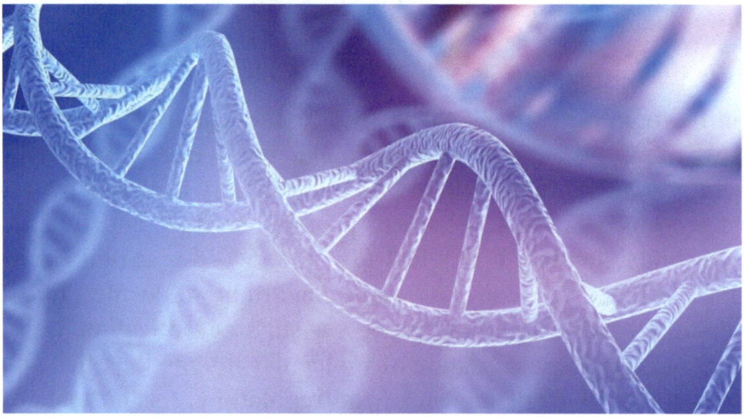

Figure 2.2. Molecular biology.

Source: https://www.news-medical.net/image-handler/ts/20220926055406/ri/950/src/images/news/ImageForNews_726082_16641860447993001.jpg.

Historical Background and Key Discoveries

The history of molecular biology has been dominated by key milestones that have shaped its fundamental concepts. At the midpoint in the 20th century, James Watson and Francis Crick's identification of the double helix structure of DNA, supplemented with Rosalind Franklin's X-ray diffraction images, provided significant insights into the molecular structure of genetic material. Such a ground-breaking discovery enlightened the mechanism of base pairing involved in the replication of DNA and transcription, providing the foundation for present-day genetics. Developments in molecular techniques, such as the invention of recombinant DNA technology in the 1970s, also fueled the field further. The innovations enabled scientists to cut and paste genes from one organism into the genome of another and thus construct GMOs and usher in an era of biotechnology. The interplay among earlier breakthroughs and technological developments has thus irrevocably shaped the discipline of molecular biology. The second key moment of this historical narrative is the beginning of epigenetics, a field of study of heritable epigenetic alterations in gene expression not encoded in the sequence of the DNA itself. Beginning as early as the early 20th century, researchers such as Conrad Hal Waddington began challenging the determinist nature of genetic inheritance. His work laid the groundwork for interpreting advanced developmental interactions that could influence phenotype outside of the genetic code. This paradigm was further energized by the work of researchers like Mary Lyon and controversy at key conferences that reiterated the value of extranuclear developmental processes (Costa and Frezza, 2014).

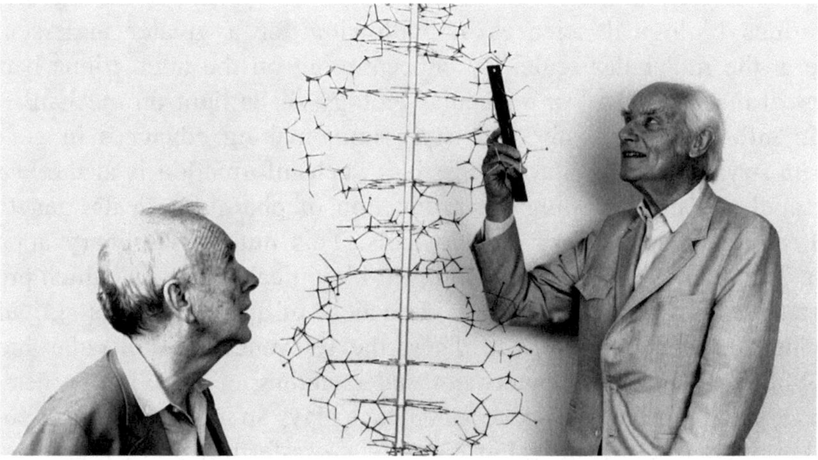

Figure 2.3. Watson and Crick posed in the early 1990s with a model like the one they made to first describe the structure of DNA.

Source: https://www.cshl.edu/wp-content/uploads/2023/03/crick_watson_DNA_model.jpg.

Such a recognition of the gene-environment interaction was a far cry from the classical models of genetics and deepened our understanding of biological regulation, describing the dynamic nature of molecular biology as an evolving science. Moreover, the philosophical implications of reductionism have caused critical debate in both biological and physical sciences, emphasizing the need to go beyond the shallow conceptualization of scientific explanation. Traditional reductionism sought to explain complex systems in less complex terminology by breaking them down into their fundamental components, a thesis not always straightforward to apply to empirical research in molecular biology.

The explanation provided by writers like Ernest Nagel has shaped current discussions, highlighting the different forms of reduction that concern genetics and molecular activity (Hüttemann and Love, 2016). This tension between holistic and reductionist philosophies is a reminder of pluralism in biological sciences and invites the pluralistic approach that can appreciate the subtle interplay between molecular events and broader biological contexts. These historical and philosophical perspectives lead to a more fertile understanding of molecular biology's conceptual foundations and its ongoing evolution (Figure 2.3).

Importance of Molecular Biology in Modern Science

Molecular biology is the core of modern science, serving as an intermediary for various biological sciences and allowing for a greater understanding of life at the molecular scale. By concentrating on the interactions between biological macromolecules, molecular biology sheds light on mechanisms of genetic information transfer and expression, making advances in genetics, biochemistry, and microbiology possible. Such information is at the heart of many applications, including the production of pharmaceuticals, genetically modified organisms, and diagnostic tools. This multi-disciplinary approach generates innovative solutions to complex biological problems, which provide important insights that impact as wide a field of applications as agriculture, medicine, and the environment. Thus, the relevance of molecular biology cannot be overstated; it enhances our understanding of basic life processes as well as applied biological sciences and hence plays an important role in solving global problems significantly. Furthermore, cross-fertilization with philosophy indicates a growing sensitivity to the meta-scientific dimensions of research activity. An attempt to understand the significance of molecular biology in the broader context of the scientific enterprise leads philosophical thinking to require serious scrutiny of the methodology and ethical concerns of experimental biology. Intellectual evolution in this field emphasizes the necessity of interdisciplinarity, from transcending empirical observation to struggling with the philosophical underpinnings of scientific inquiry. Such integration allows scientists to cultivate an enrichment of understanding of molecular interactions and their implications for life, facilitating a balanced perspective that honors both scientific thoroughness and ethical responsibility. This interaction between philosophy and molecular biology enriches the scientific discourse and creates a wide intellectual environment necessary for groundbreaking advances (Serrelli, 2017). In addition, the trajectory of the history of the philosophy of biology demonstrates that it is essential to place molecular biology in perspective within the larger context of scientific development. The discipline has witnessed significant changes, particularly with the renewed organicist thinking that emerged in the beginning of the 20th century as a response to reductive strategies that had dominated previous schemes. The renaissance invites greater engagement of interdisciplinary methods, fusing talent from various branches of science to foster a more integrated understanding of biological phenomena. Understanding of how historical school philosophies correlate with contemporary molecular biology reinforces our knowledge of the evolution of the discipline and its role in crucially influencing current scientific paradigms. According to this view, the integration of knowledge from both ancient and modern contexts ensures molecular biology remains updated on the complexity of life science, thus guiding its course to solve current bioethical challenges and scientific concerns (Gawne and Nicholson, 2015) (Figure 2.4).

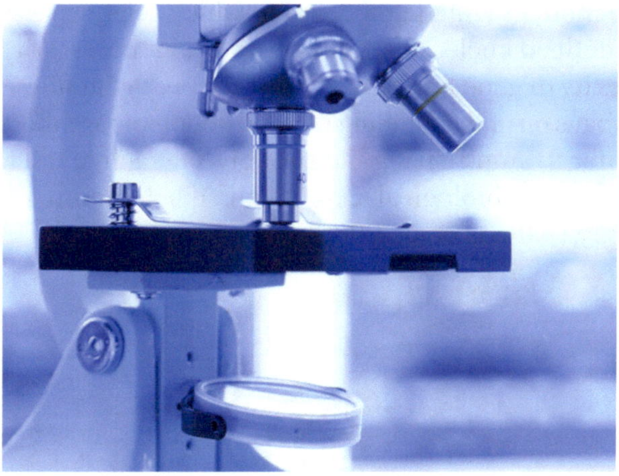

Figure 2.4. Importance of molecular biology

Source: https://d2jx2rerrg6sh3.cloudfront.net/image-handler/picture/2014/7/Molecular_Biology-620x480.jpg.

2.1. STRUCTURE OF DNA

The double helix nature of DNA is at the core of the understanding of its role in molecular biology, with its double helix formation providing the foundation for the storage and transmission of genetic information. James Watson and Francis Crick discovered the double helix structure in 1953, which describes the way two nucleotide strands are coiled over one another, each made up of a sugar-phosphate backbone with nitrogenous bases extending. The specific adenine-thymine and cytosine-guanine base pairing is necessary for the preservation of the fidelity of the genetic information. Besides stabilizing the DNA structure, it facilitates accurate replication upon cell division. The DNA structure is not merely for structural purposes; it enables the genome to be coiled and housed within the cell nucleus limits while having room for the specific interaction required for transcription, replication, and repair activities that are a part of all aspects of life processes. The knowledge of the complexity of DNA carries over to knowledge of its function and how it aids the overall concepts of genetics. The instrumental gene concept highlights the important roles of specific nucleic acid sequences in linking phenotypic characteristics to genotypes that underlie them. It is a unifying concept for experimental approaches that are dissecting gene expression and inheritance patterns as an operational entity that can be systematically manipulated within laboratory assays. In contrast, the nominal gene provides a more pragmatic model with which scientists can communicate about formally defined nucleotide sequences without necessarily having to deal with the complexity of gene function. Finally, the post-genomic vision illustrates a shift towards focus on understanding how the genome's architecture determines its functional effects, emphasizing the strongly interactive interaction between DNA architecture and biological expression, the leitmotif of contemporary molecular biology (Alberts et al., 2006).

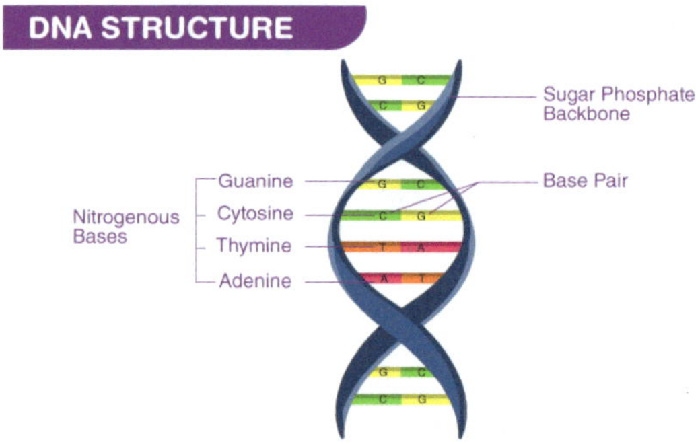

Figure 2.5. Structure of DNA.

Source: https://cdn1.byjus.com/wp-content/uploads/2023/07/DNA-Structure-Updated-1.png.

Progress in computational biology has made it possible for researchers to investigate deeper into DNA structure and function, paving the way for interdisciplinary investigations that combine biology and technology. Bioinformatics is the key development in genomic sequence analysis, where computational systems are employed to interpret complex data generated from sequencing technology. With molecular biology increasingly reliant on these computational resources, the ability to read enormous datasets – from individual nucleotides to entire genomes – is critical. Python, for instance, is now a focal programming language in this field, allowing sequence and structure analysis. With the simple-to-read syntax and powerful libraries of Python, it is possible for biologists to survey genomic databases and run computationally intensive analyses, thereby clarifying our understanding of DNA structure and its implications for evolutionary biology and genetic diversity. These interactions underscore the ways in which computational breakthroughs are revolutionizing our comprehension of molecular systems and placing increasing focus on the importance of DNA structure in the context of life sciences (Ekmekci et al., 2016) (Figure 2.5).

2.1.1. Double Helix Model and Its Significance

The double helix model of DNA, discovered by James Watson and Francis Crick in 1953, is a landmark in the field of molecular biology. The model describes the molecular structure of DNA as a double helix, where two strands are in opposite directions and held together due to the complementarity of bases. The consequences of this structure run deep: it not only provides a mechanism for the duplication of genetic information but also underlies the laws of inheritance through the specific transmission of genetic information. As demonstrated by advancements in genomic science, the specificity of base pairing ensures fidelity in DNA replication, which is necessary for genome integrity across generations. The double helix structure thus provides the foundation for a system within which the field of molecular genetics and most biotechnology enterprises exist, allowing us to comprehend heredity, mutation, and evolution. Aside from structural data regarding DNA, the double helix implications go beyond. This model gives the foundation for understanding a number of biological processes, including gene expression and gene regulation. The structural fluidity of DNA, as demonstrated in processes such as transcription, is heavily reliant upon the double helix structure, which enables uncoiling and re-coiling during RNA synthesis. The interaction between DNA and its epigenetic modifications that accompany it, including that which can be observed in the case of DNA methylation, demonstrates how the double helix is responsible for the intricate web of regulatory mechanisms that determine cellular functions and development (Mattei et al., 2022). The double helix model, therefore, not only symbolizes a groundbreaking scientific milestone but also marks a conceptual model that deepens our understanding of molecular biology. Additional research also explores the consequences of the double helix model for disease and drug development. Conditions such as cancer are correlated with alterations in DNA structure and function, underscoring the significance of the double helix in health and disease. Our knowledge of its three-dimensional structure and rules of base pairing provides us with the means to control the DNA model through techniques like CRISPR-Cas9 gene editing. With more understanding of the double helix, so too does its potential for the therapy of genetic disease and the development of target-specific therapies. The intersection of genetic sequencing technologies and computational biology allows researchers to map mutations and define their impact on phenotypes to a degree unprecedented before (Brabletz et al., 2021). Therefore, the double helix remains at the center of molecular biology, representing a landmark of the past and a vital tool for future scientific achievements (Figure 2.6).

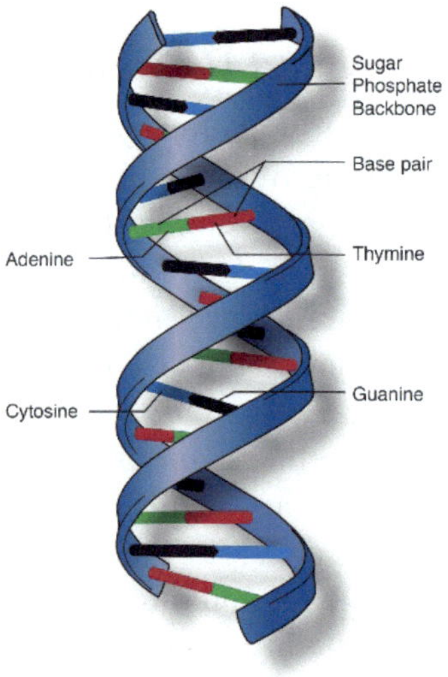

Figure 2.6. Double helix.

Source: https://cdn.storymd.com/optimized/ZoBxNMIvqM/original.jpg.

2.1.2. Nucleotide Composition and Base Pairing

The intricate relationship between nucleotide conformation and base pairing is the basis of molecular biology, giving the rationale for DNA and RNA function. Nucleotides, composed of a phosphate group, a sugar unit, and a nitrogen-containing base, are the building blocks of nucleic acids. In DNA, these components together form the adenine, thymine, cytosine, and guanine, which hybridize specifically through hydrogen bonding: adenine and thymine, and cytosine and guanine. This process of specific pairing is not a structural nicety, but it serves a significant function in enabling proper replication and transcription processes, thus ensuring genetic fidelity from generation to generation. The distinctive nucleotide content gives rise to differences in structural features, influencing the overall stability and functionality of the nucleic acids, as stressed in pioneering works stressing the indispensable role of base pairing in molecular interactions. RNA contains a more diverse nucleotide content due to the presence of uracil instead of thymine compared with DNA. This quirk, alongside modifying the forms of base pairing, which allows for the formation of more than one secondary structure, also plays a significant role in RNA's functional diversity. RNA's ability for participation in enzymatic activity and regulation of gene expression is significantly dependent on its structure, which, in turn, is determined by the exact combination of its nucleotides and their interaction through base pairing.

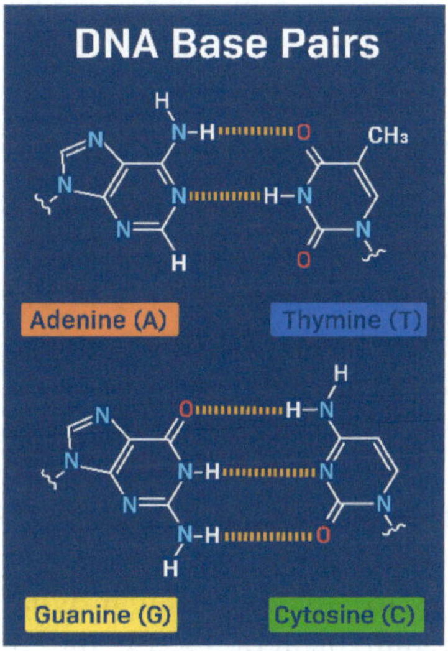

Figure 2.7. Base pairing in DNA. The 4 nitrogenous bases found in DNA are adenine (A), thymine (T), guanine (G), and cytosine (C).

Source: https://api.www.labxchange.org/api/v1/xblocks/lb:LabXchange:d2bb9765:lx_image:1/storage/slide-2-b4941df530b6fb840befcc5adde0f9fa.png.

It has been shown in recent research that RNA can also fold into complex three-dimensional motifs that determine its functional roles in the cell, explaining how nucleotide composition determines RNA dynamics. These observations contribute to an emerging understanding of molecular biology, revealing how fundamental properties of nucleotides organize biological functions at a cellular level, further highlighting the intricacies of base pairing strategies. The nucleotide structure and the base pairing interaction are not only a mere hereditary process;

they influence evolution and the origin of life on the cosmic scale. Prebiotic chemistry research has revealed that nucleotide-like compounds may have been spontaneously generated under specific environmental conditions to serve as a foundation for the evolution of replicating RNA molecules. These early-life forms took advantage of the inherent characteristics of nucleotide form to allow evolutionary modification, possibly to higher-level organisms. As explained in the research of NASA's exobiology, the hunt for biogenic molecules raises unequivocal questions about life's origin and molecular basis (Devincenzi, 2025). Further, ongoing research into the flexibility of RNAs and structural dynamics can potentially illuminate our understanding of evolution and functionality that will support our understanding of life's molecular architecture and its evolutionary importance (Aviran et al., 2018). Thus, research into nucleotide composition and base pairing is revealing important principles uniting science, life, and the universe (Figure 2.7).

2.1.3. Chromatin Structure and Organization

The intricate structure and organization of chromatin are necessary for the regulation of genetic information within the nucleus of eukaryotic cells. Chromatin, composed of DNA wrapped around histone proteins, occurs in two major forms: euchromatin and heterochromatin. Active transcription is typically associated with euchromatin because it facilitates gene expression and protein synthesis, while heterochromatin is densely packed and typically transcriptionally inactive. The dynamic aspect of chromatin structure plays a role in the regulation of gene expression, DNA replication, and repair processes. With post-translational modifications like histone acetylation and methylation, chromatin is able to respond to numerous cellular signals, modifying its structure as per the physiological requirement of the cell, thus affecting cellular function at more than one level. Recent advances in molecular biology have allowed one to understand chromatin structure and its effects on gene regulation much more critically. In particular, high-throughput methods enabled the mapping of chromatin states in different cell types and under different conditions. The usage of such advanced methods proves the dynamics of chromatin structure, demonstrating that its organization is not a mere passive framework but an extremely dynamic molecule capable of modification upon internal as well as external stimuli. For example, chromatin remodeling during cell differentiation is a critical event that establishes unique gene expression patterns, enabling the development of different cell types. Such evidence assures us of the fact that chromatin structure is at the core of comprehending larger biological paradigms like developmental biology and cellular response to environmental cues. In genomic medicine and cancer biology, information regarding chromatin structure and organization has long-lasting consequences. Chromatin element misregulation is involved in numerous diseases, including cancer, in which alterations to the chromatin landscape have the potential to lead to pathologic gene expression profiles. The World Health Organization Classification of Haematolymphoid Tumours, 5th edition, for example, encompasses numerous advances in the understanding of lymphoid neoplasms, with the pattern of chromatin modification having a featured role in tumorigenesis. Furthermore, ongoing endeavors like the Human Cell Atlas Project aim to establish the differing molecular profiles of cells in humans for a better understanding of chromatins role in both healthy and disease states. Through accurate characterization of chromatin structure, researchers are able to devise specific treatments aimed at the genetic and epigenetic modifications leading to malignancy, pointing to the utility of chromatin organization in molecular biological study (Figure 2.8).

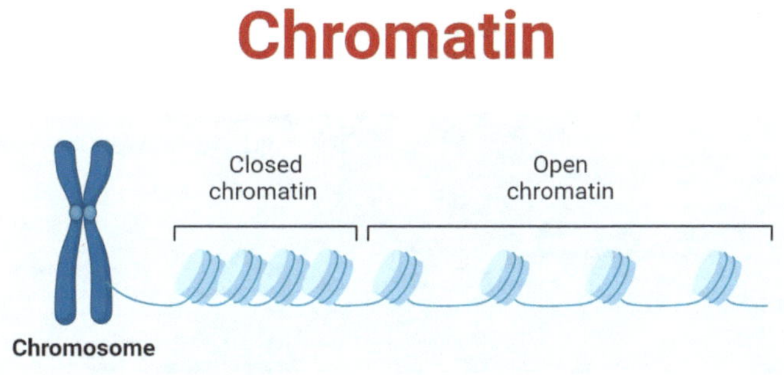

Figure 2.8. Chromatin.

Source: https://microbenotes.com/wp-content/uploads/2022/06/Chromatin.jpeg.

2.2. DNA REPLICATION

Replication of DNA is one of the central pillars of molecular biology since it involves the exact duplication of an organism's genetic information in readiness for cell division. Starting at specific sites referred to as origins of replication, the process involves unwinding the double helix structure of DNA using the assistance of helicase enzymes. This relaxation gives rise to single-stranded DNA templates that are required for the synthesis of new complementary strands. DNA polymerases are central to this, adding nucleotides to the growing strand in a complementary manner, adhering to strictly base-pairing rules. Accuracy is crucial here, particularly in early development stages, when genomic integrity must be maintained to facilitate cell differentiation and growth. Stability of DNA replication is exemplified by evidence that shows a compromise between genome size and complexity, adding the extreme limits for minimizing replication errors as species evolve (Mamun et al., 2020). The second key characteristic of DNA replication is the proofreading mechanism, which enhances fidelity in replication. DNA polymerases possess intrinsic proofreading capability, allowing them to identify and correct nucleotide-mismatched bases. If the wrong nucleotide is inserted, the enzyme has the ability to excise the incorrect base and insert the correct one before proceeding. This repair mechanism is vital, as errors in replication may cause mutations that can be transmitted to future generations and potentially create diseases, like cancer.

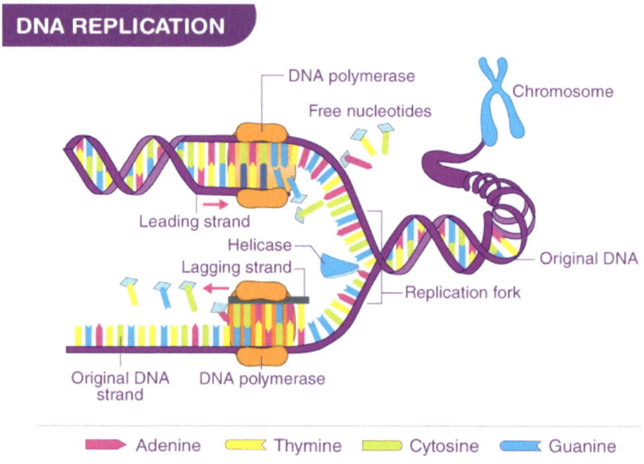

Figure 2.9. DNA replication.

Source: https://cdn1.byjus.com/wp-content/uploads/2023/07/DNA-Replication-Process-Machinery-And-Enzymes-Updated-2.png.

The importance of genomic stability cannot be overstated, especially in multicellular organisms whose growth and tissue maintenance depend on cell division. Such mechanisms are a starting point for further investigation of the mechanisms by which errors in DNA replication influence evolutionary processes, shaping adaptation and species diversity in the long term (Bandea, 2009). In addition to intrinsic components, there exist extrinsic factors that can influence the fidelity and efficiency of DNA replication. Environmental stressors such as radiation, chemical exposure, and biologic agents induce DNA lesions that block replication. These obstacles must be overcome by cellular repair mechanisms in order to ensure genomic integrity and faithful replication. Of special interest, cellular repair-DNA replication interactions form a network of intricate interactions that prevent genome aberrations. Scholars have speculated that such pressures to evolve have shaped processes of DNA repair and replication, driving adaptations in cellular organisms (Figure 2.9). The intricate dance of replication fidelity and environmental appropriateness not only accounts for the mechanisms in molecular biology but also highlights the dynamic state of evolutionary processes regarding genetic information stability and heterogeneity (Mamun et al., 2020).

2.2.1. Mechanisms of DNA Replication

DNA replication is a highly complex and coordinated mechanism that accurately duplicates the genome prior to cell division. This process is initiated with the unwinding of the double helix by the enzyme helicase, which straightens out the two strands of DNA. Single-strand binding proteins then stabilize the unwound strands, preventing them from re-annealing and allowing replication to proceed. After the strands are unbundled, primase places short RNA primers that serve as the beginning of new strands' synthesis. DNA polymerase then continues the new synthesis by adding complementary nucleotides. This stringently regulated sequence of events provides for the preservation of genetic information and its proper transmission to daughter cells as the foundation for cellular function and continuity in all living organisms. Fidelity of DNA replication is most stringent, since replicative mistakes generate mutations potentially with deleterious effects to organismal health and development. As a consequence, several proofreading activities have evolved to provide for replication fidelity. DNA polymerases possess exonuclease activity inherent to them, in that they can recognize and excise mismatched nucleotides during synthesis. This proofreading is crucial, particularly since a single base pair of misincorporation can have serious implications for cellular processes. Apart from this intrinsic proof-reading, there are other repair mechanisms that are activated after replication to correct any errors that have escaped initial detection. These complex mechanisms collectively contribute to an overall replication fidelity rate of approximately 1 error per billion nucleotides, highlighting the evolutionary importance of maintaining genome integrity across generations, particularly during rapid development when fidelity is critical (Mamun et al., 2020). Moreover, the importance of DNA replication mechanisms also traverses evolutionary timelines. The evolutionary trajectory of the organisms is determined by the mechanism of replication, especially genome size and genetic complexity. In adaptation to the organisms' environment, mutations during the process of DNA replication may be a substrate for natural selection. The balance between genome size and architectural complexity, with expression in terms of an inherent constant, reflects a trade-off that influences evolutionary fitness. As the complexity of the organism increases, it is essential that the mechanisms of DNA replication are robust enough to facilitate adaptations that promote survival. Here, the knowledge of DNA replication mechanisms not only illuminates cell functioning but also symbolizes the broader implications of genetic precision and evolution in systems across nature (Bandea, 2009) (Figure 2.10).

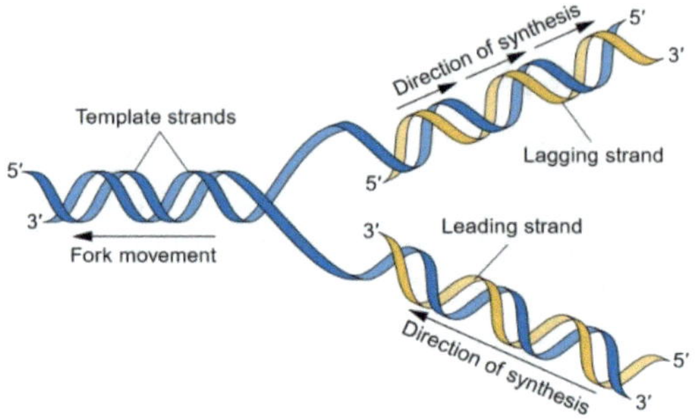

Figure 2.10. DNA replication at the growing fork.

Source: https://istudy.pk/wp-content/uploads/2020/04/mecha.jpg.

2.2.2. Role of Enzymes in Replication

Enzymes play a central role in DNA replication, a fundamental process in molecular biology ensuring that genetic information is properly duplicated and transferred to offspring. At the center of the process lies DNA polymerase, an enzyme responsible for building new strands of DNA through the addition of complementary nucleotides to the template strand. This enzyme acts in a highly coordinated manner, proofreading and fixing errors to maintain fidelity. There are other enzymes, such as helicase and ligase, that also facilitate replication. Helicase unwinds the double helix, yielding two replication single strands, while ligase closes gaps between Okazaki fragments on the lagging strand to form a complete DNA strand. The process of concerted interoperation among such enzymes is a model for the fidelity of cellular replication and a testament to the critical role played by enzymatic activity in maintaining genetic integrity through the generations. Furthermore, dynamic enzyme function in DNA replication reflects general principles of molecular biology. Thus, coordination of multiple enzymatic activities within a multi-enzyme complex optimizes efficiency as well as accuracy. Enzymes such as primase, which synthesizes short RNA primers necessary for DNA polymerase to initiate synthesis, illustrate how specialized function contributes to the replication process. The role of these enzymes is not simply catalysis; they also sense cellular signals and environmental messages. This adaptability signifies that enzymes are far more than passive catalysts, but participants actively engaged in the regulation of replication, and as such symbolize the fine balance between structure and function at the very essence of biological systems. Hence, acquiring a knowledge of the activities of these enzymes increases our awareness of appreciation of replication mechanisms as well as molecular biology in general. The emergence of enzymatic machinery for DNA replication also emphasizes some of the dominant themes in biology and the evolutionary origins of life. Paring with some theoretical predictions, such as those examining the evolution of cellular compartments, the emergence of a complex replication apparatus might be an expression of evolutionary compromises to environmental pressures and demands for efficient genetic fidelity in early life forms. This adaptive framework implies an underlying linkage between the co-evolution of various forms of life, whether viruses and cellular life, and implies that replication accuracy was central to life diversity. These notions represent debates that are presently relevant in molecular biology regarding how fundamental processes like replication contribute to our understanding of evolutionary history and the interrelatedness of life across time. Against this background, replication enzymes not only demonstrate the breathtaking sophistication of molecular biology but also remind us of their primacy in the organization of life itself, as evidenced in controversy over theories of life's evolution and the origins of cellular domains (Bandea, 2009; G et al., 2004) (Figure 2.11).

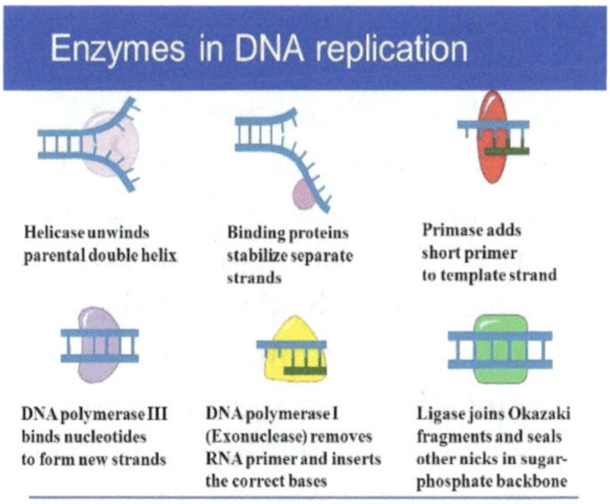

Figure 2.11. Enzymes in DNA replication.

Source: https://i.ytimg.com/vi/zpdpE3DKMro/maxresdefault.jpg.

2.2.3. Importance of Fidelity and Error Correction

The fidelity of molecular processes is absolutely essential in biology, serving as the foundation for the proper transmission of genetic information from parent to offspring. A replication, transcription, or translation mistake can cause drastic alteration in protein structure and function and therefore have pathological effects on cellular metabolism and organism development. Error-correction mechanisms are thus critical to ensuring the stability of genetic information in the presence of the sheer number of potential errors that can be created in a variety of molecular processes. These mechanisms ensure fidelity in the genetic code, enabling organisms to thrive despite environmental pressures and mutations. The presence of mechanisms to enhance fidelity and correct errors is essential to cellular operations for robustness, allowing organisms to evolve and adapt to change and minimize the risks of genetic mutation. In molecular biology, error correction is more than just straightforward processes of repair; it is a staggeringly intricate set of interactions that preserve life's complexities. For instance, the evolutionary development of error correction processes can be linked to Shannon's rate distortion theory that investigates trade-offs between information fidelity and energy cost of transmitting genetic information. This view allows scientists to visualize the stability of genetic networks as a function of their capacity for coping with uncertainties of genetic transmission. Through the use of error-correction mechanisms, cells not only maintain genetic stability but also leverage the variation that emerges as a driver of innovation. As such, information on these interactions is what sheds light on how organisms balance stability and flexibility to survive in constantly evolving environments (Glazebrook and Wallace, 2011). Furthermore, the lesson from processes of error correction also extends to complex systems theory, where genetic networks and quantum models of information can be likened. Developments in the research on complex networks have shown some of the similarities between molecular transmission of information and quantum entanglement, and it has been suggested that our biological theory of error correction may also be informed by lessons from quantum physics. As the domains of these diverse fields overlap, scientists look into how information transmission changes with conditions, shedding new insights into the evolutionary advantage of robust error correction mechanisms. Interdisciplinary in nature, the focus here is that fidelity and error correction are not separate phenomena but a part of a larger biological system, needed to uncover the primary principles that govern molecular biology and evolutionary dynamics (Biamonte et al., 2019).

2.3. TRANSCRIPTION PROCESS

The transcription process is the fundamental process in molecular biology by which the genetic information contained in the DNA is transcribed into messenger RNA (mRNA). The process of transcription begins with the attachment of RNA polymerase to a DNA region known as the promoter, causing the DNA double helix to unwind. The transcription apparatus proceeds to synthesize a single strand of RNA complementary to the template DNA strand. This is a vital process because this is where the genetic code gets translated from a stable storage form to a dynamic transcript that can then be processed further. Transcription not only facilitates gene expression but also fills the gap between the static form of genetic material and useful proteins synthesized in the cell, and therefore plays a central function in cellular operation and the regulation of biological processes. During the ongoing process of transcription, several regulatory elements and transcription factors regulate the fidelity and effectiveness of RNA synthesis.

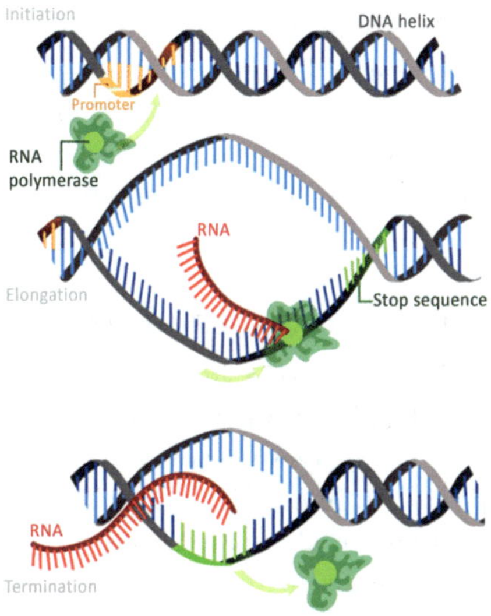

Figure 2.12. Transcription occurs in the three steps—initiation, elongation, and termination—all shown here.

Source: https://s3-us-west-2.amazonaws.com/courses-images/wp-content/uploads/sites/110/2016/06/13153312/steps.png.

These include enhancers and silencers that influence transcription rate, as well as general transcription factors that help associate the transcription complex with the promoter. This interaction is necessary to guarantee that the correct genes are being turned on and off at the correct times and quantities in response to various stimuli and developmental signals. The regulatory intricacy of transcription has significant implications in developmental biology and genetics domains since alterations in transcriptional regulation can result in variations in phenotype or even disease. Consequently, the transcriptional regulation becomes the center of much attention in present research on gene expression and its applicability to organismal biology. Further elaboration of gene dynamics in transcription shows the complexity of its information-bearing capability. Whereas genes used to be viewed as simple hereditary units, contemporary perspectives contradict this by introducing the complication of transcription and its position in a more comprehensive informational matrix. As explained in biosemiotics, genes are not simply sequences of nucleotides; they are signs reading complex interactions in a biological system (El-Hani et al., 2006). This semiotic structure demands a critical understanding of the process of transcription as an information transmitter of the genetic code, noting how the process of transcription intertwines structure and function within the genome. Further, the ongoing evolution of the gene concept—ranging from instrumental, to post-genomic definitions—also makes our understanding of gene expression and regulation ever more elaborate, correlating structural aspects of DNA with their respective functional implications in organisms (Alberts et al., 2006) (Figure 2.12).

2.3.1. Overview of Transcription and Its Stages

Transcription plays an important role in gene expression as the initial step whereby some regions of DNA are translated into RNA. This process ensures that the genetic code is precisely transcribed, which allows proteins to be synthesized in order to carry out cellular functions. Transcription occurs in three primary phases: initiation, elongation, and termination. In the initiation process, RNA polymerase binds to the promoter region of a gene with the assistance of various transcription factors. This marks the initiation of the transcription process and leads to unwinding of the double helix of DNA. Once initiation is over, the elongation phase follows, during which RNA polymerase travels along the DNA template, forming an RNA complementary strand. The highly controlled process is crucial, as errors can lead to abnormal protein synthesis, influencing cell functioning and organism survival. The elongation phase of transcription is where RNA polymerase incorporates RNA by adding ribonucleotides base paired with the DNA template strand. The phase is active, and RNA polymerase facilitates phosphodiester bonding between nucleotides, forming an extended RNA strand. The importance of precision at this level is underlined by the precision with which RNA polymerase replicates RNA to preserve the genetic information. In addition, various accessory proteins, such as elongation factors, regulate the rate and processivity of transcription since they participate in stopping transcriptional stalling. This cycle ends when the RNA polymerase reaches a termination signal, which results in the shutdown of RNA synthesis. This coordination demonstrates the intricacies of molecular interactions that define transcription, underscoring its pivotal role in the broader scheme of gene regulation and expression. With the arrival at the termination stage, transcription ends with the release of the newly synthesized RNA molecule from the elongation complex. Such cleavage is typically followed by the action of 5' cap removal and is of major significance in producing mature messenger RNA (mRNA) molecules. Following cleavage, additional modifications such as 5′ capping and polyadenylation of the 3' end occur, which are crucial for mRNA stability and subsequent translation. The significance of these events cannot be overstated since they boost the nuclear export of mRNA and protect it against degradation, ultimately resulting in successful protein translation. The intricacies of these stages, such as initiation, elongation, and termination, are a continuum of the broader topic of molecular interactions that form the core of the science of molecular biology. Clarifying these mechanisms provides deeper insights into the control of gene expression and identifies potential avenues towards the understanding of numerous various genetic diseases and disorders under the domain of ongoing research in molecular genetics (Costa and Frezza, 2014; Mamun et al., 2020) (Figure 2.13).

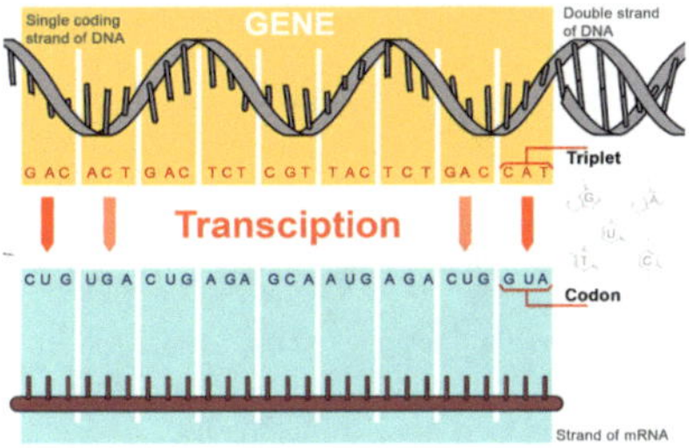

Figure 2.13. Overview of Transcription. Transcription uses the sequence of bases in a strand of DNA to make a complementary strand of mRNA. Triplets are groups of three successive nucleotide bases in DNA. Codons are complementary groups of bases in mRNA..

Source: https://s3-us-west-2.amazonaws.com/courses-images/wp-content/uploads/sites/110/2016/06/13153134/transcription.png.

2.3.2. Role of RNA Polymerase and Transcription Factors

RNA polymerase is the central activity of gene expression and cellular metabolism, and it is within this core that the essential enzyme for RNA synthesis from a DNA template is found. The enzyme initiates transcription when it binds to gene promoter sequences in a process that can be regulated by many other factors. Regulation of this process is largely performed by transcription factors, which assist in RNA polymerase binding to the distinctive DNA sequences that are needed for initiation. Such proteins can activate or inhibit RNA polymerase's binding with DNA, enhancing or inhibiting transcription within the cellular environment. Upon cells responding to internal or external cues, the interaction of diverse transcription factors offers a responsive and dynamic response, depicting the fine balance of molecular interactions governing gene expression (Locati et al., 2019). Thus, RNA polymerase and transcription factor knowledge is crucial in understanding the complexities of gene regulation. Once recruited to the promoter region by transcription factors, RNA polymerase undergoes a conformational change required for initiation of transcription.

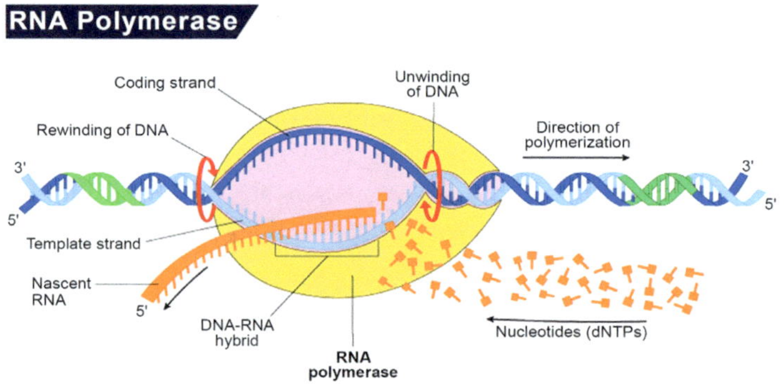

Figure 2.14. RNA polymerase.

Source: https://www.sciencefacts.net/wp-content/uploads/2023/09/RNA-Polymerase.jpg.

The transition from initiation to elongation is one in which the DNA double helix is opened, and ribonucleotides are added, basically transcribing RNA complementary to the DNA template. Transcription factors not only cause this transition but also are charged with significant functions throughout the elongation phase since they interact with the RNA polymerase complex. They can control the rate of transcription and modulate the functioning of enzymes and, therefore, impact the level and timing of mRNA formation. Regulation is carried out by various mechanisms that may respond to environmental signals, indicating that RNA polymerase and transcription factors are at the center of gene expression plasticity in terms of fluctuating cellular conditions (Cryan et al., 2019). Moreover, the association of the transcription factors with RNA polymerase does not stop at initiating transcription; it influences splicing, capping, and polyadenylation of the derived mRNA as well. Transcription factors can have other cofactors brought in that facilitate these processing reactions, thereby making the mature RNA translation-ready. This multi-functional role underscores the importance of transcription factors not just as initiation regulators but also as integral components of the entire process of gene expression. As research continues to uncover the complexities of RNA polymerase dynamics and interactions with transcription factors, their importance in various biological processes—e.g., development, cell differentiation, and stress response—grows. Detection of these interactions in health and disease can lead to therapeutic innovations, bridging basic molecular biology to translational research in biomedical research (Locati et al., 2019) (Figure 2.14).

2.3.3. Post-Transcriptional Modifications of RNA

The process of post-transcriptional RNA modifications is critical in molecular biology, serving as a primary regulatory mechanism that has an impact on gene expression. After transcription, pre-messenger RNA (pre-mRNA) is exposed to a series of processes that permit its maturation into functional messenger RNA (mRNA). These include capping at the 5 end, 3 end polyadenylation, and splicing to remove introns. The 5′ cap, a guanine nucleotide addition, protects the RNA from degradation and is required for ribosome recognition upon translation. On the other hand, polyadenylation adds a poly(A) tail, enhancing stability and helping nuclear export of the mRNA too. These operations ensure that only appropriately processed mRNA is used to make proteins, thus ensuring cellular homeostasis and fidelity of gene expression (Cavalli, 2011). Besides capping and polyadenylation, which are the essential modifications, splicing is another layer of post-transcriptional control. Alternative splicing offers a mechanism by which one gene may produce multiple versions of proteins, further increasing the complexity and diversity of the proteome. Alternative splicing requires intron removal and exons spliced together to produce mRNA isoforms with possibly distinct functional characteristics. Splicing is regulated by many splicing factors and non-coding RNAs that act in concert with the spliceosome, a molecular machine responsible for the execution of splicing. Splicing errors lead to the production of abnormal proteins and are increasingly being recognized as a causative factor in numerous diseases, including cancer and neurodegenerative diseases, and highlight the significance of post-transcriptional modification in cellular function and integrity (McKinsey et al., 2018). In addition to splicing, RNA editing is another critical post-transcriptional modification that further enhances gene expression diversity. This is achieved through enzymatic nucleotide modification of specific RNA nucleotides, leading to the production of proteins with different amino acid sequences than those synthesized by the original DNA. RNA editing may affect the stability, localization, and translational efficiency of mRNA, with downstream effects on the functional implications in cells. The role of RNA editing is not restricted to neuronal plasticity, immune function, or even evolutionary adaptation, indicating that post-transcriptional modifications are not only second-order processes but also key contributors to gene regulation. As advancements continue in molecular biology techniques and next-generation sequencing, the realm of RNA modifications continues to reveal intricate regulatory networks that compel us to re-evaluate our understanding of genetic expression and cell behavior (Cavalli, 2011) (Figure 2.15).

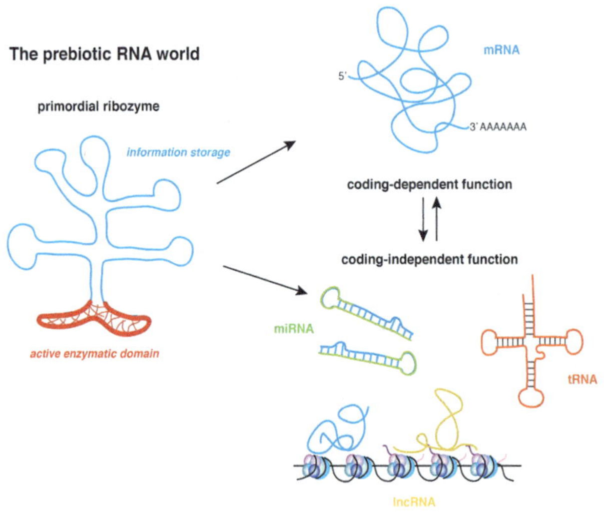

Figure 2.15. Post-transcriptional modifications of RNA.

Source: https://www.frontiersin.org/files/Articles/403882/fnmol-11-00304-HTML/image_m/fnmol-11-00304-g001.jpg

2.4. TRANSLATION MECHANISM

Within molecular biology, the translation mechanism is a crucial process of gene expression in which messenger RNA (mRNA) is being translated into proteins. This highly detailed process is located within the ribosome, a sophisticated molecular machine composed of ribosomal RNA and protein. Translation proceeds when the ribosome attaches to the mRNA molecule, enabling the corresponding of codons—three nucleotides corresponding to an amino acid—to their respective transfer RNA (tRNA) molecules. Each tRNA molecule carries a specific amino acid, allowing stepwise polypeptide chain assembly. The fidelity and efficiency of translation are controlled by a variety of factors, including the availability of mRNA, tRNA, and ribosomal subunits, and the presence of different translation factors. This poised coordination enhances the accuracy of protein synthesis, which is not only important for cellular function but also for the expression of phenotypic characteristics in multicellular organisms (Hershey et al., 2013). Along the way, there are multiple discrete phases—initiation, elongation, and termination—each playing an essential role in determining the ultimate outcome of translation. During initiation, the ribosomal subunits associate with the mRNA and position the start codon at the P site where the first tRNA is bound. Elongation follows, during which there is a sequential arrival of tRNA molecules for each codon of the mRNA. This involves forming peptide bonds between consecutive amino acids, hence growing the polypeptide chain. Next, the termination step is triggered when a stop codon is reached, releasing the finished protein and leading to ribosome breakdown. Importantly, translation complications extend beyond the mere fact of peptide formation; they involve control systems that ensure proteins are only made when needed to respond to adaptive reactions in the organism, which is a testament to the strong relationship between gene expression and complexity in biological systems (Bandea, 2009).

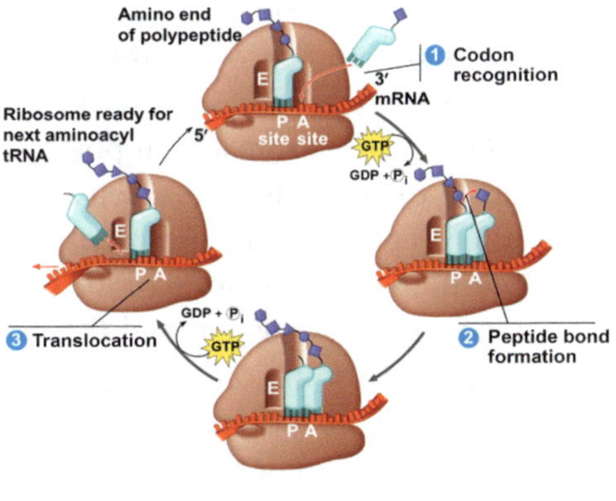

Figure 2.16. Mechanism of translation.

Source: https://www.researchgate.net/profile/Ritika-Gupta-21/publication/318422578/figure/fig1/AS:516017036431364@1500039456868/Mechanism-of-translation.png.

Recent developments underscore the relevance of translation in evolutionary processes, particularly from viral biology and parasitism perspectives. Viruses as obligate host cellular machinery dependents for replication pose an evolutionary paradox in molecular biology. Their ability to hijack the translation machinery of host cells poses a fascinating interaction between viral and cellular life forms. As has been argued in new research, viruses can be defined as molecular organisms due to the fact that they need hosts to express their genetic information as proteins (Bandea, 2009). Not only does this definition add more to our understanding of viral evolution, but it also raises critical questions about the definition of life itself. In line with this, research in translation mechanisms deepens our understanding of both the fundamental principles of molecular biology and the dynamic waltz of life at a molecular level, emphasizing the interconnection between various biological entities in evolutionary history (Hershey et al., 2013) (Figure 2.16).

2.4.1. Overview of the Translation Process

Translation is a vital and intricate process in the protein synthesis process where genetic information stored in messenger RNA (mRNA) is translated into polypeptide chains, which ultimately fold into functional proteins. Situated within the ribosome, the process starts with the binding of mRNA onto the small subunit, facilitating the correct selection of the transfer RNA (tRNA) molecules that bring in individual amino acids. Each tRNA is linked to a codon formed by three nucleotides of the mRNA sequence. Codon-anticodon specificity ensures the accuracy of protein synthesis, an essential aspect of cellular function. Misinterpretation of the genetic code leads to the production of aberrant proteins, which can contribute to numerous diseases. The effectiveness and regulation of the translation process thus ensure cellular homeostasis and coordination of the response to environmental stimuli. The elongation phase of translation is particularly important as it involves sequential addition of amino acids according to the position of the moving ribosome along the molecule of mRNA. Once a tRNA has entered the ribosome at the A site, peptidyl transferase catalyzes the formation of peptide bonds to link amino acids together in a growing chain of polypeptide. While this is happening, the ribosome itself also moves, advancing the mRNA by one codon. Not only is this an energy-demanding process in the guise of guanosine triphosphate (GTP), but it also involves several initiation and elongation factors that enhance the rate and accuracy of translation. An understanding of these operations is essential to deciphering how cells regulate protein synthesis in conformity with metabolic need and environmental stimulus. Further, current research indicates potential therapeutic uses in the correction of translation mistakes, thereby highlighting the importance of the process in high-level molecular biology. The termination of translation marks protein synthesis finality, which is triggered by the recognition of stop codons in the sequence of mRNA. As opposed to regular codons, there are stop codons without associated tRNAs that result in the recruitment of release factors, enabling the disassembly of the ribosomal complex and the release of the newly synthesized polypeptide. This step illustrates the subtleties of regulation of gene expression, with the control of translation playing a role in the availability as well as the functionality of proteins in the cell. In addition, comprehension of the termination process reflects broader paradigms within molecular biology, including the interaction between translation and transcription control. Additional philosophical examination of these interconnected cellular events may reveal new approaches for addressing generic issues of biomedical research; for instance, the understanding that enhancement of our understanding of translation could solve drug development problems, as described by Armiento et al. (2020). The interconnectedness of these molecular processes emphasizes the value of core knowledge in guiding therapeutic approaches (Figure 2.17).

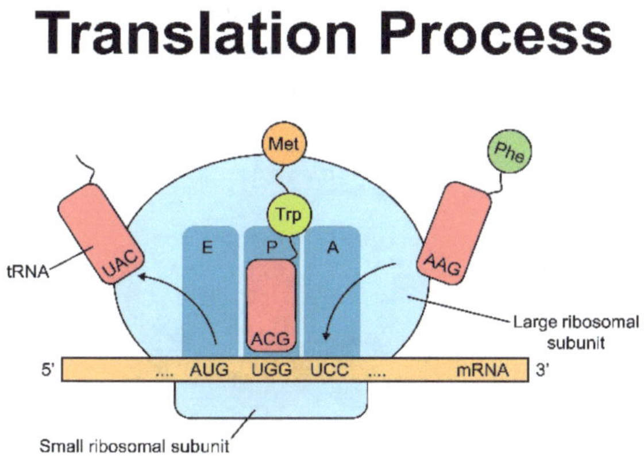

Figure 2.17. Translation process.

Source: https://ecdn.teacherspayteachers.com/cdn-cgi/image/format=avif,quality=70,onerror=redirect/thumbitem/Translation-Process-of-mRNA-Molecule--7742227-1716300075/750f-7742227-1.jpg.

2.4.2. Role of Ribosomes and Transfer RNA (tRNA)

Ribosomes are the central part that is active in translation, being an essential step in gene expression by which mRNA gets translated into a polypeptide chain. These macromolecular complexes are made up of ribosomal RNA (rRNA) and proteins, organized in two subunits: the large and small subunit. The small subunit participates in the reading of the mRNA sequence, while the large subunit participates in the catalysis of the peptide bond creation between amino acids, thus allowing protein synthesis to occur. This multifaceted process relies heavily on the accuracy and efficiency of ribosomes to ensure that the correct polypeptide is synthesized according to the genetic code. An appreciation of the significance of ribosomes in cell biology reinforces the reliance of the genetic material on the apparatus that deciphers it, reinforcing the fundamental nature of the ribosomal process in all organisms. tRNA is a major adapter molecule in the process of translation, matching single amino acids with codons for them within the mRNA chain. One tRNA molecule contains an attached amino acid at one end and an anticodon arm at the other end that is complementary to the specific mRNA codon. This formation allows tRNA to place the correct amino acid in the correct position for addition to the growing polypeptide chain. The efficiency and fidelity of the translation process largely rely on the interaction between ribosomes and tRNA. When the ribosomes function well in collaboration with tRNA, they enhance protein synthesis rates, which, in turn, influence cellular processes and growth in organisms. Such collaboration between ribosomes and tRNA reflects not only a dynamic but also tightly regulated process, central in the maintenance of cellular homeostasis and adaptation to external stimuli.

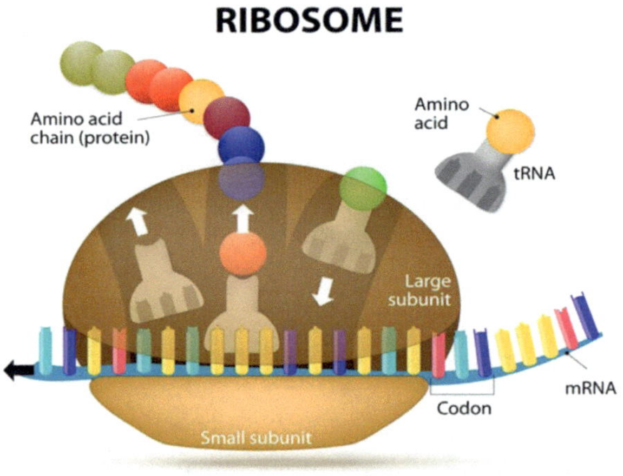

Figure 2.18. Ribosome.

Source: https://www.news-medical.net/image-handler/picture/Ribosome%20during%20protein%20synthesis.%20The%20Interaction%20of%20a%20Ribosome%20with%20mRNA.%20Process%20of%20initiation%20of%20translation-Designua_thumb.jpg.

The activities of ribosomes and tRNA extend beyond translation, engaging them in broader cellular operations like regulatory equipment and stress response mechanisms. A number of characteristics have recently been described to link polyamines with the control of translation, emphasizing the intricate world of control within cells. Polyamines are involved in the modulation of translation through the modification of ribosomal stabilizing factor function and tRNA binding, suggesting more intimate interactions among metabolism, gene regulation, and cell function. Modification of either ribosomal function or tRNA activities may have far-reaching cellular consequences, including altered transcription, translation efficiency, and cell viability. Besides, such aberrations are linked with other disease conditions, including cancer and neurodegenerative diseases, highlighting the role of ribosomes and tRNA in maintaining cellular integrity and health (Hutchison et al., 2016; Miller-Fleming et al., 2015) (Figure 2.18).

2.4.3. Post-Translational Modifications (PTMs) of Proteins

The PTM theory explains a crucial aspect of protein biology whose effects have substantial repercussions on protein function, stability, and interactions in cellular contexts. These are the post-translational modifications that occur after protein synthesis and involve a variety of chemical modifications, such as phosphorylation, acetylation, methylation, and ubiquitination. All these modifications have the potential to fundamentally alter the conformation and activity of proteins and allow cells to regulate processes such as signaling pathways, metabolic processes, and gene expression in response to different stimuli. This dynamic regulation is vital, as in vascular health, where oxidative post-translational protein modification plays a central role in the control of vascular smooth muscle cell behavior under oxidative stress and consequently modifies the pathophysiology of conditions like hypertension (Camargo et al., 2015). Additionally, the interaction between oxidative stress and post-translational modification indicates the intricacy of redox signaling in cells. During hypertensive conditions, increased production of reactive oxygen species (ROS) favors oxidative modifications that can adversely affect cellular function. For instance, cysteine residues of vascular proteins undergo modifications like S-nitrosylation and S-glutathionylation, which are able to regulate signaling pathways involved in smooth muscle cell growth and vascular remodeling. The evolution from a contractile to a proliferative phenotype in smooth muscle cells in response to such changes underscores the clinical potential that can be gleaned from an understanding of PTMs in vascular biology (Camargo et al., 2015). Thus, elucidating the mechanisms behind such changes provides insight into the cellular shifts present with stress and disease. The significance of post-translational modifications extends beyond short-term cellular reactions; it is an essential component of developmental processes and pathogenesis in other tissues, including cardiac development. More recent studies established correlations between chromatin remodeling, epigenetics, and PTMs, such that similar regulatory systems managing gene expression can influence protein behavior. For instance, signal-induced crosstalk between long non-coding RNAs and histone modifications reorganizes the chromatin architecture and contributes to cardiac hypertrophy and fibrosis pathogenesis. Such intricate interactions demonstrate that post-translational modifications are not only crucial for the regulation of individual proteins but also a component of the cellular regulation and response repertoire as a whole, further confirming their status as components of a central molecular biology concept (Figure 2.19).

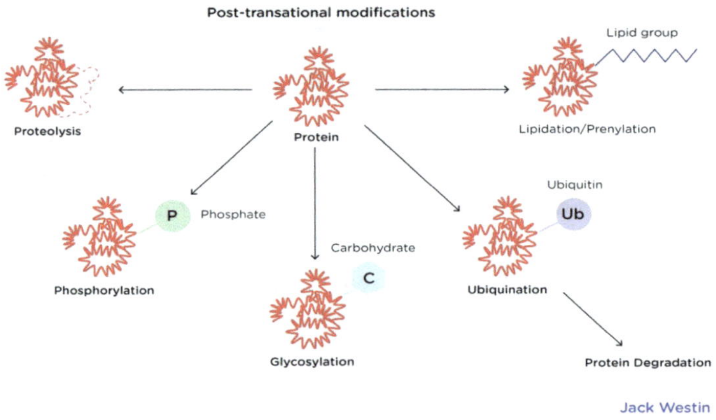

Figure 2.19. Post-translational modifications of proteins.

Source: https://i2.wp.com/cms.jackwestin.com/wp-content/uploads/2020/02/Post-Translational-Modifications.jpg?resize=1025%2C668&ssl=1.

2.5. GENE REGULATION

Gene regulation is a dynamic and complex process at the center of molecular biology, governing gene expression in response to cellular state, developmental signals, and environmental signals. A cell's ability for gene expression regulation allows for an efficient response to changing conditions, ensuring survival and function. This regulation occurs at various levels, including transcriptional control, translational modulation, and post-translational modifications. Transcription factors, for instance, will bind to specific DNA sequences to enable or inhibit the recruitment of RNA polymerase, thereby controlling the gene transcription. These processes demonstrate the interplay between biological specificity and information, where specific interactions define the output of genetic information in an environment within a cell (Griffiths et al., 2017). It is necessary to elucidate these regulatory systems to appreciate the intricacy of cellular behavior and genetic network interdependence. Gene regulation at the molecular level is closely coupled with chromatin structure and epigenetic marks that are pivotal in regulating gene access. DNA folding into chromatin affects not only transcription factor binding but also three-dimensional genome organization within the nucleus. There are a number of changes, such as methylation and acetylation, that can flip gene expression modes on and off, activating or deactivating particular genes according to what the cell requires. Such control reveals that gene expression is not static but rather a facultatively adaptive process which reacts to internal and external stimuli. The concept of the gene must therefore shift as our understanding of its function in a regulon shifts, both in terms of the sequence of DNA itself, but generally in terms of gene interaction and regulatory contexts (Alberts et al., 2006). Furthermore, advances in gene regulation are significant for biomedicine, particularly in disease treatment and diagnosis. Dysregulation of mechanisms is most commonly involved in a wide variety of disorders, ranging from cancer to genetic syndromes. For instance, mutations in transcription factors or chromatin conformational changes can lead to tumorigenesis-guided patterns of aberrant gene expression. Thus, targeting such regulatory pathways offers new therapeutic prospects, enabling the development of strategies to correct normal gene function. By studying gene regulation multi-dimensionality, scientists can determine causality between genotype and phenotype, allowing for intervention more precisely tailored to a person's genetic makeup. This is further reinforced by the necessity of studying gene regulation in molecular biology terms because this connects genetic information systematically to cellular processes and organismal health.

2.5.1. Mechanisms of Gene Expression Regulation

Gene expression regulation mechanisms are critical to the dynamic interaction between various biological processes and cellular homeostasis. Central control of transcription is the intricate regulation of transcription, where RNA polymerase must bind to the promoters in the presence of a variety of transcription factors. These transcription factors are molecular switches that initiate or suppress transcription based on cellular conditions and environmental signals. In addition, chromatin remodeling and positioning of the nucleosome are primary mechanisms to render the DNA accessible to transcription machinery. This degree of control guarantees that the genes are expressed in a timely and tissue-specific manner, translating environmental signals into respective cellular responses. These complexities underscore the simple concept that gene expression is not a linear process but rather a beautiful orchestration of various regulatory inputs, underscoring the magnificence of molecular biology in orchestrating biological processes. Post-transcriptional processes also contribute to the regulation of gene expression so that the resulting mRNA is efficiently processed and modified before translation. Splicing, editing, and polyadenylation are decisive processes that influence mRNA stability and translational effectiveness. For instance, alternative splicing allows a single gene to code multiple protein isoforms, enhancing the proteome diversity without the need for additional genetic information. MicroRNAs (miRNAs) also play a decisive role in post-transcriptional regulation by binding to target mRNAs, which results in their degradation or translation inhibition. These processes show biological specificity since they bring about finely tuned reactions to cell and environmental variations, enabling organisms to exist and be in a state of adaptation with changing conditions.

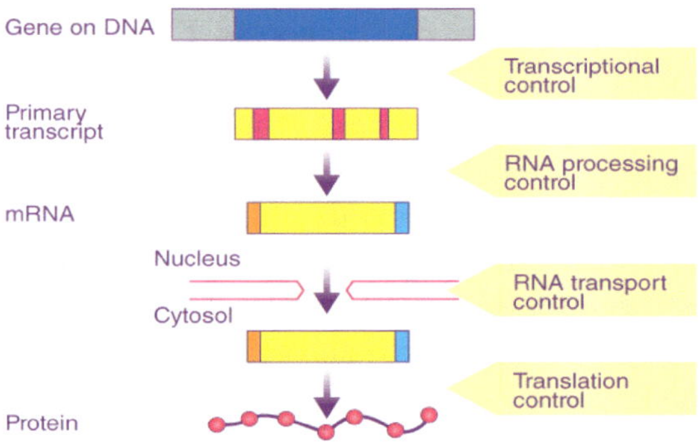

Figure 2.20. Regulation of gene expression.

Source: https://cdn1.byjus.com/wp-content/uploads/2018/01/Gene-Regulation-3.png.

As noted, biological specificity is a feature of the molecular level of organizing information, reiterating the significance of gene regulation to organismal integrity (Griffiths et al., 2017). At the translational level, gene expression regulation is still a strong driving force behind protein function and synthesis. Translation initiation is a highly regulated process involving many initiation factors that bind and recognize the mRNA 5′ cap. They facilitate the integration of the ribosomal machinery and proper translation. In addition, RNA-binding proteins, for example, have the ability to control the translation of specific mRNAs in accordance with cellular context, thus dictating which proteins are produced under a given set of conditions. Gene concept, particularly in the post-genomic context, emphasizes how genome function and structure become intertwined to generate phenotypic determinations. This observation illustrates the complexity of gene regulation, which illustrates effective gene expression as a balance of regulatory checkpoints at different levels of biology, substantiating the interrelatedness of molecular biology (Alberts et al., 2006) (Figure 2.20).

2.5.2. Role of Enhancers and Silencers

The intricate regulation of gene expression is imperative to cellular functioning, and enhancers and silencers are imperative regulators in this dynamic process. Enhancers are cis-acting regulatory DNA sequences that, after binding by specific transcription factors, enhance transcription of homologous genes, typically located at considerable distances from genes they regulate. This is facilitated by the three-dimensional organization of DNA, which brings enhancers close to promoters to facilitate effective transcription initiation. The importance of enhancers is placed in perspective by their ability to regulate genes in a tissue-specific manner, thus contributing to the diversity of cell types within multicellular organisms. The nuanced and elegant roles played by these components illustrate the precise regulation of gene expression to meet the developmental and physiological requirements of an organism. Silencers, conversely, are repressive factors that negatively regulate gene expression. These DNA sequences recruit repressive transcription factors ahead of time that occlude the assembly of the transcriptional machinery at a gene's promoter. Spatial organization of the chromatin near silencers also guarantees that transcription is sufficiently blocked, which is essential to ensure cellular identity and prevent aberrant gene expression. A good case in point is the silencing of certain heat shock proteins under non-stressful conditions, demonstrating how silencers contribute to homeostasis through regulation of the organism's response to environmental conditions. The combination of enhancers and silencers demonstrates the duality of gene control because it highlights the role of both activation and repression in the attainment of specific control of gene expression according to the organism's needs. Current studies emphasize the significance of enhancers and silencers in understanding complex diseases as well as developmental disorders. Enhancers, for instance, which are distally located in relation to gene promoters, may harbor mutations related to a range of pathologies, as suggested by their involvement in atrial fibrillation. A genomic locus at 4q25 has been suggested to be risk-increasing by virtue of its regulatory control of the PITX2 gene, crucial in cardiac development. The functional connection between such regulatory factors and target genes unravels the molecular underpinnings of disease mechanisms. As ongoing research increasingly refines our understanding of the roles played by enhancers and silencers in gene regulation, particularly by modern genomic technologies, they are not only passive onlookers but actively involved players in coordinating genetic networks, as suggested in research like that summarized in Alonso et al. (2016) and Eilbeck et al. (2021) (Figure 2.21).

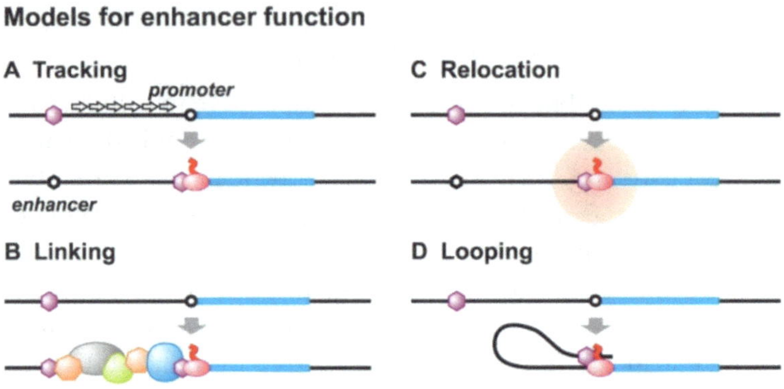

Figure 2.21. Existing models for the function of enhancers.

Source: https://biologynotesonline.com/wp-content/uploads/2024/10/image-297.png.

2.5.3. Epigenetic Factors Influencing Gene Regulation

Epigenetic mechanisms play a pivotal role in gene regulation, influencing gene expression without modifying the underlying DNA sequence. These include DNA methylation, histone modification, and RNA-associated silencing, all of which can dynamically regulate chromatin structure and accessibility. For instance, the methylation of DNA will hinder the binding of transcription factors, in effect silencing some genes. Conversely, histone acetylation tends to promote gene expression by loosening the DNA-histone interactions, making the genetic material more accessible for transcription. Such intricate regulatory mechanisms illustrate how epigenetics goes beyond traditional genetic determinism, presenting a nuanced layer of control essential for various biological processes, including development and cellular differentiation. The historical journey of epigenetics highlights its significance in understanding biological development. They originated with early 20th-century scientists and progressed with the integration of genetics and embryology, as in Waddingtons foundational ideas. This interdisciplinary beginning provided a platform where gene regulation was not just considered as a linear event dependent on genetic data, but an interaction between external and internal factors. This shift in paradigm occurs again in the works of leaders like Mary Lyon, who emphasized the impact of extranuclear factors in developmental events. Studying these elements reveals the intricacy of gene regulation, propelling the field toward novel therapeutic applications and insights into evolutionary biology (Costa and Frezza, 2014). Present studies emphasize the capacity of epigenetic modifications to be inherited by future generations, significantly impacting the following generations. This heritability is expressed through processes such as genomic imprinting and parent-of-origin effect gene silencing, which suggest that environmental influences can cause gene expression changes that are retained over generations. The implications of this are significant, as epigenetic tags could explain the inheritance of some traits, diseases, and susceptibilities in populations. Furthermore, understanding these mechanisms upholds the causal specificity theory in biological systems, in that molecular entities' interactions are inherently linked with particular biological functions. Next, epigenetic factor research not only deepens our understanding of molecular biology but also compels us to reconsider the role played by environmental influence in the regulation of genes (Griffiths et al., 2017) (Figure 2.22).

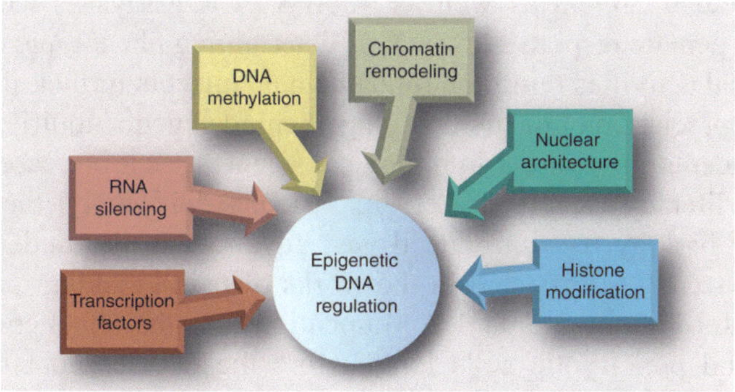

Figure 2.22. Factors influencing epigenetic regulation of DNA information.

Source: https://www.researchgate.net/profile/Quaovi-Sodji/publication/221894485/figure/fig1/AS:339972369469444@1458067137625/Factors-influencing-epigenetic-regulation-of-DNA-information.png.

2.6. MOLECULAR TECHNIQUES IN BIOLOGY

The study of molecular techniques in biology has introduced new mechanisms of understanding cell processes and genetic mechanisms and, therefore, defined the foundation of modern-day biological sciences. Methods such as polymerase chain reaction (PCR), gel electrophoresis, and DNA sequencing have revolutionized our method of analyzing genetic materials quickly and precisely. These techniques enable one to amplify specific DNA sequences, isolate nucleic acids by size, and determine nucleotide sequences, respectively. Simple information gathered from these methods enables advancements in fields ranging from genetics to biotechnology to become a reality, as their profound impacts on other scientific disciplines justify the far-reaching impact of molecular biology. For instance, the identification of genetic markers, besides aiding in our understanding of heredity, has been pivotal in the development of targeted therapy and personalized medicine, which is a paradigm shift in the practice of healthcare. The significance of molecular approaches is not only in the basic research but is interwoven with applied biological sciences as well as with educational frameworks. For example, experimental learning that engages students in molecular experimentation builds critical thinking and hands-on skills required in scientific inquiry. A good example of an educational intervention was a multi-week lab where students investigated Mendel's stem length phenotype in pea plants and linked traditional genetics to modern molecular methods. Given the circumstances, students utilized gibberellic acid treatment as well as dCAPS PCR assays as techniques for investigating genetic responses as well as determining phenotypic expressions. These practical activities reinforce fundamental concepts further, in addition to promoting appreciation of the methodology behind genetic inquiries, ultimately preparing students for a career in biological sciences, as it increases their level of scientific literacy (Kaplinsky et al., 2015). Additionally, the integration of molecular biology has introduced vast advances in the understanding of eukaryotic systems, as reflected in the works of path-breaking cell biologists like Edouard Chatton. He laid a foundation for the distinction between eukaryotic and prokaryotic cells long before the development of advanced techniques of microscopy, emphasizing the significance of cellular constituents in reproduction mechanisms and mitosis processes. Further work employing up-to-date molecular biology methods confirmed Chatton's own early hypotheses concerning chromosome organization and cell division. The development of these sciences has substantiated historical records, and the ability to possess a more comprehensive knowledge of cell and genetic behavior has become a reality. This ongoing discussion between past and present techniques enhances our understanding of fundamental biological concepts and inspires innovative research practices while highlighting the importance of technological advancement in biological science (Soyer-Gobillard, 2010).

2.6.1. Overview of Techniques Such as Polymerase Chain Reaction (PCR) and Gel Electrophoresis

The PCR is an innovative molecular biology method that facilitates the amplification of a target DNA sequence and makes it possible for scientists to create millions of copies of an infinitesimal piece of genetic material. The process involves the use of a thermocycler, which alternates temperature conditions to allow the denaturation of the DNA, annealing of the primer, and extension of the DNA polymerase. The specificity of PCR originates in the approach to primer design on either flank of the target sequence so that only the target sequence is replicated. PCR has become an essential reagent in numerous applications, ranging from cloning and sequencing to diagnosis and forensic analysis. For instance, its use in laboratory experiments allows students to connect theoretical concepts to experimental activity, as illustrated by its utilization in the determination of Mendelian characters in plants and emphasizing the significance of modern molecular techniques in teaching (Kaplinsky et al., 2015). Gel electrophoresis is a simple technique of separating and analyzing proteins and nucleic acids based on size and charge. This method uses a polyacrylamide or agarose gel medium into which charged molecules can migrate under the force of an electric field. Small DNA fragments will travel through the gel more easily than large ones, separating the samples into distinct bands. Visualization is then commonly achieved through intercalating dyes that insert between DNA base pairs, so that researchers can observe and examine the product under UV light. Not only is gel electrophoresis the foundational technique for PCR product confirmation, but it also acts as a central part of other applications such as genotyping, mutational analysis, and protein research. By connecting the separation of genetic material with broader scientific inquiry, gel electrophoresis allows for greater understanding of molecular variation and genetic expression. New developments, such as microfluidics, further extend the scope of capability of traditional techniques like PCR and gel electrophoresis. Microfluidic systems use tiny channels to manipulate minute amounts of fluid, which simplifies processes that were previously restricted to large laboratory environments. The integration of PCR in microfluidic platforms provides quick and effective amplification of DNA with minimal reagent expenses and sample amounts. The combination of PCR and gel electrophoresis in microfluidic devices enhances precision and throughput in various analyses, particularly in complex applications such as the surveillance of cancer (Azad et al., 2016). As scientists step into these cutting-edge technologies, the art of molecular biology evolves, closing gaps between theoretical education and innovative techniques, ultimately expanding the boundaries of genetic research and diagnosis.

2.6.2. Applications of Molecular Cloning

The applications of molecular cloning play an important role in the invention of various domains of molecular biology, particularly biotechnology. One of them is the production of recombinant proteins, e.g., insulin, which is vital for diabetes management. Molecular cloning techniques enable splicing of the human insulin genes into a bacterial plasmid, permitting mass production of this lifesaving hormone. This biotechnology application not only boosts the production of insulin but also reduces the reliance on animal sources, hence curbing the ethical concerns and production cost. Also, molecular cloning for designing biological systems uncovers avenues for the design of synthetic biology applications wherein intricate biological entities are customized for particular uses, such as drug discovery and simulation of diseases, illustrating the capabilities of molecular cloning to revolutionize healthcare and pharmaceutical industries. Molecular cloning has also been pivotal in agricultural biotechnology, especially in the development of genetically modified organisms (GMOs). By using cloning techniques, scientists can design crop traits such as resistance against insects, drought, and nutritional content. For instance, introducing bacterial genes into crops makes them resistant to certain insects and reduces the need for chemical insecticides, thereby promoting sustainable agriculture. Additionally, the applications of molecular cloning also include epigenomics because the technique can be utilized to engineer gene expression profiles in transgenic plants in such a manner that scientists can uncover concealed mechanisms of plant development and adaptation to environmental stresses. These advancements underscore the broad implications of molecular cloning in addressing global food security problems and improving agricultural productivity for a growing population. In medicine, molecular cloning facilitates the development of new therapeutic technologies, such as vaccines and gene therapies. Through cloning, researchers can isolate targeted disease-causing genes, thus deriving targeted vaccines with high immune responses. An example is the manufacturing of recombinant vaccines from cloned viral genes that have revolutionized preventive health regimens. Additionally, molecular cloning plays a very important part in gene therapy by providing the means to deliver normal genes into patients' cells to correct genetic flaws. With continued advancement in synthetic biology, the application of engineered biological systems has the potential to create advanced treatments tailored for each patient, ultimately changing the paradigm of personalized medicine. As highlighted in the growing body of literature, applications of molecular cloning are poised to transform agricultural and medical biotechnology, highlighting its inherent relevance to modern biology (Baianu et al., 2004; Heinemann et al., 2006).

2.6.3. Importance of CRISPR Technology in Gene Editing

The discovery of CRISPR technology revolutionized the field of gene editing, taking molecular biology to a new era of precision and affordability. Originally derived from a natural defense mechanism present in bacteria, CRISPR-Cas systems enable researchers to make specific DNA sequences for the first time (N/A, 2017). Not only does such capability enable gene editing on a targeted scale, but it also enables us to gain more insights into gene function and regulation. As a general-purpose technology, CRISPR can be employed across diverse fields, ranging from agriculture, where it is applied to generate crops with improved traits like disease resistance or drought tolerance. This application points to the significance of the technology in addressing the issue of global food security, highlighting its dual purpose in promoting scientific wisdom and practical problem-solving in contemporary biology. The ethical considerations of CRISPR technology are equally profound and necessitate serious examination. With the enhancement of the potential of gene editing, so does the potential for misuse or unanticipated consequences. The difference between traditional breeding and CRISPR-mediated modification raises questions regarding regulatory frameworks and public support for genetically modified organisms (GMOs) (N/A, 2017). Such nuances matter to the community of organic farming, which is concerned with balancing innovation with preservation of organic integrity. The stakeholders must engage in discussions regarding the permissible gene editing practices in harmony with organic methods to ensure transparency and consumer trust while leveraging the potential provided by CRISPR technology. Therefore, the social implications of CRISPR must be examined alongside its scientific advancement to facilitate responsible usage in various fields. The multiplicity of CRISPR in usage extends to the therapeutic frontiers, charting potential avenues for the cure of genetic disorders. By enabling precise mutation corrections causing illnesses, such as cystic fibrosis or sickle cell anemia, CRISPR can reshape the future of medicine significantly (Bonomo et al., 2018). Additional studies are still ongoing to refine delivery methods and eliminate off-target effects, which are all vital to ensure patient safety and treatment effectiveness. The development of CRISPR therapies holds not only the potential to broaden our understanding of genetic disease but also arms both patients and doctors with new instruments for effective intervention. So, the value of CRISPR technology in gene editing is more than mere research, advancing to actual health treatments that realize the real essence of molecular biology's potential to advance human health and welfare.

SUMMARY

In summarizing an examination of the fundamentals in molecular biology, it is more apparent that these concepts not only defend the complexity of life but also hold vital information regarding the machinery behind cellular function and interaction. Through this discussion, molecular biology has been made to show that genetic control, protein synthesis, and cellular structure organization are intricately linked to molecular biology. Myofibrillogenesis, for example, illustrates that the complex phenomena are controlled at the molecular level to maintain muscle function. This type of information is not just important in grasping normal biological processes but also significantly contributes towards unraveling pathologies resulting from disruptions of these fundamental processes, upholding the significance of research in this field (Kobeza, 2025). In addition, the interdisciplinary aspect of molecular biology, in which aspects of biology, chemistry, and physics are brought together, cannot be emphasized enough. Methods such as nuclear magnetic resonance (NMR) and electron paramagnetic resonance (EPR) are prime examples of this fusion, demonstrating how physical principles can be used to explain biological processes at the molecular level. The understanding of hyperfine interactions, for instance, elucidates complex behaviors and performances of molecular systems that enable one to explore further the interactions that sustain life itself. As research continues, such methods are sure to be expanded, opening new horizons and uses in both fundamental research and clinical usage (Novotný et al., 2024). In brief, molecular biology is a broad subject that encompasses a vast array of ideas that come together to form the basis of life that we have. The intricate dynamics of molecular structure, genetic material, and cellular processes attest to the necessity of ongoing research and progress within this discipline. As we continue to make progress in our understanding of processes such as myofibrillogenesis and the mechanisms of molecular interaction, we are all the better equipped to advance treatment and eventually maximize health outcome. Thus, the foundational principles of molecular biology not only lay the groundwork for academic inquiry but also hold the key to revolutionary practical applications in medicine and beyond, underscoring the long-term relevance of this field in shaping our understanding of life itself.

Molecular biology relies upon its ability to explain the unseen processes of life at the molecular level, namely through the study of nucleic acids and proteins. Biological philosophy of information, which entails storage, transmission, and processing of genetic information, represents the foundation of the discipline. One key advance in understanding this concept is recognizing that biological specificity is directly related to the causal principles underlying molecular interactions. By examining where terms dealing with information in molecular biosciences originate, scientists have mapped how biological specificity directs causation interpretation in biological systems. Specifically, this way of thinking maintains that highly specific causal relationships are inextricably informational and thus bridge molecular biology to more abstract philosophical controversies surrounding causation and specificity in biological systems, which is elaborated in Griffiths et al. (2017). The interaction between genetics and protein synthesis highlights the centrality of the central dogma of molecular biology, which embodies the one-way flow of genetic information from DNA to RNA to protein. This paradigm not only outlines the general processes of replication, transcription, and translation but also names the intricate systems of regulation that manage gene expression. Molecular biology emphasizes that gene regulation is required to offer cellular responsiveness and adaptability to environmental cues. Furthermore, recognizing threshold concepts in areas such as genetics and biochemical pathways becomes essential in shaping educational frameworks. For example, identification and accentuation of these threshold concepts can facilitate deeper understanding of them by students and help bridge intrinsic challenges in comprehending complex biological themes, as observed in a recent pilot study where subjects reflected upon their learning experiences in various areas of study, including biological sciences, as discussed by Jamieson et al. (2011). Additionally, the emerging field of molecular biology recognizes the implications of biotechnological advancements in agriculture, medicine, and ecology, which promote a broader debate regarding ethical considerations. Genetic engineering, for example, CRISPR technology, is the epitome of the wide-ranging impact of molecular biology on applied sciences, enabling the precise modification of genetic sequences in organisms. But as the examination of these technologies takes place, so also does the raising of ethical questions, such as long-term environmental effects and implications for genetic diversity. As advances occur, the inclusion of ethical discussion in the study of molecular biology ensures that scientific advancement is balanced by careful consideration of its effects upon society. Thus, the development of a well-rounded understanding of both the scientific and ethical aspects of molecular biology is essential for the creation of appropriate stewardship of biotechnology innovation.

REFERENCES

1. Al Mamun, M., Albergante, L., Blow, J. J., Newman, T., et al. (2020). 3 tera-base pairs as a fundamental limit for robust DNA replication. https://core.ac.uk/download/322476058.pdf (accessed on 3rd October 2025).
2. Alaggio, R., Amador, C., Anagnostopoulos, I., Attygalle, A. D., Araújo, I., Berti, E., Bhagat, G., et al. (2022). The 5th edition of the World Health Organization Classification of Hematolymphoid Tumors: Lymphoid neoplasms. *Leukemia, 36*, 1720–1748. https://doi.org/10.1038/s41375-022-01620-2 (accessed on 3rd October 2025).
3. Alberts, B., Finta, C., Waters, C. K., Communi, D., et al. (2006). Genes in the postgenomic era. https://core.ac.uk/download/11921232.pdf (accessed on 3rd October 2025).
4. Alonso, M. E., Fernández-Miñán, A., Gómez-Skarmeta, J. L., Manzanares, M., et al. (2016). Long-range regulatory interactions at the 4q25 atrial fibrillation risk locus involve PITX2c and ENPEP. https://core.ac.uk/download/45446838.pdf (accessed on 3rd October 2025).
5. Anikeeva, P., Bao, Z., Benfenati, F., Bezanilla, F., et al. (2018). Roadmap on semiconductor-cell bio interfaces. https://core.ac.uk/download/287623639.pdf (accessed on 3rd October 2025).
6. Armiento, A., Gare, A., Hilfinger, A., Karmon, A., Lin, A., Ross-Gillespie, A., Sekikawa, A., et al. (2020). Analytic philosophy for biomedical research: The imperative of applying yesterday's timeless messages to today's impasses. https://core.ac.uk/download/237182315.pdf (accessed on 3rd October 2025).
7. Aviran, S., & Li, H. (2018). Statistical modeling of RNA structure profiling experiments enables parsimonious reconstruction of structure landscapes. https://core.ac.uk/download/323069317.pdf (accessed on 3rd October 2025).
8. Azad, M., Goudarzi, M., Mobarra, N., Sahmani, M., et al. (2016). Microchips and their significance in isolation of circulating tumor cells and monitoring of cancers. https://core.ac.uk/download/52205560.pdf (accessed on 3rd October 2025).
9. Baianu, I. C., & Glazebrook, J. F. (2010). Categorical ontology of complex systems, meta-systems and theory of levels: The emergence of life, human consciousness and society. https://core.ac.uk/download/pdf/332869.pdf (accessed on 3rd October 2025).
10. Baianu, I. C., Lin, H. C., Lozano, P., Prisecaru, V., et al. (2004). Novel techniques and their applications to health foods, agricultural and medical

biotechnology: Functional genomics and basic epigenetic controls in plant and animal cells. https://core.ac.uk/download/pdf/85962.pdf (accessed on 3rd October 2025).

11. Bandea, C. I. (2009). A unifying scenario on the origin and evolution of cellular and viral domains. https://core.ac.uk/download/pdf/288874.pdf (accessed on 3rd October 2025).
12. Bandea, C. I. (2009). The origin and evolution of viruses as molecular organisms. https://core.ac.uk/download/pdf/288872.pdf (accessed on 3rd October 2025).
13. Biamonte, J., De Domenico, M., & Faccin, M. (2019). Complex networks from classical to quantum. http://arxiv.org/abs/1702.08459 (accessed on 3rd October 2025).
14. Bich, L., & Green, S. (2016). Is defining life pointless? Operational definitions at the frontiers of biology. https://core.ac.uk/download/131211791.pdf (accessed on 3rd October 2025).
15. Bonomo, M. E., & Deem, M. W. (2018). The physicist's guide to one of biotechnology's hottest new topics: CRISPR-Cas. http://arxiv.org/abs/1712.09865 (accessed on 3rd October 2025).
16. Brabletz, S., Schuhwerk, H., Brabletz, T., & Stemmler, M. P. (2021). Dynamic EMT: A multi-tool for tumor progression. *The EMBO Journal, 40*. https://doi.org/10.15252/embj.2021108647 (accessed on 3rd October 2025).
17. Cavalli, G. (2011). From linear genes to epigenetic inheritance of three-dimensional epigenomes. https://core.ac.uk/download/30436052.pdf (accessed on 3rd October 2025).
18. Costa, R., & Frezza, G. (2014). Crossovers between epigenesis and epigenetics: A multicenter approach to the history of epigenetics (1901–1975). https://core.ac.uk/download/144472860.pdf (accessed on 3rd October 2025).
19. Cryan, J. F., O'Riordan, K. J., Cowan, C. S. M., Sandhu, K. V., Bastiaanssen, T. F. S., Boehme, M., Codagnone, M. G., et al. (2019). The microbiota-gut-brain axis. *Physiological Reviews, 99*, 1877–2013. https://doi.org/10.1152/physrev.00018.2018 (accessed on 3rd October 2025).
20. De Lucca Camargo, L., Dulak-Lis, M., Harvey, A., Montezano, A. C., et al. (2015). Redox signaling, Nox5 and vascular remodeling in hypertension. https://core.ac.uk/download/42367463.pdf (accessed on 3rd October 2025).
21. Devincenzi, D. L. (2025). Second symposium on chemical evolution and

the origin of life. https://core.ac.uk/download/pdf/42840896.pdf (accessed on 3rd October 2025).

22. Eilbeck, K., Logie, C., Lovering, R. C., Mungall, C., et al. (2021). Sequence ontology terminology for gene regulation. https://core.ac.uk/download/477691687.pdf (accessed on 3rd October 2025).
23. Ekmekci, B., McAnany, C. E., & Mura, C. (2016). An introduction to programming for bio scientists: A Python-based primer. http://arxiv.org/abs/1605.05419 (accessed on 3rd October 2025).
24. El-Hani, C., Emmeche, C., & Queiroz, J. (2006). A semiotic analysis of the genetic information. https://core.ac.uk/download/146502284.pdf (accessed on 3rd October 2025).
25. Gawne, R., & Nicholson, D. J. (2015). Neither logical empiricism nor vitalism, but organicism: What the philosophy of biology was. https://core.ac.uk/download/228135446.pdf (accessed on 3rd October 2025).
26. Glazebrook, J. F., & Wallace, R. (2011). "The frozen accident" as an evolutionary adaptation: A rate distortion theory perspective on the dynamics and symmetries of genetic coding mechanisms. https://core.ac.uk/download/pdf/289861.pdf (accessed on 3rd October 2025).
27. Griffiths, P. E., & Stotz, K. (2017). Biological information, causality and specificity – An intimate relationship. https://core.ac.uk/download/153326381.pdf (accessed on 3rd October 2025).
28. Heinemann, M., & Panke, S. (2006). Synthetic biology—Putting engineering into biology. https://core.ac.uk/download/pdf/12944386.pdf (accessed on 3rd October 2025).
29. Hershey, J. W. B., Polunovsky, V. A., & Sonenberg, N. (2013). Introduction to translation. https://core.ac.uk/download/323075444.pdf (accessed on 3rd October 2025).
30. Hutchison, C. A., Chuang, R. Y., Noskov, V. N., Assad-Garcia, N., Deerinck, T. J., Ellisman, M. H., Gill, J., et al. (2016). Design and synthesis of a minimal bacterial genome. *Science, 351*. https://doi.org/10.1126/science.aad6253 (accessed on 3rd October 2025).
31. Hüttemann, A., & Love, A. (2016). Reduction. https://core.ac.uk/download/199235464.pdf (accessed on 3rd October 2025).
32. Jamieson, S., MacAndrew, S. B. G., & Spedding, N. (2011). How was it for you? A cross-disciplinary study of 'troublesome knowledge' as identified by undergraduate students and lecturers in geography, medical science and psychology. https://core.ac.uk/download/228176308.pdf (accessed on 3rd October 2025).

33. Jarrah, A. S., Veliz-Cuba, A., Jarrah, A. S., Murrugarra, D., Castiglione, F., Hinkelmann, F., Pe'er, G. U. Y., et al. (2010). A mathematical framework for agent-based models of complex biological networks. http://arxiv.org/abs/1006.0408 (accessed on 3rd October 2025).
34. Kaplinsky, N. J., Kudish, P., & Schlag, E. (2015). An inquiry-infused introductory biology laboratory that integrates Mendel's pea phenotypes with molecular mechanisms. https://core.ac.uk/download/84122222.pdf (accessed on 3rd October 2025).
35. Kobeza, P. (2025). Myofibrillogenesis: Formation of the contractile apparatus of muscle cells (Literature review. Issues of integration between research disciplines). *Morphologic.* https://www.semanticscholar.org/paper/fef6035b9c386b0d05a41b95f678209510a98bdf (accessed on 3rd October 2025).
36. Locati, M., Curtale, G., & Mantovani, A. (2019). Diversity, mechanisms, and significance of macrophage plasticity. *Annual Review of Pathology: Mechanisms of Disease, 15*, 123–147. https://doi.org/10.1146/annurev-pathmechdis-012418-012718 (accessed on 3rd October 2025).
37. Mattei, A. L., Bailly, N., & Meissner, A. (2022). DNA methylation: A historical perspective. *Trends in Genetics, 38*, 676–707. https://doi.org/10.1016/j.tig.2022.03.010 (accessed on 3rd October 2025).
38. McKinsey, T. A., Morimoto, T., Poller, W., Turner, B. M., & Wang, Y. (2018). Epigenomes in cardiovascular disease. https://core.ac.uk/download/323078188.pdf (accessed on 3rd October 2025).
39. McKinsey, T. A., Vondriska, T. M., & Wang, Y. (2018). Epigenomic regulation ofheart failure: Integrating histone marks, long noncoding RNAs, and chromatin architecture. https://core.ac.uk/download/323077345.pdf (accessed on 3rd October 2025).
40. Miller-Fleming, L., Olín-Sandoval, V., Campbell, K., & Ralser, M. (2015). Remaining mysteries of molecular biology: The role of polyamines in the cell. *Journal of Molecular Biology, 427*, 3389–3406. https://doi.org/10.1016/j.jmb.2015.06.020 (accessed on 3rd October 2025).
41. N/A. (2017). Compatibility of breeding techniques in organic systems. https://core.ac.uk/download/161988586.pdf (accessed on 3rd October 2025).
42. Novotný, J., Komorovsky, S., & Marek, R. (2024). Paramagnetic effects in NMR spectroscopy of transition-metal complexes: Principles and chemical concepts. *Accounts of Chemical Research, 57*, 1467–1477. https://www.semanticscholar.org/paper/1774c3b6400d9a8266a2f951d1012cc14b3ac713 (accessed on 3rd October 2025).

43. Regev, A., Teichmann, S. A., Lander, E. S., Amit, I., Benoist, C., Birney, E., Bodenmiller, B., et al. (2017). The Human Cell Atlas. *eLife, 6*. https://doi.org/10.7554/elife.27041 (accessed on 3rd October 2025).
44. Serrelli, E. (2017). Meta scientific views: Challenge and opportunity for philosophy of biology in practice. https://core.ac.uk/download/131214315.pdf (accessed on 3rd October 2025).
45. Soyer-Gobillard, M. O. (2010). Edouard Chatton (1883–1947) and the dinoflagellate protists: Concepts and models. https://core.ac.uk/download/159084167.pdf (accessed on 3rd October 2025).

CHAPTER 3

BASIC PRINCIPLES OF CHEMISTRY FOR SYNTHETIC BIOLOGY

Contents

INTRODUCTION TO SYNTHETIC BIOLOGY

Synthetic biology involves the convergence of multiple scientific disciplines, engineering complex biological systems using engineering principles. It is an interdisciplinary field that integrates biology, chemistry, and engineering to control and redesign organisms for specific purposes. By harnessing the fundamental principles of chemistry, synthetic biology seeks to generate novel biological parts, devices, and systems, essentially blurring the line between life and designed life. The rigorous application of chemical concepts provides the ability to explain metabolic pathways and molecular interactions, facilitating specifically tailored changes in microbial cultures or plant systems. This way, scientists can develop remedies that solve principal world problems, such as biofuel production, waste treatment, and disease protection, thus demonstrating the discipline's transformative capacity in preserving the harmony of nature and promoting human health and trade. Among the foundational components of synthetic biology are DNA sequence manipulation, founded on sound chemical principles. DNA, which consists of nucleotides linked by phosphodiester bonds, is assembled by different biochemical techniques. These include polymerase chain reaction and ligation to construct or replicate DNA sequences. Moreover, chemical properties also play a key role in the synthesis of genetic circuits for use as control mechanisms within cells. These genetic circuits are able to react to external stimuli, thus creating organisms that can adapt to changing environments. Directed evolution, too, employs high-level chemistry to induce mutations and select desired characteristics, further enhancing the effectiveness of synthetic application. This sophisticated interweaving of chemistry and biology leads to innovation in biomolecule synthesis and manipulation of metabolism that is necessary for cutting-edge drug design and bioremediation uses. The practical relevance of synthetic biology goes far beyond mere theoretical investigation; it offers actual, significant benefits in medicine and environmental management. Developments like the production of extracellular vehicles (EVs) are novel diagnostic and therapeutic agents, an embodiment of the intricate dance of chemistry and biology in the nanoscale. With their targeting of EV cargo, researchers are beginning to study their use as biomarkers of disease and as drug-delivery vehicles more specifically (Welsh et al., 2024). Moreover, the transition towards green synthesis processes signals a shift in the paradigm of material production where bioinspired processes reduce environmental impact while increasing material efficiency (Singh et al., 2018). All these, put together, not only demonstrate the essence of synthetic biology but also its potential to drive sustainable solutions to address the intricate demands of our world. The future of synthetic biology lies in the capability to integrate rigorous chemical principles with biological processes, to allow for innovations appropriate to the multidimensional challenges that society is now confronted with.

Definition and Scope of Synthetic Biology

Synthetic biology's definition is inherently complex, as it crosses a variety of disciplines like biology, engineering, and chemistry. Synthetic biology, in its very essence, seeks to design and construct novel biological devices, parts, and systems and to reengineer existing, naturally occurring biological systems for useful purposes. To achieve this entails a sophisticated understanding of the underlying principles of chemistry, where manipulation of biological systems is fundamentally reliant upon chemical interactions. Utilizing these rules, scientists are able to design synthetic pathways replicating or complementing nature. Besides, the growing interest in understanding undirected self-assembly within biological systems further indicates the close alliance between chemistry and synthetic biology, since most self-assembled structures rely on high-energy transient conditions to perform work, hence the necessity for a robust chemical basis in this multidisciplinary discipline (Desai et al., 2018). The application of synthetic biology is not just to simple genetic alteration but also the creation of completely new organisms with engineered functions.

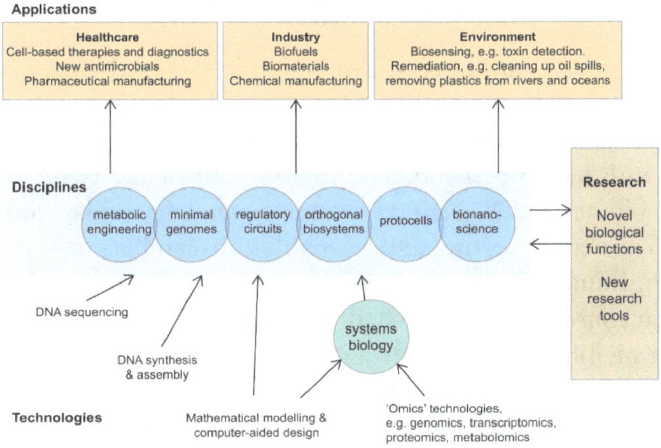

Figure 3.1. Overview of synthetic biology.

Source: https://www.researchgate.net/figure/Overview-of-synthetic-biology-Technologies-underpinning-synthetic-biology-bottom_fig1_355613736

This perspective calls for a holistic comprehension of biochemical processes that govern life. While scientists in synthetic biology explore the potential of genetically engineered organisms for applications such as biofuels, pharmaceuticals, and environmental cleanup, they are also confronted with the ethical and philosophical aspects of their research. The challenge is not so much in the technical implementation of these designs but in determining what constitutes "life." A lack of definitions for life complicates the field. Rather than abandoning definitions, scientists are encouraged to probe deeper into operational definitions that can support practical experimentation and facilitate novel ways of thinking (Bich and Green, 2016). To be able to understand the implications of synthetic biology, it is necessary to realize how it came to be and how it has developed. By bridging chemical principles into the functional framework of synthetic biology, there can be an enhanced understanding of life itself. In attempts at engineering life-like systems, synthetic biologists will resort to methodologies that mirror the processes constituting the natural biological realm. By investigating how chemical-driven dissipative processes may have contributed to the evolution of living systems, researchers can learn about the evolutionary pathways that led to life today. This research not only enhances the scientific community in synthetic biology but also informs the design of advanced materials and systems that can mimic biological functions, ultimately resulting in innovative applications in medicine, energy, and environmental sustainability (Desai et al., 2018) (Figure 3.1).

Importance of Chemistry in Synthetic Biology

The intricate relationship between synthetic biology and chemistry is central to understanding the design and engineering of biological systems. Synthetic biology relies heavily on chemical principles to manipulate biological components with high precision. The intersection of the two emphasizes the synthesis of biomolecules such as proteins, nucleic acids, and lipids. The ability to construct these molecules enables researchers to create artificial cells, a central interest in synthetic biology. Interestingly, innovations in microfluidic technology have pushed the field, allowing one to regulate the fabrication of membrane-based artificial cells with particular parameters, such as size and composition. Consequently, the innovations bear witness to the importance of chemistry in facilitating complex biological functions, unveiling understandings not only of how biological systems function but also of how they may be engineered to serve specific purposes, hence reasserting the central position of chemistry in synthetic biology (Elani, 2016). In addition, part of the rationale for the progress of synthetic biology stems from the open-sharing culture of scientific research, particularly that of chemistry. The shared pursuit for knowledge has brought to the fore the importance of accessible information to drive innovations in the field. Open access advocates see that open-sharing of chemical information makes research more efficient and effective, enabling an environment conducive to developments in synthetic biology. The intersection of cheminformatics—chemistry and data science—allows for the management and interpretation of big data sets, a key element in the design of synthetic systems. By promoting transparency in sharing methodology and outcomes, including negative outcomes, researchers can maximize experimental designs and synthetic pathways, consequently enhancing research quality in general and accelerating the advancement of new technologies in synthetic biology (Bird and Frey, 2013). Finally, the core foundation in chemistry serves as an anchor bulwark for the broadened series of synthetic biology applications, such as biotechnology, medicine, and environmental science. Chemical manipulation of interactions allows scientists to design new avenues to biosynthesis and metabolic engineering. Such a capability assists in the creation of biofuels, pharmaceuticals, and biodegradable materials, illustrating chemistry concepts in response to basic challenges of our world. The collaboration of chemistry and synthetic biology not only advances our understanding of nature but also enables researchers to respond to complex problems by discovering innovative solutions. As the field develops, ongoing research of chemical principles will undoubtedly shape the future of synthetic biology and create new methods and approaches that can revolutionize how we tackle sustainability and health (Elani, 2016).

3.1. FUNDAMENTAL CONCEPTS OF CHEMISTRY

The fundamental principles of chemistry are important in understanding synthetic biology because they contain the fundamentals that allow us to make biological changes at the molecular level. At the heart of this discipline are the laws of chemical reactions, which dictate how molecules interact and change in biological processes. An example is the concept of enzymes as catalysts that explains how certain proteins speed up biochemical reactions without being utilized in the process. This knowledge enables scientists to engineer artificial environments in which these reactions can be optimized for the production of desired biological compounds. Also, the rules of thermodynamics and equilibrium apply, enabling scientists to predict the performance and stability of biosynthetic processes. This chemical basis opens doors for new uses, from biosensors to biofuels, exemplifying the union of chemistry and biology in synthetic biology. Finally, the concept of self-organization in chemical systems is inherent in abiogenesis as well as synthetic biology. Autopoiesis, or the capacity for a system to reproduce and sustain itself independently, is a powerful theoretical framework for the study of life's origin and synthetic counterparts. Autopoiesis focuses on the mechanism of self-organization in order for complex biological systems to arise from simpler chemical precursors, thus implying the potential for artificial forms of life. Understandings of self-organization also better inform the development of synthetic biology applications, in which scientists engineer cells to take certain actions independently of external inputs. In accordance with newer research, the inclusion of self-organization theory is particularly useful in addressing the challenge posed by the very complexity of biological systems. By employing such ideas, synthetic biologists can innovate more and bridge gaps between fields, resulting in further advancement in the field. Finally, the integration of semiconductor-based materials into the domain of synthetic biology is a paradigm shift to link chemistry to biophysics. These materials make monitoring and control of biological entities possible, with the potential to open up new areas of research and application. Semiconductor devices, for example, are capable of creating dynamic interfaces to monitor cellular signaling processes in real time, and researchers can gain insights into the dynamics of cells that were previously not possible. The agenda for the use of these semiconductor-based technologies is on their power to promote biointerface design, brokering the cross-talk between electronic engineering and living systems. With effective use of these materials, scientists are capable of not only advancing the understanding of rudimentary biochemical dynamics but also revolutionizing the assembly of next-generation devices to be uniquely optimized for applications in bioengineering, validating the importance of fundamental chemistry principles in the advancement of synthetic biology (Figure 3.2).

Figure 3.2. Understanding Chemistry

*Source:*https://nailknowledge.org/wp-content/uploads/2021/04/chemical-1536x1024.jpg.

3.1.1. Atoms and Molecules: Building Blocks of Matter

The intricate universe of atoms and molecules constitutes the defining character of matter, which lays the basis for synthetic biology. Atoms are the tiny particles of chemical elements, which consist of a nucleus packed with protons and neutrons, with electrons surrounding them. Their atomic structure determines the properties and reactivity of materials. Molecules, formed by the joining of two or more atoms by covalent bonding, have diverse functionalities based on their structures. The ability to do this is due to the manipulation of these simple building blocks to form new biological systems and enhance existing organisms. Understanding these basic concepts is crucial, since the intricate interaction of atoms and molecules leads to the design of new biomaterials and drugs that can mimic or modify biological function. Further, the behavior of atoms and molecules relies extensively on how they interact and what they surround. Chemical bonding—the force holding atoms together—is varied, from ionic, covalent, and metallic bonds, among others. The character of the bond formed governs the reactivity and stability of the molecules. For instance, in the synthesis of pyrazoles, a class of compounds prevalently used in medicinal chemistry, structural nuances regulated by different bonding interactions can give rise to varied biological activities and reactivity patterns (Alam et al., 2019). This implies that the area of atomic and molecular interaction is not only abstract; it has practical applications, particularly in the formulation of targeted drugs and biochemical pathways that can effectively address complex health challenges in synthetic biology. Additionally, advances in understanding metal sulfides, particularly transition-metal sulfides, reveal the potential for large-scale applications of different atomic structures. These materials possess various chemical features as compared to their oxide counterparts, which render them more beneficial in applications such as catalysis and bioimaging (Dengo et al., 2017). Stabilized colloidal suspensions of such materials are crucial, as their unique surface chemistry defies the predictability observed in metal oxide behavior. Consequently, ongoing exploration of the reactivity and functionalization of metal sulfides confirms that the right interpretation of atomic and molecular principles is important to develop novel synthetic biology applications. These principles, alongside the optimization of existing methodologies, will also help to identify new compounds with the capability to further the field (Figure 3.3).

Figure 3.3. Atoms and molecules.

Source: https://i.ytimg.com/vi/V6-CkPAqmlc/maxresdefault.jpg.

3.1.2. Chemical Bonds: Ionic, Covalent, and Hydrogen Bonds

Synthetic biology provides the umbrella for understanding the fundamental ways of chemical bonding—ionic, covalent, and hydrogen bonds—to affect the discovery and design of biological molecules. Ionic bonds result from electrostatic attraction between charged ions so that complex biomolecule association occurs in cell structure and function. For instance, ionic bonding between chloride and sodium ions results in the formation of table salt, which is essential to osmotic balance in living things. Ionic association not just regulates biomolecule stability but also their biological activity.

The importance of ionic bonds in synthetic biology is also underscored by their use in designing biosensors, whose sensitivity and accuracy in recognizing biological signals through electrochemical interactions depend on the charge properties of molecules, as such challenges are addressed by employing the interfacing of sensing technology with nanomaterials (Naresh et al., 2021). Covalent bonds, in which pairs of electrons are shared between atoms, are fundamentally central to biological macromolecules, including proteins, nucleic acids, and polysaccharides, structure and function. The ability of carbon to form four covalent bonds makes a vast array of complex organic molecules for life possible. Covalent modifications, i.e., the addition of functional groups or the creation of cross-links, are used in synthetic biology to enhance biomaterial properties. For example, the change in polymer backbone structures by covalent bonding allows the creation of biocompatible materials that can efficiently interact with biological systems, resulting in better performance in drug delivery and biosensor applications.

This manipulation of covalent bonding also allows scientists to stabilize biomolecules in artificial pathways such that more stable and uniform chemical reactions are attained (Bullock et al., 2020). Hydrogen bonds, as weak as they are, play a crucial role in the structural formation of biological molecules, particularly nucleic acids and proteins. The accurate base pairing in DNA, stabilized by hydrogen bonds, is responsible for storing and transferring genetic information. Through synthetic biology, the utilization of hydrogen bonds can facilitate the design of smart biomaterials that respond to environmental stimuli and enhance their performance in drug delivery and molecular sensors, for instance. Furthermore, the dynamic nature of hydrogen bonds offers the flexibility and responsiveness of biomolecular interactions that govern the signaling pathways and metabolic pathways. Therefore, ionic, covalent, and hydrogen bond interactions are at the core not only to comprehend biological processes but also to design synthetic biological devices that tap into these laws for novel biotechnological applications (Naresh et al., 2021) (Figure 3.4).

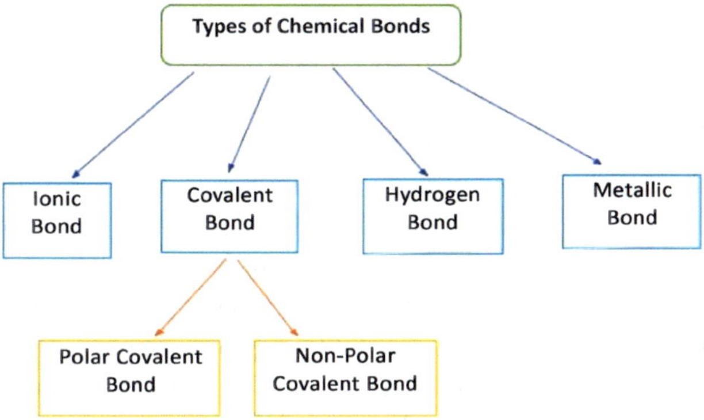

Figure 3.4. Types of chemical bonds.

Source: https://alevelchemistry.co.uk/wp-content/uploads/2019/11/Types-of-Chemical-Bonds.jpg.

3.1.3. States of Matter and Phase Transitions

The chemistry of state and phase transitions is at the heart of appreciating the fundamentals of chemistry that inform synthetic biology. Matter exists in various states—solid, liquid, gas, and plasma—each with specific molecular structure and energy levels. Transition between them involves drastic changes in structure and energy under conditions of temperature and pressure. For instance, melting transforms a solid into a liquid by the action of thermal energy that disturbs the ordered arrangement of the molecules. Cooling the liquid allows for crystallization, which reveals an ordered state of the solid. These transformations are not hypothetical but have significant consequences in synthetic biological systems, namely, how biological molecules will behave under varied states of their surroundings. Phase transition effects have consequences in the realm of microscopic interactions, particularly in biological organisms where molecular topologies play determinant roles. Glycan-lectin interaction insights suggest reaction punctuations as the outcome of experimental concentration gradients between reacting partners. Applying a biological renormalization technique derived from statistical physics to investigate phase transitions enables researchers to highlight a range of universality-class behaviors that are qualitatively distinct from those of conventional physical systems. It is apparent that synthetic biology is able to avail itself of these unique biochemical properties, developing logic gates in terms of molecular switches that enable complex biological processes. These discoveries not only merge chemistry and biology but also reveal fresh synthetic potential capable of copying intricate natural phenomena, thus perfecting our capabilities in biotechnology. Besides, the amorphous pathway-driven transition processes are also an example of the dynamic nature of biological systems that often tend to favor non-equilibrium states to enhance functionality. For instance, crystallization processes can be manipulated to enable organisms to regulate activation energies critical to activities such as morphogenesis. The templated ordering of amorphous phases at heterogeneous interfaces illustrates how living systems skillfully exploit phase transitions to achieve desired outcomes. This action is replicated within the encapsulation of compounds like TiO_2 in polymer nanocages, where non-equilibrium conditions are highly effective in promoting adaptability and guest accommodation to facilitate disorder-to-order processes. Such a mechanism replicates concepts observed in biological polymerization and suggests that the employment of synthetic biomimicry may lead to innovative methodologies in material handling at the nano-scale and highlights complex communications between states of matter and directional principles for synthetic biology (Wallace, 2012) (Figure 3.5).

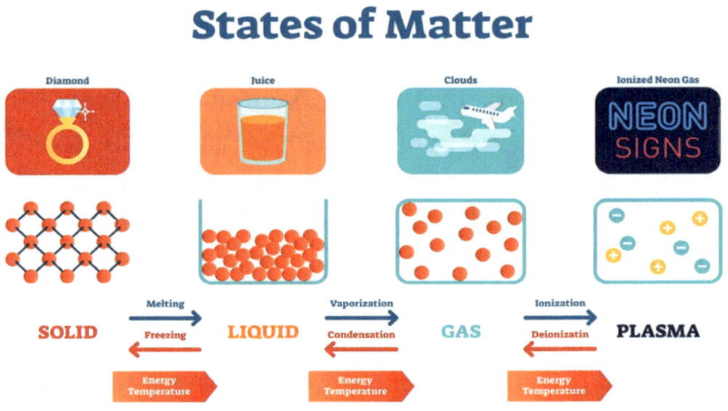

Figure 3.5. States of matter.

*Source:*https://www.thoughtco.com/thmb/sXdRXMvUrRydV9FwIlw-dzBvKHA=/1500x0/filters:no_upscale():max_bytes(150000):strip_icc()/GettyImages-947148218-1742a84786ae46b0b229da848348fb0f.jpg.

3.2. BIOCHEMICAL FOUNDATIONS

One should know the biochemical foundations of synthetic biology because these are the principles that govern the interaction and activities of the biological molecules that are used within biotechnological processes. Operational definitions of life are central to this field, serving as major frameworks for scientists. As explicated, over and above classification, these definitions facilitate scientists in dealing with the applied aspect of life's elements in experimental analysis and design. Operational definitions, for instance, can be used to clarify what constitutes a living system in synthetic biology, conditioning research agenda design as well as the interpretation of experimental outcomes. The ability to cross these definitions determines not only the theoretical models but also the methodologies employed in synthesizing organisms, gradually bringing us closer to a grasp of life itself (Bich and Green, 2016). Artificial cells are an exemplary case of the biochemical principles that form the crux of synthetic biology, having been a synthesis of biochemistry, engineering, and biology. These constructs are intended to replicate the fundamental properties of natural cells but provide a simpler system for manipulation and examination. A common artificial cell is composed of a biological membrane, translation and transcriptional machinery, and genetic modules that trigger cellular processes. The simplification by design not only facilitates the study of the core cellular processes but also demonstrates the current discrepancies between artificial and natural systems. As researchers work to narrow these gaps, they confront fundamental questions about the minimum requirements for function and complexity in these artificial systems, which reflect broader questions about the essence of life (Ding et al., 2014). Moreover, progress in understanding biochemical pathways is critical to expanding the functionality of synthetic organisms. By demystifying complex interactions in natural cells, scientists can engineer and construct artificial cells with more advanced functionalities, enabling them to perform specific functions such as drug delivery or environmental sensing. Pursuing these directions makes possible the construction of genetic circuits that adaptively respond to environmental signals, thereby bridging the gap between man-made systems and natural systems. Continued development in this area emphasizes the value of interdisciplinary approaches, combining knowledge from biochemistry, molecular biology, and engineering to create new synthetic systems. As synthetic biology advances, understanding molecules derived from these fundamental principles will have widespread implications for medicine, environmental science, and beyond, expanding our comprehension of life itself and our ability to manipulate it (Ding et al., 2014).

3.2.1. Structure and Function of Proteins

Proteins, the structural macromolecules of biological organisms, are linear polymers of amino acids coiled into specific three-dimensional conformations. The function and shape of a protein are defined by its specific amino acid sequence, or primary structure. Secondary structures, including alpha helices and beta sheets, arise due to hydrogen bonding among the backbone units of the peptide chain, while tertiary structures are set up as more complex interactions, such as hydrophobic interactions and ionic bonds, further stabilizing the overall conformation. Quaternary structure, which can include multiple protein subunits, adds a level of complexity that impacts the proteins function. Understanding these hierarchies of structure is crucial in proceeding with synthetic biology, as knowing how structure relates to function is required to be able to design proteins for desired function. In synthetic biology, protein structure manipulation enables novel functionalities not possible in nature to be developed.

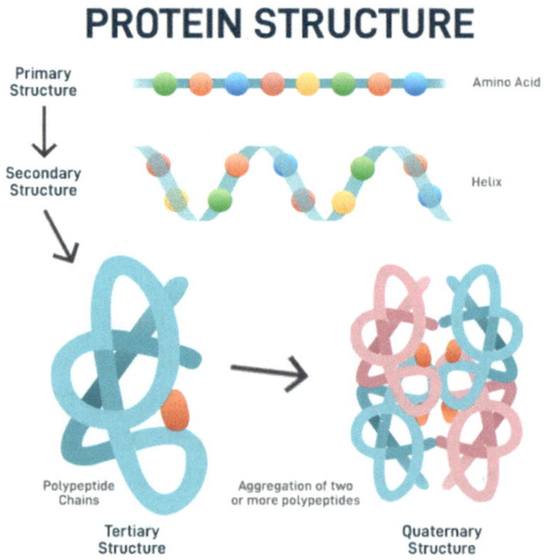

Figure 3.6. Protein structure.

Source: https://media.geeksforgeeks.org/wp-content/uploads/20220823180008/ProteinStructure.png.

For instance, by altering the amino acid sequence, researchers can create proteins that are more stable, specific, or catalytic. This has profound implications, particularly in the creation of artificial cells. These are simple things that try to mimic natural cellular processes and are constructed using biological membranes, transcriptional apparatus, and genetic modules. As noted in research, artificial models like these can be designed to possess equivalent properties as biological cells and thus offer cellular dynamics insights with lower complexity. Effective design and use of the artificial cells require a deep understanding of protein structure and its interface with functionally important outcomes that bridge gaps between natural and synthetic systems, as identified in the case of synthetic biology (Ding et al., 2014). Advancements in protein engineering are significantly boosted by methods such as microfluidics, which enable the controlled fabrication of artificial cells. Scientists can construct membrane-based artificial cells with size and biomolecular content-defined parameters by taking advantage of microfluidics. This is fundamental in understanding protein interaction and dynamics at the cellular level. The high-throughput approach enables robust interrogation of the impact of protein structure alterations on the behavior of cells as a whole. As the field advances, it is critical to answer basic questions about what the minimal conditions for artificially functioning cells are and whether they can communicate with the environment. It is important to comprehend the intricate relationship between protein structure and function because it forms the foundation on which these sophisticated systems are developed, subsequently leading to application in biology and medicine (Elani, 2016) (Figure 3.6).

3.2.2. Nucleic Acids: DNA and RNA Roles in Synthetic Biology

The roles of nucleic acids, primarily DNA and RNA, are self-evident in synthetic biology, a discipline that seeks to construct biological systems on the principles of chemistry and biology. The DNA is the prevailing genetic material that provides the blueprint required for synthesizing proteins and regulating the roles of cells. Through synthetic constructs, scientists can design DNA sequences to create new organisms with desirable traits, enhancing abilities like metabolic efficiency or stress tolerance. The ability to work on DNA through techniques such as CRISPR-Cas9 has revolutionized the field, allowing for targeted changes with profound implications in agriculture, medicine, and industrial biotechnology. Therefore, it is necessary to understand the chemical fundamentals underlying such processes to further advance synthetic biology and its applications. In addition to DNA, RNA plays a crucial role in the expression of genetic information and is a very useful tool for synthetic biology as well. mRNA, in particular, takes a pivotal role in translating the genetic information into functional proteins, whereas other forms of RNA, including ribozymes and siRNA, are utilized to regulate gene expression. The recent developments in mRNA technology, with the example of its application towards the manufacture of vaccines, demonstrate the great potential of RNA as a biotech tool. Moreover, constructing RNA molecules to perform specific functions such as metabolite pathway control or serving as synthetic gene circuits shows the subtle balance between chemistry and biological outcomes. Thus, a close understanding of RNA chemistry plays a central role in revealing new methods in synthetic biology. The incorporation of both RNA and DNA into synthetic biology platforms is part of an ongoing trend toward the exploitation of nucleic acids for technological and therapeutic applications. This strategy not only enhances our understanding of genetic and biochemical processes but also enables the development of advanced biosensors, efficient drug-delivery vehicles, and individualized therapeutic agents. The application of nucleic acids in nanotechnology, particularly in the development of smart delivery systems for targeted therapy, demonstrates the potential of the coming together of biology and chemistry to improve health and environmental challenges through novel solutions. With the progress of synthetic biology, surmounting the intrinsic issues, along with the ethical issues involved in the manipulation of genetic materials, is central to securing strong and effective innovations in the field (Chehelgerdi et al., 2023; Luo et al., 2023) (Figure 3.7).

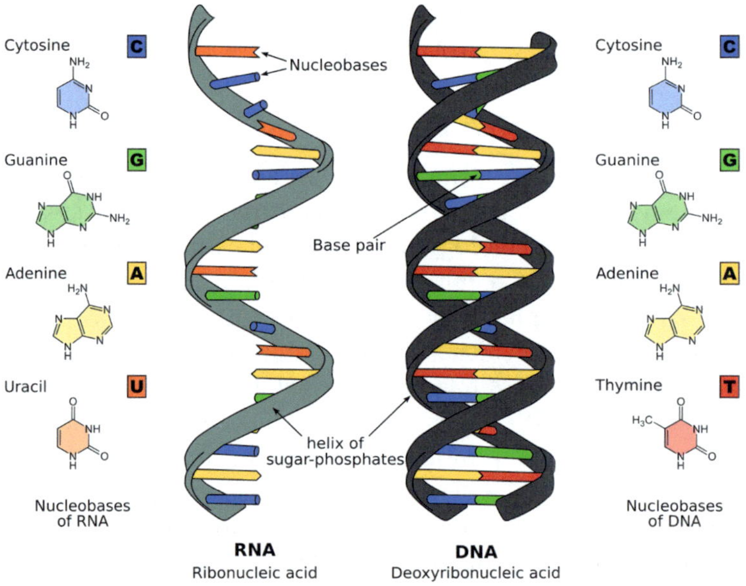

Figure 3.7. Nucleic acids RNA (left) and DNA (right).

Source: https://upload.wikimedia.org/wikipedia/commons/3/37/Difference_DNA_RNA-EN.svg.

3.2.3. Metabolism and Energy Transfer in Biological Systems

The intricate metabolic processes are the foundation of energy exchange in living organisms necessary for maintaining life. Metabolism consists of two general pathways: catabolism and anabolism. Catabolic processes break molecules down to release energy, while anabolic processes form complex molecules, typically requiring energy input. This balance decides cellular functions, impacting growth in response to the environment. In synthetic biology, an understanding of these metabolic pathways is important since it can be utilized in engineering organisms to perform specific functions. For instance, the modification of metabolic pathways allows us to make biofuels or drugs. As researchers explore such processes, the focus on energy conservation in metabolism emphasizes the need to optimize these reactions for yield and sustainability, often drawing on theories of autopoiesis and self-organization to maximize biological processes for designated purposes (Gradoff, 2017). In examining energy transfer, an examination of the work of adenosine triphosphate (ATP) as the primary energy currency of the cell is crucial. ATP is produced largely through cellular respiration and photosynthesis, mechanisms that illustrate the remarkable efficiency of energy conversion in biological processes.

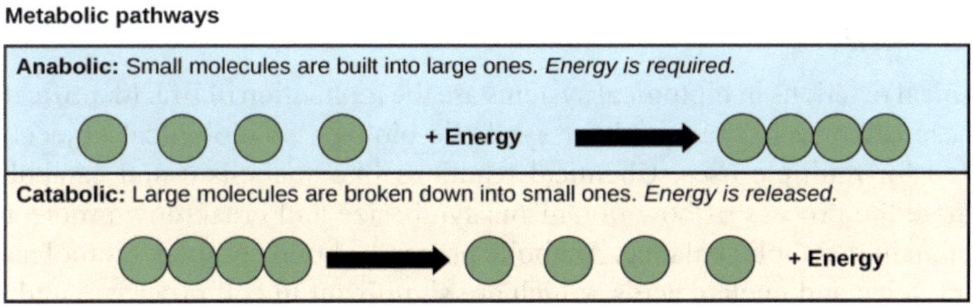

Figure 3.8. Catabolic pathways are those that generate energy by breaking down larger molecules. Anabolic pathways are those that require energy to synthesize larger molecules. Both types of pathways are required for maintaining the cell's energy balance.

Source: https://opentextbc.ca/biology/wp-content/uploads/sites/96/2015/02/Figure_04_01_02.jpg.

The individual steps of these pathways are tightly regulated to maximize the overall metabolic efficiency required for cellular function. In addition, ATP is a critical mediator of signaling pathways, linking energy availability to cellular response. Understanding these dynamics is essential in synthetic biology, where researchers focus on creating engineered systems that mimic or optimize natural efficiency. Having a thorough understanding of ATP production and consumption provides information that can be quantified in terms of synthetic circuits to perform complex operations, ultimately leading to biotechnology and bioengineering applications (Anikeeva et al., 2018). Furthermore, synthetic biology research is constructed around new approaches that examine how biological materials and systems interact at the molecular level. Materials derived from semiconductors, for instance, hold vast potential to integrate with biological components and consequently uncover metabolic processes across different scales. This integration not only enhances our understanding of biophysical dynamics but also leads to the development of next-generation devices enabling the monitoring and regulation of cellular processes. Using such technologies, scientists can explore new signal processes, potentially unveiling new aspects of cellular metabolism and energy transfer. Through continued advancement in the areas of chemistry, biology, and materials science, these fields will be integral to creating synthetic biology, enabling new avenues towards revolutionary applications that address scientific and societal concerns (Anikeeva et al., 2018) (Figure 3.8).

3.3. CHEMICAL REACTIONS IN BIOLOGICAL SYSTEMS

Chemical reactions in biological systems are the foundation of life. Identification of these responses is required for synthetic biology, as biological objects are created for multiple uses. Chemical reactions like anabolism and catabolism illustrate the process of how organisms synthesize and break down molecules to maintain metabolic balance. Anabolic processes build complex biomolecules like proteins and nucleic acids, which are significant in cell processes and cell growth. Catabolic processes enable the breakdown of these biomolecules to release energy, typically in the form of ATP. The fine balance between these reactions is a demonstration of life's dynamic character, with equilibrium repeatedly being pursued and optimized in adaptation to environmental and physiological change. The function of enzymes as biological catalysts is perhaps the most common thread throughout the investigation of chemical reactions in biological systems. Enzymes catalyze reactions by decreasing the activation energy needed, thereby optimizing the responsiveness and efficiency of metabolic pathways. This is especially crucial in synthetic biology, where the engineered pathways usually demand extremely controlled rates of reaction to achieve maximum product output. Enzyme specificity also serves to direct substrates into desired products while suppressing unwanted side reactions that are otherwise possible in complex biological systems. Therefore, knowledge of enzyme kinetics and the conditions of these reactions, i.e., temperature and pH, is of utmost significance for the synthesis of artificial biological circuits, enabling researchers to achieve high-precision processes. Also, the intersection of chemical reaction and biological information represents the existing controversy surrounding life and work parameters of life. In synthetic biology, life is defined as more than just categorization to include functional models that direct agendas for investigation. Working definitions like these allow researchers to build systems in which chemical reactions can be fashioned such that living systems or processes that resemble life can be designed. For example, by examining underlying metabolic pathways and how they chemically interact, scientists are able to develop novel means of re-engineering life forms for biotechnology applications. Such a search for understanding of the chemical nature that underlies life also raises philosophical questions, e.g., what life is, and illustrates the sophistication and depth of synthetic biology as seen in the recent controversy over definitions of life and their meaning in science (Bich and Green, 2016).

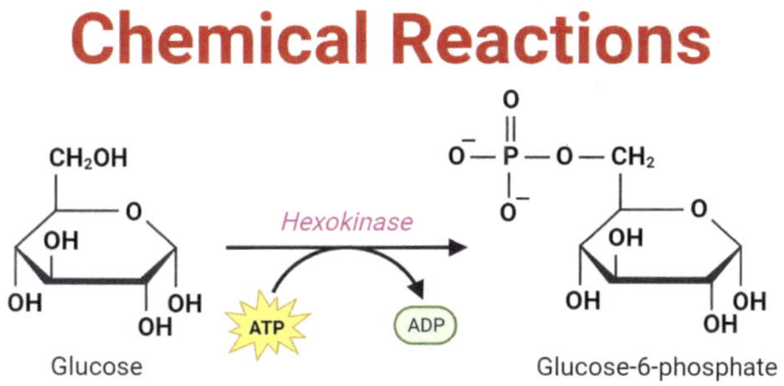

Figure 3.9. Chemical reactions.

Source: https://microbenotes.com/wp-content/uploads/2023/12/Chemical-Reactions.jpg.

3.3.1. Types of Chemical Reactions: Synthesis, Decomposition, and Exchange

Chemical reactions form the essence of synthetic biology concepts, dictating the interaction and modification that lead to the formation of new substances. Synthesis reactions are particularly pertinent among all chemical reactions, as they involve combining simpler compounds to form more complex molecules. This reaction is crucial in the development of biomolecules such as proteins, nucleic acids, and polysaccharides, forming a foundation for cellular processes and metabolic pathways. In synthetic biology, such reactions are designed to construct new biological systems, making it possible to produce bioproducts for addressing environmental or human health concerns. These skills are essential in developing new products such as biofuels and bioplastics from renewable resources, as shown in present advancements revolving around agricultural residues as feedstock for bioproducts (Mujtaba et al., 2023). Decomposition reactions, on the other hand, break down complex substances into simple compounds, playing a vital role in recycling materials in biological systems. These answers are ubiquitous in metabolism, facilitating the breakdown of carbohydrates, lipids, and proteins to provide the energy and building blocks an organism requires to grow and maintain itself. In synthetic biology, scientists can design microbial systems capable of breaking down contaminants effectively or taking waste streams and generating valuable resources if they understand decomposition pathways. By applying these natural processes, researchers can optimize bioremediation and develop bioconversion strategies towards sustainability goals. Ongoing exploration of decomposition processes uncovers the possibility of optimizing metabolic pathways for the production of environmentally safe bioproducts as well as minimizing the impact of industrial effluent (Mujtaba et al., 2023).

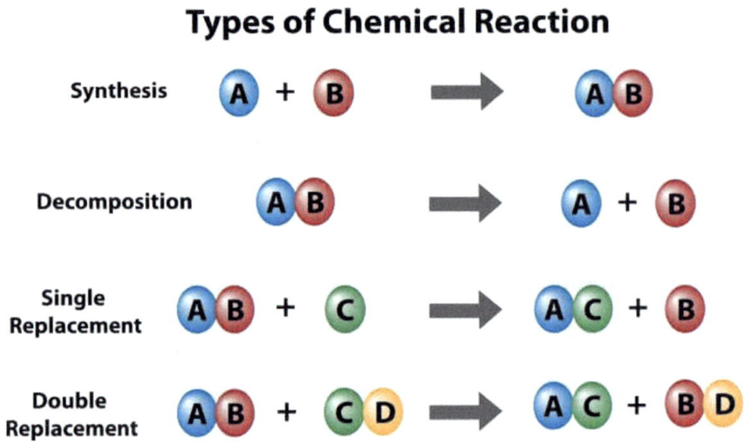

Figure 3.10. Types of chemical reactions.

Source: https://www.reagent.co.uk/wp-content/uploads/Diff/re/Different-types-of-chemical-reactions-853x506-1.jpg.

Exchange reactions, or substitution reactions, involve the exchange of components within molecules and are central to most biochemical processes. These types of reactions permit atom, ion, or functional group exchange to produce new compounds with an identical overall molecular design. In synthetic biology, exchange reactions are central to probing mechanisms of gene editing, such as CRISPR-Cas systems, which rely on accurate interactions between nucleic acids to alter genetic content. By employing such reactions, researchers can successfully manipulate genomes, enabling organisms to be engineered with specific traits. As the technology improves, an understanding of the kinetics and mechanisms of exchange reactions contributes to the progress of genetic engineering and synthetic synthesis of intricate biomolecules. Their application to practice is a support for the imperative to critically analyze chemical reactions in their pursuit of sustainable solutions in synthetic biology (Figure 3.10).

3.3.2. Enzymatic Reactions and Catalysis

Enzymatic reactions are at the core of the biochemical processes that are the foundation of synthetic biology, functioning as catalysts that speed up chemical reactions without being consumed in the process. These biological catalysts are site-specific in nature, allowing exact control of reaction pathways, which is of utmost importance in application to synthetic biology. Enzyme reactions, as they are investigated, bring together chemical and biological concepts to understand how enzymes such as cellulases and β-glucosidases function at the molecular level. The insight attained on these mechanisms shows how such enzymes can be manipulated for better performance in industrial bioprocesses ranging from the production of biofuels to synthesis in the pharmaceutical industry. Increased catalytic efficiency ensures cost-effectiveness and eco-friendliness, showing the relevance of enzymatic reactions to green applications in synthetic biology (N/A, 2025). Enzyme structure and function dictate their catalytic function. Enzymes are typically globular proteins with optimally shaped active sites for binding substrates, allowing reactants to be converted to products. The interaction of the enzyme with the substrate can usually be explained in terms of the lock-and-key model or the induced fit model, both of which emphasize molecular compatibility. In addition to this, temperature, pH, and whether or not cofactors and inhibitors are present can also have a significant impact on enzymatic activity, emphasizing the requirement for controlled conditions in applications in synthetic biology very strongly. For example, the cellulase enzyme system, endoglucanase, exoglucanase, and β-glucosidase work synergistically to degrade cellulose to glucose, demonstrating the functional diversity of enzyme systems (Ahmed et al., 2017). This equilibrium in enzymatic action is important for optimizing biotechnological processes. Enzyme catalysis not only facilitates biochemical reactions but also depends on the specificity and efficiency mechanism that can be harnessed and engineered for synthetic biology. Enzyme design for operation under various industrial conditions can be achieved through methods such as molecular modeling and directed evolution. Consequently, scientists can design enzymes for specific applications, in the hopes of creating lower-cost and greener processes in industries including textiles, biofuels, and food processing. Moreover, the increasing demand for enzymes such as β-glucosidase, driven by the critical role in breaking down complicated carbohydrates, highlights the feasibility of biotechnological innovation meeting the needs of industry while being in line with sustainable processes (Ahmed et al., 2017). This ongoing research priority serves to underscore the significance of enzymatic reactions and catalysis within the context of synthetic biology and its long-term implications for the future of biotechnology.

3.3.3. Thermodynamics and Kinetics in Biological Reactions

Thermodynamic concepts are central to the understanding of biological reactions, especially in synthetic biology. Thermodynamics dictates how energy is converted and transferred in biological systems, with a significant influence on metabolic processes. The first law of thermodynamics, stating that energy cannot be created or destroyed, would imply that all biological processes must have an initial input of energy. For example, in cellular respiration, glucose is broken down to free energy that is used to build ATP, the cellular energy currency. The second law offers the concept of entropy, which underscores the systems' trend towards greater disorder. This is especially relevant in synthetic biology, where the design of efficient biochemical pathways must be accounted for in terms of energy dissipation and product stability. Understanding these thermodynamic models allows scientists to predict and manipulate the behavior of biological reactions better (Booth et al., 2014). In addition to thermodynamics, kinetics also offers a vital viewpoint under which the rates of biological reactions can be examined. A kinetic study examines the rate at which reactions occur and what regulates those rates, such as enzyme levels, the presence of substrates, and temperature. Enzymes, the biological catalysts, substantially lower the activation energy for reactions and cause metabolic processes to run at much higher rates than they would under spontaneous conditions. In synthetic biology, enzyme kinetics optimization is capable of producing more efficient biochemical pathways and improved product yield. For instance, optimization of substrate and cofactor concentrations could enhance the activity of enzymes. Furthermore, understanding the kinetics of reaction cascades facilitates the construction of complex biological circuits, such that it is now feasible to engineer systems with natural process mimicry or novel functions. The link between kinetics and synthetic biology emphasizes the necessity of manipulation in synthetic biological systems (Andrianantoandro et al., 2009). As scientists integrate thermodynamic and kinetic concepts into synthetic biology, opportunities for creative applications spread exponentially. With the application of thermodynamic principles of stability, biologists can design molecular systems that favor certain paths or products over others, reducing side reactions and maximizing efficiency. The integration of kinetic models with thermodynamic data enables the development of more insightful predictive models of biological behavior. This can lead to breakthroughs in engineered organisms for biotechnological applications, such as the production of biofuels or the synthesis of drugs. The addition of these principles not only enhances the understanding of biological systems but also enhances the range of tools available to synthetic biologists for designing more sophisticated molecular architectures that mirror or outclass natural processes (Booth et al., 2014) (Figure 3.11).

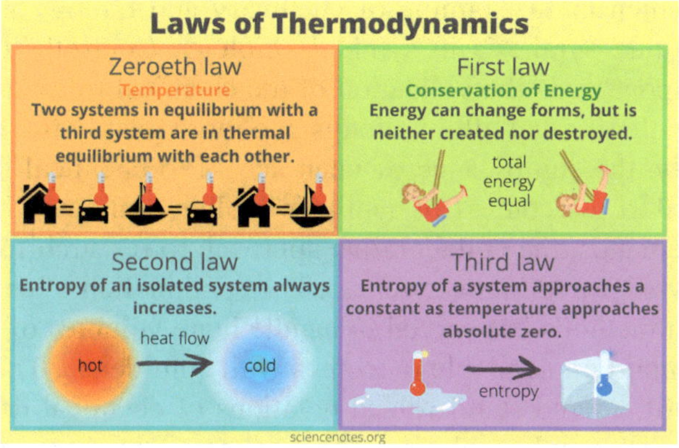

Figure 3.11. Laws of thermodynamics.

Source: https://sciencenotes.org/wp-content/uploads/2021/12/Laws-of-Thermodynamics.png.

3.4. MOLECULAR TECHNIQUES IN SYNTHETIC BIOLOGY

Molecular approaches in synthetic biology are foundational tools used to manipulate and understand biological systems at the molecular level. Through borrowing ideas from chemistry, researchers can design and construct proteins, RNA, and DNA, which form the backbone of biological function. Techniques such as CRISPR-Cas9 have revolutionized gene editing by allowing for accurate alterations in genetic material and the manipulation of gene function through the investigation of probes and producing genetically altered organisms. The control behind this is a firm grasp of molecular interactions, with knowledge of chemical bonding and molecular shape playing a huge role in effective experimentation. The marriage of such molecular methods not only adds to our understanding of biological processes but also opens up novel applications in the fields of medicine, agriculture, and environmental science. Advances in molecular methods are closely tied to advances in chemical methodologies, specifically those that deal with DNA and protein synthesis and analysis. For example, advances in next-generation sequencing technologies have allowed rapid and deep analysis of genomes that has yielded decisive information regarding genetic variabilities and interactions. Unveiling these molecular interactions, as recent research has shown, underscores the importance of semiconductor-based materials for observing biological processes at various scales. The application of such materials allows for the development of interface devices with biological components, which is an example of chemistry and biology integration that is pivotal to future synthesis in synthetic biology. Controllability over such molecular components is the validation of the chemical rules controlling such advanced technologies that allow scientists to more readily traverse biological space. Secondly, the significance of open access to chemical data cannot be overemphasized in the context of synthetic biology. As noted by recent open data-sharing demands, the collaborative approach to research fosters a culture of shared knowledge and innovativeness. The combination of cheminformatics and e-Research methods has served to handle huge volumes of data, enabling researchers to build on available outcomes with confidence. The development of unique identifiers and standardized notations of chemical data makes data use and access easier and enables scientists to cooperate. Acquiring knowledge about the sharing of positive and negative results can develop the field while being resource-efficient. As synthetic biology continues to progress, the integration of molecular methods and open science paradigms will grow more significant in addressing grand challenges and ultimately shift our approach to living systems and their applications in society (Anikeeva et al., 2018; Bird and Frey, 2013).

3.4.1. DNA Synthesis and Manipulation

DNA manipulation is the focus of synthetic biology, enabling the redesign and construction of biological entities to meet specific needs. DNA synthesis can be performed using a variety of methodologies that fulfill unique roles within the field. Techniques such as polymerase chain reaction (PCR) and solid-phase synthesis enable the construction of synthetic constructs that can be utilized to lead to functional proteins or entire pathways. Such advances are of primary significance in developing artificial cells, which are constructed by assembling DNA sequences that code for essential biological functions. This involves the coordinated assembly of building blocks whereby the specificity and quality of the synthesized DNA directly influence the robustness of the resulting cellular behavior. In efforts to emulate natural cellular dynamics, these underlying methods are essential to bridge the gap between artificial and natural cellular designs (Elani, 2016). Encapsulation and structure are the largest obstacles in trying to engineer advanced artificial cells. Effective DNA manipulation includes not just precision synthesis but also precise incorporation within membranous structures mimicking the protective barriers of native cells. Microfluidic technologies have revolutionized this field, making high-throughput production of artificial cells with set parameters such as size and biomolecular content possible. Such interventions make it possible to study how alterations in these parameters affect cellular communication and function, hence the possibility of targeted experimentation. By encapsulating transcription and translation machinery with genetic modules, scientists aim to mimic cellular processes within a simplified framework. It has far-reaching implications for decoding the rules of cellular life while simultaneously driving innovation in synthetic biology applications from therapeutic development to bioengineering (Ding et al., 2014). Besides, DNA manipulation is more than just synthesis in that it aims at dealing with a whole system that includes regulation and evolution in synthetic constructs. As scientists package more and more information into artificial cells, they eventually face significant sustainability and environmental interaction issues. Through the incorporation of feedback loops and genetic networks, artificial cells can exhibit dynamic behavior that replicates their biological counterparts. But how low can the level of complexity go before these cells are able to function independently? It will take innovative methods and interdisciplinarity, a climate of discovery, to bridge the gap between synthetic devices and their biological counterparts. As this research progresses, the potential for using artificial cells for beneficial purposes in medicine, environmental science, and biotechnology becomes increasingly real, mirroring the significance of manipulating DNA in synthetic biology (Ding et al., 2014).

3.4.2. Protein Engineering and Design

Protein engineering and design is a significant branch of synthetic biology that enables the construction and creation of new proteins with desired functions. By applying chemistry and principles of molecular biology, scientists can manipulate amino acid sequences to design proteins with specific biochemical properties. This spans across a range of applications from enzyme engineering for industrial catalysis to the design of therapeutic proteins for pharmaceuticals. Protein function and structure prediction has been radically boosted by computational modeling and bioinformatics resources that now offer the capability to design more stable, specific, and active proteins. As this area continues to evolve, the integration of synthetic techniques and instruments will close the gap between the intended protein and its natural counterpart even further, allowing unprecedented control and flexibility in biological systems. Besides, the concept of artificial cells further emphasizes the importance of protein engineering in yielding biologically relevant constructs. Artificial cells, which aim to mimic the cellular functions of natural cells, rely heavily on the integration of engineered proteins capable of conducting essential biochemical processes. These miniaturized systems consist of biological membranes, transcription and translation devices, and genetic modules, stressing the importance of engineered proteins in cell dynamics. As scientists go about encoding increasingly complex functionalities into artificial cells, there has been a pressing need to query not only the design of such proteins but even how they interact within the cell framework. Asking the question of the minimum number of factors for optimal operation of artificial cells highlights the subtle relationship between protein design and system function, inviting a higher-level use in biomedical research and synthetic biology (Ding et al., 2014). Building biointerfaces using semiconductor-based materials continues to emphasize the need for protein engineering. By creating devices that can probe and manipulate biological components, scientists explore upcoming electronic and optoelectronic interfaces with a greater understanding of complex biophysical dynamics. Since proteins act as an intermediary between semiconductor engineering and biology, the potential of finding new signaling processes is vast. As engineered interfaces advance, they offer deeper insights into cellular mechanics and open the door to novel therapeutic strategies. Therefore, the meeting of protein engineering, design, and new technologies brings the intersection of organic chemistry with synthetic functions to the forefront, which places protein engineering at the foundation of driving the field of synthetic biology in general (Anikeeva et al., 2018) (Figure 3.12).

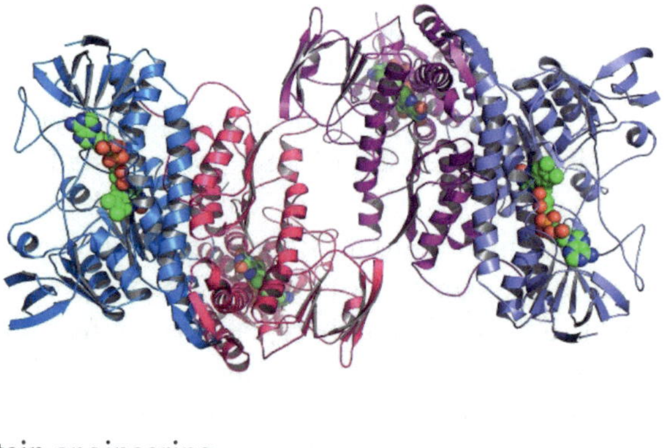

Figure 3.12. Protein engineering.

Source: https://www.creativebiomart.net/images/Protein%20Design%20for%20Protein%20Engineering.jpg.

3.4.3. Metabolic Engineering for Biosynthesis

Metabolic engineering is a key intersection of synthetic biology and biochemistry that aims to rewire cellular metabolism for the optimal production of valuable compounds. Through the regulation of metabolic pathways through gene editing and pathway reassembly, researchers can enhance the biosynthesis of specific metabolites, including drugs, biofuels, and specialty chemicals. This designed approach allows microorganisms to assume novel functions as effective, green factories that optimize yield and reduce the environmental footprint compared to traditional manufacturing operations. Such developments emphasize that there is a necessity of comprehensive knowledge of metabolic networks and the associated biochemical mechanisms, offering a general framework for developments in synthetic biology that leverage engineered metabolic networks.

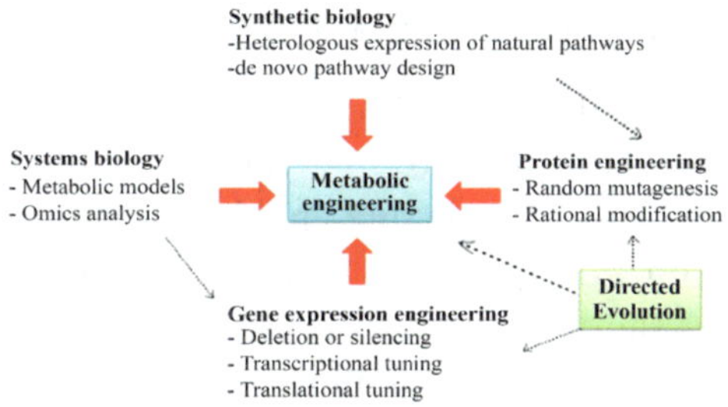

Figure 3.13. Overview of the metabolic engineering process.

Source: https://ars.els-cdn.com/content/image/3-s2.0-B9780128121603000027-f02-09-9780128121603.jpg

At the forefront of metabolic engineering innovation is the application of synthetic biology philosophies in terms of design, construction, and optimization of biological systems. Having standard genetic components to work with enables modular pathway assembly to provide researchers with the tool they require to design complex metabolic circuits for particular biosynthetic purposes. For instance, through the modular design of artificial gene networks, researchers can achieve toggling or oscillatory control of metabolic flux, enabling dynamic control over biochemical production. Furthermore, the improved ability in DNA de novo synthesis enables the incorporation of synthetic regulatory elements, enhancing the precision of metabolic engineering. By accomplishing that, this multi-dimensional strategy forms a solid basis for generating novel biotechnological applications far beyond traditional metabolic pathways, representing a huge leap in biosynthetic capability. The study of hybrid technologies that combine biological catalysts and inorganic materials represents the future of metabolic engineering in biosynthesis. These technologies take advantage of the strengths of both synthetic and biological elements to form higher energy efficiencies and novel production strategies. Specifically, the combination of semiconductor light harvesters with microbial systems presents routes for solar-to-chemical conversions that readily exceed the efficiency of natural photosynthesis. However, realization of these hybrid systems in practice necessitates detailed insight into the materials-microbe synergies, which underscores the necessity of bio- and chemocompatibility. While scientists encounter these challenges, the intersection of chemistry with biological engineering is likely to introduce revolutionizing technologies that reorder the biosynthesis landscape (Cestellos-Blanco et al., 2018; Heinemann et al., 2006) (Figure 3.13).

3.5. TOOLS AND TECHNOLOGIES IN CHEMISTRY

The integration of advanced tools and technologies in chemistry plays a vital role in the developments exhibited in synthetic biology. Semiconductor-based materials are specialized tools that enable researchers to study complex biophysical dynamics on various scales. Semiconductor materials are essential to the conception and creation of innovative electronic devices that engage with biology. By leveraging semiconductor building block properties, researchers can enhance the capability to track and regulate the actions of biological components, leading to breakthrough information in cellular signaling processes. This discovery not only expands the boundaries of chemical exploration but also enhances the possibility of creating novel biointerfaces, thus creating a significant interface between biology and technology (Anikeeva et al., 2018). The development of artificial cells sealed by lipid membranes is a significant milestone in synthetic biology. These cell mimics are especially well worth their while through their biocompatibility, along with the capability of reconstituting biological materials therein. However, the challenge of achieving an accurate control over the formation of these membrane-based systems remains a focal point. Recent innovations highlight the importance of exercising fine control over important parameters such as size, biomolecular composition, and spatial organization. Microfluidic techniques are now essential tools for overcoming such challenges and producing artificial cells at high throughput. Such technologies not only streamlined the production process but also offered avenues for exploration of the functional capabilities of engineered membranes (Elani, 2016). Such progress in tools and technologies is not merely confined to constructing artificial systems but also translates into a deeper understanding of the governing biochemical principles. Techniques such as mass spectrometry and high-performance liquid chromatography play a vital role in providing an understanding of the molecular underpinnings of synthetic biology. Additionally, improvements in microscopy encourage cell-cell interactions to be closely viewed at historically unprecedented resolutions, allowing scientists to visualize and quantify biochemical reactions in real-time. These exacting examinations are essential to the furtherance of synthetic biology because they allow scientists to test engineered biological pathways and circuits with unprecedented accuracy. As these methods continue to evolve, their applications not only clarify elementary chemical principles but also lead to the design of better synthetic systems, driving future innovation in the field (Anikeeva et al., 2018) (Figure 3.14).

Figure 3.14. Chemical technology

Source: https://advancedchemtech.com/wp-content/uploads/2021/07/pexels-Chokniti-Khongchum-3938023.jpeg

3.5.1. Spectroscopy and Its Applications in Synthetic Biology

Spectroscopy is a central technique in synthetic biology, enabling the explanation of biochemical protocols at the molecular level. Using various spectroscopic methods, researchers can analyze biomolecular interactions, monitor metabolic routes, and assess the performance of designed systems. For instance, techniques such as UV-Vis, fluorescence, and nuclear magnetic resonance (NMR) spectroscopy unveil details regarding protein folding kinetics, ligand binding rates, and nucleic acid structure. Adoption of these methods facilitates optimization of synthetic biological constructs, which in turn lead to better applications like metabolic engineering and biofuel manufacturing. Additionally, the real-time monitoring benefit offered by spectroscopic methods enables researchers to conduct slight changes to the reaction, enhancing the synthetic biology process efficiency under varied experimental conditions. The integration of spectroscopy and synthetic biology also includes the development of hybrid systems, which involve both biological and inorganic components, with the goal of maximizing the optimization of light-harvesting and energy conversion. Such systems have huge potential for solar-to-chemical production because they can allow energy and charge transfer from renewable materials to desired chemicals. One issue with these approaches is understanding how energy and charge transfer occur within microorganisms. Spectroscopy plays a crucial role in answering such questions, as it provides information about the interface interactions between the material and the microorganism.

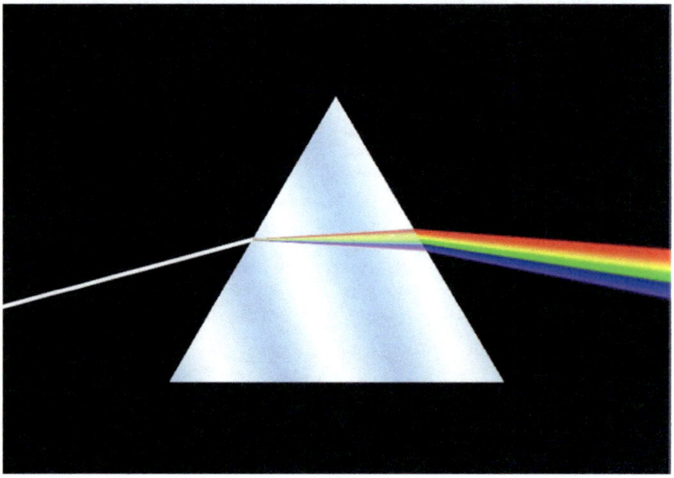

Figure 3.15. Spectroscopy.

Source: https://www.thoughtco.com/thmb/fbIZQ4rsnadGOYRFgJokZ8N0CFQ=/1500x0/filters:no_upscale():max_bytes(150000):strip_icc()/Dispersion_prism-1--5897aa7b5f9b5874ee77da71.jpg.

With the use of techniques that describe the electronic properties and behavior of these hybrid materials, researchers can work towards producing more effective biocompatible materials and improving energy efficiencies higher than those achieved by natural photosynthesis, and thus propel the field (Cestellos-Blanco et al., 2018). Moreover, the application of spectroscopy in synthetic biology is not only committed to the examination of proteins and nucleic acids; it also includes the exploration of glycan interactions involving complex carbohydrates. Recent findings have emphasized the significance of glycosylation patterns and how they influence cellular behaviors. The sugar code—a complex array of information encoded within glycan structures—is to blame for controlling various biological processes, including cell signaling, adhesion, and recognition. By applying spectroscopic techniques such as NMR and molecular modeling, researchers can distinguish glycoprotein structural architectures and interaction behaviors. Such data not only sheds light on biological functions but also informs the construction of synthetic routes for glycosylated compound assembly, paving the way for new therapeutic paradigms and biotechnological uses (Gabius et al., 2001). Through these multidimensional strategies, spectroscopy remains a valuable asset that closes the gap between fundamental research and applied uses in synthetic biology (Figure 3.15).

3.5.2. Chromatography Techniques for Separation and Analysis

In synthetic biology, chromatography techniques are of utmost importance for the separation and analysis of biomolecules. These procedures make use of the principle of disentanglement of compounds between a stationary and mobile phase to achieve effective separation using different physical and chemical characteristics. For example, HPLC is widely employed to examine metabolites and proteins due to its high resolution and versatility. In synthetic biology, the ability to isolate certain proteins using chromatography not only helps define their functions but also identifies potential applications within metabolic pathways. The scientific efficacy with which HPLC separates complex mixtures, therefore, serves as a supporting role in important studies for designing engineered organisms with the ability to produce desirable bio-based products. In addition to these techniques, advanced chromatographic techniques such as gas chromatography (GC) and affinity chromatography are also involved in applications of synthetic biology. GC is particularly suitable for volatile compounds, allowing scientists to analyze small organic compounds relevant to biosynthetic processes. Affinity chromatography, on the other hand, provides a particular technique for protein purification using specific binding interactions, which is crucial when analyzing the functions of particular enzymes in synthetic pathways. This specificity of separation ensures significantly high throughput of proteomic analysis, thereby improving the worldwide understanding of cellular functions in synthetic systems. With continued development of synthetic biology, the integration of such chromatographic techniques will never be replaceable for exact characterization and control of biological molecules. Moreover, chromatography method optimization cannot be overstated, especially when applied in processes such as folate manufacturing through yeast, as discussed in ongoing research. Application of optimized HPLC and LC-MS methods enables efficient quantitation and comparison of folate yield in various yeast strains, and this illustrates the flexibility and sensitivity of cultivation conditions on metabolite profiles. As described in investigations concerned with yeast-based folate production, both techniques must be individually set to ensure maximum yield and stability, exemplifying an undeniable intersection of analytical chemistry and synthetic biology. Coupling these techniques with what is known about biological systems, scientists can create new processes for enhancing nutrient content in food and advance the art of metabolic engineering's potential (Aebersold et al., 2009; Patring, 2007).

3.5.3. Computational Chemistry and Modeling in Synthetic Biology

The union of modeling and computational chemistry with synthetic biology is pivotal to the creation of better insight into biological systems, as well as aiding the design of novel synthetic entities. Computational chemistry provides the tools for simulating how molecules interact with each other and predicting the behavior of the constituents of biological systems at the atomic level. This capability allows scientists to model complex biochemical phenomena, thus leading to more effective design techniques for synthetic biological systems. By applying computational techniques, it is possible to investigate the dynamics of artificial cells, which are, in essence, reduced-complexity analogs of natural cells that replicate their functions. Through simulations, scientists have the ability to test the way synthetic cells react in their environments, thus answering fundamental questions regarding their potential for communication and evolution, an area highly linked to the emergence of work in synthetic biology (Ding et al., 2014). Computational methods also play a vital role in informing engineering principles embedded in synthetic biology, promoting an understanding of the hierarchical organization of biological systems. Computational programs allow for the simplification of complex biological processes into manageable fragments, which can be subjected to organized scrutiny and redesign. For instance, the use of engineering-based operations, such as toggle switches and oscillators within DNA, is facilitated by computational models, which predict interactions and responses of synthetic genetic circuits. This foundational knowledge is fundamental in the development of uniform components that can be well-handled to create the foundation for synthesizing advanced biological systems. With the growth of the field, the synergy between synthetic biology and computational modeling will bring forward new biomanufacturing and pharmaceutical innovations (Heinemann et al., 2006). Besides, the continued advancement of computational techniques is fully in harmony with the increasing capacity for the de novo synthesis of DNA, which is ideally suited to the assembly of complex synthetic systems. This harmony of design and implementation is at the heart of achieving the goals of synthetic biology, i.e., the construction of functional artificial cells with characteristics analogous to those of biological cells. With the boosting computing power, scientists can rationally design proteins and predict new functions by orchestrating domain architectures like modular assembly, as with Legos. With synthetic biology embracing such computing powers, aspirations for useful applications become increasingly tantalizing. Thus, computational chemistry is both a foundation and driving element of innovation, driving the development towards the realization of functional and efficient synthetic biological constructs (Ding et al., 2014; Heinemann et al., 2006).

3.6. ETHICAL CONSIDERATIONS IN SYNTHETIC BIOLOGY

The rapid development of synthetic biology has raised ethical concerns regarding responsible innovation. One of the central ethical concerns involves the implications of creating synthetic life forms that are possibly capable of displaying features akin to consciousness. With the advent of machines that exhibit cognitive sophistication and social complexity, as the morphospace model based on complexity explores, ethical concerns regarding the potential status of such entities become pertinent. Meeting with autonomy and cognitive potential obfuscates the moral landscape by making issues of rights and welfare exist for such artificial entities, particularly where they could infringe on or intersect with human social systems. Hence, the ethical dialogue must be wary of the undertows present in synthetic biology, challenging stakeholders to consider not only the technological possibilities but also the humanitarian implications of what they are making. Additionally, the scientists and engineers must be trained and educated in synthetic biology with the incorporation of ethical training that is targeted at the special challenges posed by the field. As cited in recent controversy over the maturity of synthetic biology as a field, the application of traditional engineering principles is not adequate to the treatment of ethics. Employing an engineering design cycle approach, as advocated for in calls for synthetic biology pedagogy, ensures that future professionals will be educated to possess a robust ethical paradigm alongside scientific and technical competencies. Such an integrated approach can impart a sense of moral compass to emergent professionals, guiding their choices as they engage with powerful technologies that could remake life itself. Encouraging the interdisciplinary dialogue between ethicists, biologists, and engineers can give rise to a richer debate, placing responsible innovation at the center. Moreover, it is important to acknowledge that synthetic biology is within a sophisticated socio-political landscape that can influence the public acceptance and perception of its advancements. With issues surrounding the safety, security, and environmental implications of engineered organisms, ethical questions need to be considered not just the narrower technical challenges but also the broader societal impacts of synthetic biology. It is necessary that there be stakeholder involvement since community involvement demystifies scientific advancement and can reverse public misconceptions. Open lines of communication not only create trust but also offer an opportunity to consider diverse opinions in decision-making. Through the involvement of various stakeholders in advance—e.g., policymakers, businesses, and the wider public—scientists can direct the ethical issues of synthetic biology in such a manner that technological gains are aligned with societal values and interests, ultimately with the goal of balanced innovation.

3.6.1. Safety and Risk Assessment of Synthetic Organisms

Synthetic organism development is a significant step in science, namely synthetic biology, which entails strict safety and risk assessment. Because these organisms are created to perform specific functions, their potential impact on ecosystems and human health is a topic of interest. The type of system in question typically introduces unknown factors; therefore, risk assessment is essential to setting parameters for safe use. Since they are man-made, synthetic organisms may interact with natural biological systems in unforeseen ways and may create ecological impacts that may not be immediately apparent. Hence, conducting detailed evaluations is required to find potential risks, quantify exposure levels, and examine the overall risks of releasing synthetic organisms into various environments so that ethical considerations can be merged with scientific procedure and policy development (Chaturvedi et al., 2015). Another key aspect of synthetic organism risk assessment and safety arises through the incorporation of public attitudes and ethical considerations in policy development. Understanding what the public perceives regarding synthetic biology has a major influence on the regulatory processes and acceptance of these technologies. Cultural environments, socio-economic status, and history can all influence attitudes toward synthetic organisms and how they are integrated into society. Hence, public participation is vital in formulating open policies that not only address the safety concerns involving synthetic organisms but also the ethical dilemmas involved in their use. An inclusive approach that engages diverse perceptions can help to build trust and acceptance and ensure that advances in synthetic biology are ethically and socially integrated with values (Boyen et al., 2025). Moreover, global cooperation in regulating synthetic biology is important in establishing uniform safety standards and risk assessment procedures. The approach to synthetic biology and its risks varies significantly all over the world, based on different cultural norms and regulatory contexts. Building comprehensive governance frameworks that encompass ethical codes, risk analysis methods, and public involvement can help in facilitating the effective control of synthetic organisms. Cooperative frameworks, as proposed in global science governance studies, illustrate potential convergences that can be utilized to craft an effective and peaceful international response (Chaturvedi et al., 2015). Such cooperation not only provides assurance for responsible innovation but also allows for a context where the potential of synthetic organisms can be maximized while providing an assurance for the minimization of potential harmful effects on ecosystems and public health.

3.6.2. Environmental Impact of Synthetic Biology Applications

The application of synthetic biology has evoked intense controversy regarding its environmental impact, necessitating cautious research into the intricate interaction between engineered life and the environment. Synthetic organisms are a source of particular concern in that they have the potential to disrupt established ecological balances. The very reason for creating such organisms to play certain roles—bioremediation or waste reduction, say—is that they can inadvertently outcompete native species or create unintended biochemical interactions. The complexity of biological systems ensures that having a solid grasp of such interactions is imperative; as the literature points out, the unbracketed complexity of biological systems requires care in applying synthetic biology tools without extensive ecological testing (Gradoff, 2017). Furthermore, the moral issues of ecosystem manipulation for human convenience bring up the question of whether synthetic biology will be sustainable in the long term. Besides, while synthetic biology can contribute to environmental sustainability—e.g., by advancing the development of biofuels and biodegradable plastics—it is important to critically assess the lifecycle studies of such products. Manufacturing procedures involved in synthetic biology can be resource-intensive, which might form huge carbon footprints. For instance, energy consumption in cultivating engineered microbes and converting their outputs might counteract the environmental benefits they provide. It is crucial to understand these trade-offs, particularly as we seek to transition towards cleaner technologies. The roadmap presenting the role of semiconductor-based materials indicates a path of more efficient biointerfaces, where plans can be taken towards reduced energy use and increased maximum ecological benefit from synthetic biology (Anikeeva et al., 2018). A holistic approach interlinking chemical concepts with environmental science is thus necessary. Lastly, regulatory regimes and public opinion also play overarching roles in deciding the environmental impact of synthetic biology technologies. The significance of open dialogue on the strengths and weaknesses of engineered biologies cannot be overemphasized in creating public trust. Misconceptions surrounding synthetic biology fuel resistance against its applications, slowing progress towards agriculture, medicine, and pollution clean-up. With synthetic biology increasingly becoming the site to fight climate change and resource limitation, there is a necessity to build strong regulatory standards that maintain ecological integrity. Such standards must synthesize knowledge from interdisciplinary-informed studies that uncover complex interdependencies between synthetic and natural ecosystems, aligning scientific advancement with public values. Lastly, the safe and prudent use of synthetic biology relies on a collaborative effort among scientists, policymakers, and society to make sure that these technologies lead to environmental sustainability while defending natural systems.

3.6.3. Ethical Implications of Genetic Modification

The advent of genetic modification in the framework of synthetic biology raises profound ethical concerns, particularly regarding the manipulation of living organisms at the molecular level. As gene engineering technologies continue to develop, the ability to modify genes to increase farm output, fight disease, or extend the lifespan of humans is no longer a fantasy but a real possibility. Such interventions do push society to deal with ethical issues about the possible dangers of unforeseen ecological impacts and the modification of natural species. These considerations prompt policymakers and researchers to tread a fine line between innovation and ethics to guarantee that scientific advancement does not compromise environmental integrity or exacerbate social inequalities, especially in regard to access to new biotechnology and biobased resources (N/A, 2010). Furthermore, the ethical consequences of genetic modification trickle down into public perception, which profoundly influences policy-making and research funding. In examining how societies perceive genetic engineering, broad disparities become apparent between cultures and indicate variation in ethical norms and systems of value. This is especially true in regions such as Europe, India, and China, where ethics in scientific policy is discussed with regard to historic, cultural, and social constructions (Chaturvedi et al., 2015). For instance, European controversies regarding GMOs focus on precautionary principles, based on appeals to opinion expressed through environmentalism and safety. In contrast, in rapidly expanding developing nations, the focus may be brought to bear more on societal benefits through GMOs as food security and economic solutions. This variation underlines the imperative need for discussion and education attuned to differing values but advocating cautious genetic innovation. Lastly, ethical considerations need to be integrated into the regulation of genetic modification technologies to provide a basis for building societal trust and acceptance. As new applications of synthetic biology are developed, there needs to be a system that allows openness in scientific practice to increase accountability among researchers and businesses. Ethical controls need to be designed that ensure public engagement, guaranteeing extensive stakeholder representation in risk assessment and policy-making. Such democratic integration assists in boosting public engagement and aids in preventing questions of unethical behavior, thus promoting a responsible culture of innovation. Finally, the issue is to develop an integrative approach that promotes scientific advancement but addresses the ethics surrounding modifying life at the genetic level, safeguarding sustainable steps toward the future of biotechnology (Chaturvedi et al., 2015).

SUMMARY

As synthetic biology continues to evolve, the basic concepts of chemistry remain key to comprehending and innovating in this cross-disciplinary area. Chemical understanding is integral to building complex biomolecular systems, especially as scientists work toward synthesizing artificial cells with lipid membrane encapsulation. These cell mimics are not mere theoretical constructs; they constitute a significant advance in biological engineering, pushing the boundaries of what is possible in cell design. Microfluidic technology, as the enabling technology in achieving this, facilitates high-throughput manufacture of these artificial cells with defined parameters. A clear understanding of the chemical underpinnings of such synthetic biological constructs thus underpins the effectiveness and optimality of such constructs, paving the way for future studies and use in synthetic biology (Elani, 2016). Furthermore, sharing and collaboration on chemical information are critical to driving further advances in synthetic biology. The chemistry community has been generally slow to embrace open access and data sharing, slowing overall advances. However, contemporary calls for openness and collaboration, such as those made in high-profile reports from the Royal Society and the European Commission, highlight the importance of constructing on previously laid research foundations. In open science, the researcher can facilitate more informed decision-making regarding synthetic routes and procedures, thereby optimizing innovation in the field. The exponential development of e-Research technologies and cheminformatics underscores the need for the chemistry community to effectively curate and manage high-dimensional data sets to make significant information available to scientists in subsequent generations (Bird and Frey, 2013). Overall, the intersection of chemistry and synthetic biology is offering hopeful avenues for scientific advancement. When scientists attempt to create new biological systems, the principles of chemistry provide the framework necessary to achieve this. Challenges will still persist, particularly those related to data sharing and reproducibility of results. Overcoming these challenges requires a concerted effort by the scientific community to embrace collaborative approaches and open access as a priority. Finally, with the synergy of robust chemical principles and emerging synthetic biology approaches, researchers are not just capable of expanding the frontiers of biological inquiry but are also closer to realizing mega challenges that face us in the realm of artificial cell construction and beyond (Bird and Frey, 2013; Elani, 2016).

One of the main ideas in synthetic biology is the design of artificial cells, typically enclosed in lipid membranes. Such constructs possess many advantages, including biocompatibility and a reconstitutable platform for biological machinery. Yet, one of the main constraints in the advancement of this field is the generation of membrane-based artificial cells in a controlled manner to facilitate high-throughput production. Key parameters such as size, biomolecular make-up, and spatial arrangement need to be precisely outlined to achieve functional outputs. Novel advances in microfluidic generation techniques are a breakthrough in addressing such challenges, allowing precision in the construction of artificial cell systems and new areas of opportunity in synthetic biology applications. Thus, knowledge of these principles is paramount to any venture that aims to make use of chemistry within synthetic biology (Elani, 2016). Another important principle of chemistry relevant to synthetic biology is the use of semiconductor materials. These materials are crucial in dictating biophysical dynamics at various length scales, which is significant for effective biointerface design. The plan for integrating semiconductor building blocks emphasizes their ability to interface and process biological components. These technologies hold the promise of enhanced monitoring and interaction with living organisms and gaining insight into complex biological signaling mechanisms. Building next-generation electronic and optoelectronic devices with seamless interaction between biological components and technology interfaces is also a necessity. Implications of such integration signal not only advances in synthetic biology but also overhauls our understanding of fundamental biological processes (Anikeeva et al., 2018). Lastly, the integration of artificial cells and the incorporation of semiconductor materials into biological architecture highlighted the interdisciplinary scope of synthetic biology. Such endeavors bridge chemistry, biology, and materials science, highlighting concerted efforts in untangling biological complexity. With advances in research, the fundamental principles of artificial cell synthesis and biointerface engineering are not only enhancing our engineering capabilities but also redefining the paths towards understanding life at a molecular level. It is this convergence of fundamental chemical principles with new engineering solutions that propels synthetic biology and compels researchers to follow new lines of biotechnological application. Emphasizing such interdisciplinarity will yield unprecedented progress, extending our capability to engineer, control, and understand biological systems chemically (Anikeeva et al., 2018; Elani, 2016).

REFERENCES

1. Aebersold, R., Andersen, J., Anderson, B. L., Anderson, G. M., Anderson, L., et al. (2009). Power and limitations of electrophoretic separations in proteomics strategies. https://core.ac.uk/download/52696251.pdf (accessed on 3rd October 2025).
2. Ahmed, A., Batool, K., Bibi, A., Nasim, F., et al. (2017). Microbial β-glucosidase: Sources, production and applications. https://core.ac.uk/download/188825639.pdf (accessed on 3rd October 2025).
3. Alam, A., Secrieru, A., Alkorta, I., Andes, D., Anwar, A., Behr, J., Brown, S., et al. (2019). Revisiting the structure and chemistry of 3(5)-substituted pyrazoles. https://core.ac.uk/download/288958529.pdf (accessed on 3rd October 2025).
4. Andrianantoandro, E., Arkin, A., Ashkenasy, G., Bath, J., Benner, S., Soloveichik, D., Winfree, E., et al. (2009). DNA as a universal substrate for chemical kinetics. https://core.ac.uk/download/4884366.pdf (accessed on 3rd October 2025).
5. Anikeeva, P., Bao, Z., Benfenati, F., Bezanilla, F., et al. (2018). Roadmap on semiconductor-cell bio interfaces. https://core.ac.uk/download/287623639.pdf (accessed on 3rd October 2025).
6. Arsiwalla, X. D., Herreros, I., Moulin-Frier, C., Sanchez-Fibla, M., et al. (2017). The morphospace of consciousness. http://arxiv.org/abs/1705.11190 (accessed on 3rd October 2025).
7. Bich, L., & Green, S. (2016). Is defining life pointless? Operational definitions at the frontiers of biology. https://core.ac.uk/download/131211791.pdf (accessed on 3rd October 2025).
8. Bird, C., & Frey, J. G. (2013). Chemical information matters: An e-research perspective on information and data sharing in the chemical sciences.
9. Booth, C., Brecher, D., Lowe, M. S., Stone, A., et al. (2014). Chemistry curriculum map 2013–2014. https://core.ac.uk/download/70981645.pdf (accessed on 3rd October 2025).
10. Boyen, C., Børresen, T., Dobson, A., Höfle, M., et al. (2025). *Marine Biotechnology: A New Vision and Strategy for Europe*. https://core.ac.uk/download/pdf/29241634.pdf (accessed on 3rd October 2025).
11. Bullock, R. M., Chen, J. G., Gagliardi, L., Chirik, P. J., Farha, O. K., Hendon, C. H., Jones, C. W., et al. (2020). Using nature's blueprint to expand catalysis with Earth-abundant metals. *Science, 369*. https://doi.org/10.1126/science.abc3183 (accessed on 3rd October 2025).
12. Cestellos-Blanco, S., Kornienko, N., Lim, J., Liu, C., et al. (2018).

Physical biology of the materials-microorganism interface. https://core.ac.uk/download/287623640.pdf (accessed on 3rd October 2025).

13. Chaturvedi, S., Ladikas, M., Stemerding, D., Zhao, Y., et al. (2015). *Science and Technology Governance and Ethics–A Global Perspective from Europe, India and China.* https://core.ac.uk/download/42136809.pdf (accessed on 3rd October 2025).
14. Chehelgerdi, M., Chehelgerdi, M., Allela, O. Q. B., Pecho, R. D. C., Narayanan, J., Rao, D. P., Thamaraikani, T., et al. (2023). Progressing nanotechnology to improve targeted cancer treatment: Overcoming hurdles in its clinical implementation. *Molecular Cancer, 22*. https://doi.org/10.1186/s12943-023-01865-0 (accessed on 3rd October 2025).
15. Dengo, N., Gross, S., & Vittadini, A. (2017). Functionalization of colloidal transition metal sulphides nanocrystals: A fascinating and challenging playground for the chemist. https://core.ac.uk/download/141699569.pdf (accessed on 3rd October 2025).
16. Desai, A., Mishra, A., Sorrenti, A., Dambenieks, A. K., Grzybowski, B. A., Ravensteijn van, B. G. P., et al. (2018). Energy consumption in chemical fuel-driven self-assembly. https://core.ac.uk/download/195758473.pdf (accessed on 3rd October 2025).
17. Ding, Y., Tan, C., & Wu, F. (2014). Synthetic biology: A bridge between artificial and natural cells. https://core.ac.uk/download/323068426.pdf (accessed on 3rd October 2025).
18. Elani, Y. (2016). Construction of membrane-bound artificial cells using microfluidics: A new frontier in bottom-up synthetic biology. https://core.ac.uk/download/77011265.pdf (accessed on 3rd October 2025).
19. Ettlinger, R., Lächelt, U., Gref, R., Horcajada, P., Lammers, T., Serre, C., Couvreur, P., et al. (2022). Toxicity of metal–organic framework nanoparticles: From essential analyses to potential applications. *Chemical Society Reviews, 51*, 464–484. https://doi.org/10.1039/d1cs00918d (accessed on 3rd October 2025).
20. Gabius, H. J., Jiménez-Barbero, J., Kaltner, H., Lieth, C., et al. (2001). Towards defining the role of glycans as hardware in information storage and transfer: Basic principles, experimental approaches and recent progress. https://core.ac.uk/download/16432899.pdf (accessed on 3rd October 2025).
21. Gradoff, O. (2017). Autopoiesis concepts for chemical origins of life and synthetic biology [Stenogram of the popular lecture on the foreign bibliographic seminar]. https://core.ac.uk/download/186331807.pdf (accessed on 3rd October 2025).

22. Hall, G. M., & Howe, J. (2012). The impact of synthetic biology in chemical engineering – Educational issues. https://core.ac.uk/download/6112945.pdf (accessed on 3rd October 2025).

23. Heinemann, M., & Panke, S. (2006). Synthetic biology—Putting engineering into biology. https://core.ac.uk/download/pdf/12944386.pdf (accessed on 3rd October 2025).

24. Luo, Y., Abidian, M. R., Ahn, J. H., Akinwande, D., Andrews, A. M., Antonietti, M., Bao, Z., et al. (2023). Technology roadmap for flexible sensors. *ACS Nano, 17*, 5211–5295. https://doi.org/10.1021/acsnano.2c12606 (accessed on 3rd October 2025).

25. Mujtaba, M., Fraceto, L. F., Fazeli, M., Mukherjee, S., Savassa, S. M., de Medeiros, G. A., do Espírito Santo Pereira, A., et al. (2023). Lignocellulosic biomass from agricultural waste to the circular economy: A review with focus on biofuels, bio composites and bioplastics. *Journal of Cleaner Production, 402*, 136815. https://doi.org/10.1016/j.jclepro.2023.136815 (accessed on 3rd October 2025).

26. N/A. (2010). Criteria for the diploma qualifications in science at advanced level: Principal learning. https://core.ac.uk/download/4151080.pdf (accessed on 3rd October 2025).

27. N/A. (2025). ECUT (Energy Conversion and Utilization Technologies) program: Biocatalysis project. https://core.ac.uk/download/pdf/42832304.pdf (accessed on 3rd October 2025).

28. Naresh, V., & Lee, N. (2021). A review on biosensors and recent development of nanostructured materials-enabled biosensors. *Sensors, 21*, 1109. https://doi.org/10.3390/s21041109 (accessed on 3rd October 2025).

29. Patring, J. (2007). Development and validation of chromatographic methods to study folate derivatives produced by yeasts. https://core.ac.uk/download/11695153.pdf (accessed on 3rd October 2025).

30. Singh, J., Dutta, T., Kim, K. H., Rawat, M., Samddar, P., & Kumar, P. (2018). 'Green' synthesis of metals and their oxide nanoparticles: Applications for environmental remediation. *Journal of Nanobiotechnology, 16*. https://doi.org/10.1186/s12951-018-0408-4 (accessed on 3rd October 2025).

31. Wallace, R. (2012). Beyond the fuzzy lock-and-key: Spontaneous symmetry shifts and glycan/lectin logic gates. https://core.ac.uk/download/pdf/290598.pdf (accessed on 3rd October 2025).

32. Welsh, J. A., Goberdhan, D. C. I., O'Driscoll, L., Buzás, E. I., Blenkiron, C., Bussolati, B., Cai, H., et al. (2024). Minimal information for studies of extracellular vesicles (MISEV2023): From basic to advanced approaches. *Journal of Extracellular Vesicles, 13*. https://doi.org/10.1002/jev2.12404.

CHAPTER 4

GENETIC ENGINEERING TOOLS

Contents

INTRODUCTION

Genetic engineering has rapidly become one of the most promising areas of modern biotechnology. As we venture into the intricacies of genetic engineering tools, it is imperative to realize their value potential in a wide array of fields, from agriculture and medicine to environmental sciences. These tools enable scientists to meddle with genetic material with unprecedented precision, which has led to advances in genetically modified organisms (GMOs) and precision treatments. The tools of genetic engineering not only enables us to promote desired characteristics in plants but also opens doors to novel therapies for genetic ailments. In going through this discovery, it is important to have an idea of the fundamental principles and methods that define genetic engineering in order to appreciate the promise and shortfalls embedded in these technologies. At the forefront of the genetic engineering pack is a series of pioneering techniques that have taken the science to where it stands, with some of those including CRISPR-Cas9, gene cloning, and recombinant DNA. All these technologies operate on the premise of interfering with the DNA sequence to achieve intended outcomes, be it adding, removing, or altering genes. CRISPR-Cas9 has generated significant buzz because it is easy to use and highly efficient, enabling scientists to edit genomes with extremely high accuracy.

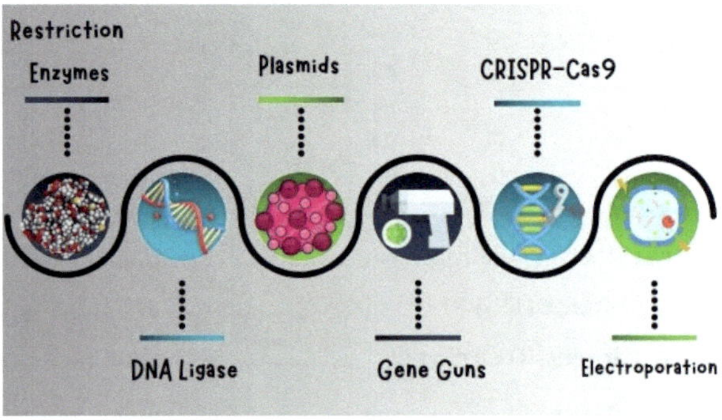

Figure 4.1. Types of tools used in genetic engineering.

Source: https://files.codingninjas.in/article_images/custom-upload-1685435011-ff372690.png.

Furthermore, the historical context made available by innovations in synthetic biology and gene therapy demonstrates a trajectory of continuous progress, leading to current high-level methodologies. Since genetic engineering is growing, proper analysis of these tools will provide an overall understanding of their operations in shaping the future of the application of biotechnology. The consequences of genetic engineering technology extend beyond the act of scientific inquiry in itself; they present ethical and regulative problems that need to be addressed by the public. As advances are made in the manipulation of genetic material, concerns about bioethics, safety, and the potential unintended consequences of such manipulation become more imminent. Policymakers are required to ensure that policies are developed to ensure that the technology is responsibly used while promoting innovation within the sector. In addition, societal attitudes also play a critical role in the adoption of genetically modified products, influencing market forces and regulatory policies. It thus becomes essential to have a balance between the development of science and the ethical concerns surrounding such advanced sciences (Figure 4.1).

Definition of Genetic Engineering

Genetic engineering in the field of genetics is the deliberate alteration of an organism's genome with the assistance of biotechnological tools. The alteration may be accomplished by inserting, removing, or replacing specific DNA sequences to achieve specific characteristics, from increased tolerance to drought in plants to the production of insulin in bacteria. The underlying assumption of genetic engineering relies on the understanding of DNA structure and function, thus allowing scientists to play with specific genes responsible for particular traits. By applying highly specific techniques, scientists can create genetically altered organisms (GMOs), which could one day revolutionize agriculture, medicine, and even bioenergy production. Therefore, genetic engineering is a key to addressing world issues such as food security and disease control. Beyond the most basic definition, the tools and techniques that facilitate genetic engineering are diverse and continually advancing. Among the most prevalent techniques is CRISPR-Cas9, an innovative gene-editing method that allows for precise modifications at various sites in the genome. It leverages bacteria's innate defense mechanisms to enable scientists to target and manipulate DNA sequences with unprecedented precision and efficiency. More traditional methods, such as recombinant DNA technology and gene splicing, are still largely employed. These tools have collectively contributed to a broader understanding of genetics and molecular biology. Furthermore, they have encouraged new techniques in medical interventions, including the use of genetic engineering in gene therapy, which can treat genetic diseases in humans (Company et al., 2013). The use of genetic engineering transcends the technique and definition; it raises ethical issues and social debates concerning biodiversity, food security, and genetic privacy. Since organisms are genetically transformed to express traits that appeal to humans, environmental consequences might be tremendous. Concerns about potential allergenicity, pest resistance, and long-term environmental impacts need rigorous examination of genetic engineering applications. Another area, the ethical controversy of genetic manipulation, has also raised concerns about ownership of genetic resources and potential eugenic applications in human genetics. Therefore, the debate on the definition and use of genetic engineering must extend beyond its scientific to include its ethical considerations, public attitudes, and the regulatory frameworks that govern its deployment (Brunetti et al., 2015) (Figure 4.2).

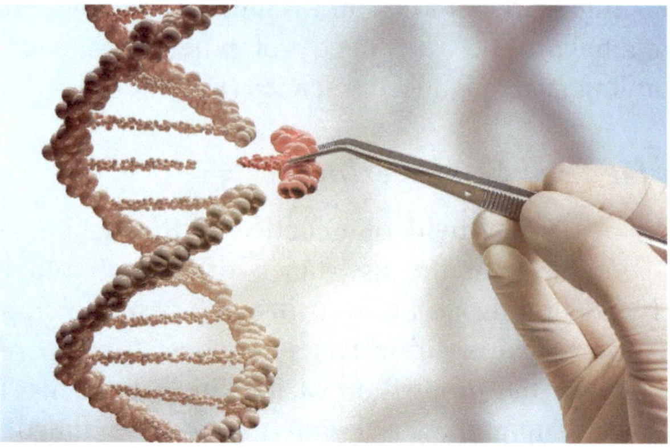

Figure 4.2. Definition of genetic engineering.

Source: https://www.pirbright.ac.uk/sites/default/files/styles/large/public/quick_media/gene-editing-genetics-genomics-pirbright-research.jpg?itok=k0p27DM1.

Importance of Genetic Engineering in Modern Science

The significance of genetic engineering cannot be overstated in modern science as it has direct effects on various fields of study like medicine, agriculture, and environmental science. Genetic engineering is one of the key advantages that assist in enhancing crop resistance and yield, which is crucial in fighting food security challenges posed by climate change and population growth. With techniques like CRISPR and base editing, scientists can engineer GMOs not only more nutritious but also resistant to disease and pests. Such capability is instrumental in ensuring that agriculture can keep pace with the demand for food in a sustainable manner while reducing its impact on the environment. Thus, the equipment developed for genetic engineering serves not only science, but society as a whole by contributing to food security and sustainable agriculture. In medicine, genetic engineering has the potential to revolutionize its application, particularly in the development of targeted therapies against previously recalcitrant genetic diseases. Breakthroughs in gene-editing technologies allow for precise modifications to be made at the individual loci within the genome, which can cure diseases such as cystic fibrosis or sickle cell disease. This recent population genetics breakthrough emphasizes the necessity for harnessing genome editing technologies with traditional genetic research to explore the implications of genetic variation in populations. As pointed out in recent debates, a working study of population-specific variation with the aid of cutting-edge genome editing can assist in confirming evolutionary hypotheses, pushing medical research and evolutionary biology forward. This interface between the two fields is a clear demonstration of the pivotal role genetic engineering serves in deepening our knowledge of genetic disorders and improving the effectiveness of therapeutic strategies (Fox et al., 2020). Furthermore, the ethical aspects of genetic engineering cannot be ignored as technology advances. Scientific inquiry must be balanced with ethical concerns about tampering with life at the genetic level. The controversy on the use of gene editing technologies, particularly on humans, involves issues of grave concern regarding long-term consequences and potential unexpected outcomes. The imperative for new conceptual and theoretical thinking, highlighted in recent literature, underscores the need for responsible approaches to genetic engineering that take into account ecological and societal implications (Arnold et al., 2013). While researchers test boundaries of what is scientifically possible, they also have to discuss matters of morality, equity, and potential for abuse so that products developed serve humanity for the common good while remaining safe, just, and ethically responsible in scientific advancement (Figure 4.3).

Figure 4.3. Importance of genetic engineering

Source: https://www.news-medical.net/images/Article_Images/ImageForArticle_22413_16508965894669618.jpg.

Overview of Genetic Engineering Tools

The landscape of genetic engineering has rapidly transformed with the emergence of powerful molecular tools to enable high-precision manipulation of genetic material. CRISPR/Cas9 technology is one such tool among these that has revolutionized gene editing. CRISPR/Cas9 exploits the bacterial mechanism of immunity against viral attack and enables targeted genome modification with unprecedented precision and efficiency. This system uses a guide RNA to direct the Cas9 enzyme to the target DNA sequence, making cuts and enabling subsequent modifications. The use of CRISPR in various organisms like plants and microbes in recent research has opened up prospects for enhanced features and metabolic processes, enhancing the discovery rate and commercialization of biotechnology. Continued refinement of CRISPR approaches further highlights the tremendous potential of these technologies for numerous applications in medicine and agriculture, a new paradigm of genetic engineering. In addition to CRISPR/Cas9, other genetic engineering reagents such as TALENs (Transcription Activator-Like Effector Nucleases) and ZFNs (Zinc Finger Nucleases) provide alternative approaches to targeted genome editing. TALENs employ TALE proteins for DNA sequence binding, while ZFNs employ zinc finger motifs for recognition. Both technologies enable researchers to introduce double-strand breaks at predetermined locations in the genome, leading to precise alterations via cellular repair mechanisms. The process, however, is supplemented by intricacies in designing such nucleases over CRISPR systems. As noted in recent research, the development of these alternative tools provides a broader set of tools for researchers, particularly when CRISPR is inaccessible. The widening toolkit of genetic engineering tools enables the exploration and manipulation of a larger set of genetic targets, thereby sparking innovation in synthetic biology and biomedicine. Other technologies in genetic engineering that are advancing include advances in synthetic biology, where engineering principles are brought together with biological systems to create new functions and capabilities. Among such techniques is the use of engineered genetic circuits that drive gene expression in response to specific environmental stimuli. This technology is gaining momentum, particularly in the engineering of microbial strains for biochemical synthesis and bioremediation. Recent studies of fast-growing species and non-model organisms in cyanobacterial research hold potential for novel methods in transformation and regulation. Moreover, technologies for next-generation metabolic engineering, including those from current CRISPR studies, are unlocking opportunities for new uses in pharmaceuticals and biofuels. As the genetic engineering toolkit continues to be enhanced, it promotes innovation in addressing global problems like sustainable manufacturing and health issues, ultimately shaping the future of environmental conservation and biotechnology.

4.1. CRISPR-CAS9 TECHNOLOGY

The discovery of CRISPR-Cas9 technology has been a groundbreaking advancement in the area of genetic engineering, revolutionizing the entire molecular biology and genetics landscape. This process, based as it is on the immune mechanisms of bacteria, provides researchers with the opportunity to make specific edits in the genome with unprecedented accuracy. Unlike previous genetic editing techniques, which made random changes, CRISPR-Cas9 allows for specific changes, such as gene knockouts or insertions, thus making the creation of cellular and animal models of human diseases possible. The implications for biomedical research are considerable since the technology enables researchers to study the causes of various disorders at the genetic level and offers a pathway toward new therapeutic approaches, particularly in the case of gene therapy, where the repair of patient-specific mutations offers a pathway toward more effective treatments (Ferreira et al., 2020). Although its potential is considerable, the application of the CRISPR-Cas9 technology in a clinical setting faces hurdles to its use.

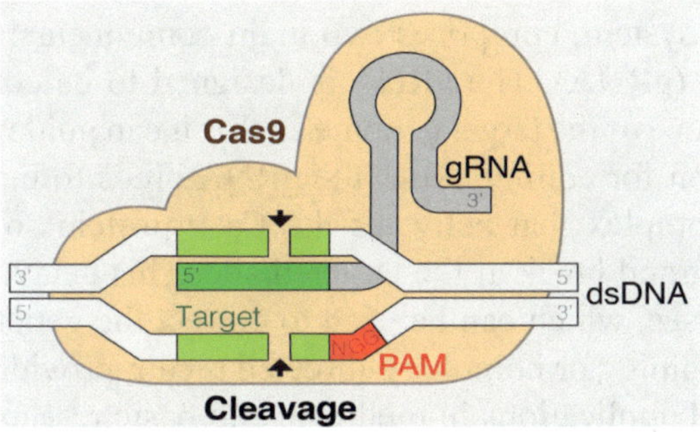

Figure 4.4. The CRISPR-Cas9 system.

Source: https://uskudar.edu.tr/assets/uploads/sayfa/693/images/crispr-021.png

As scientists continue to translate these laboratory discoveries towards therapy, they will also have to overcome the hurdles of efficiency, specificity, and delivery of CRISPR components into the target cells of the human body. The danger of off-target effects, where the wrong genetic changes are introduced, presents additional ethical and safety issues that must be extremely carefully resolved before there can be widespread clinical utilization. Besides, the development of effective means of delivering CRISPR-Cas9 into target tissues is a significant challenge. Overcoming these challenges is crucial to harnessing the full potential of CRISPR-Cas9 so that its application can extend beyond research to become effective, therapeutic treatments for genetic diseases (Ferreira et al., 2020). CRISPR-Cas9 not only edits human genomes; it is also an effective tool in treating viral infections. Such work has shown its potential in breaking latent infections caused by herpesviruses, long a problem of difficult treatment due to their ability to evade immune response. Researchers can induce specific mutations in the viral genome using CRISPR-Cas9 that incapacitate the virus during the latent phase, something that current antiviral strategies cannot do effectively. This innovative approach not only moves the understanding of viral genetics forward but also puts CRISPR-Cas9 as a probable contender in developing adjunctive therapies to the conventional treatments, thereby increasing its utility in translational medicine. As more work is underway, CRISPR-Cas9 can potentially transform how we perceive and treat viral infections by completely shifting the gameplay of viral latency and persistence (Figure 4.4).

4.1.1. Mechanism of CRISPR-Cas9

The CRISPR-Cas9 system is now a revolutionary genetic engineering tool due to its specificity and versatility. The system, initially derived from the bacterial adaptive immune system, comprises two main components: the Cas9 enzyme and a guide RNA (gRNA). The gRNA is designed to be complementary to a target DNA sequence of the target genome so that it can guide the Cas9 nuclease to the very location for editing. After the gRNA binds to the target DNA, the gRNA forms a complex that activates the Cas9 protein so that it is able to induce a double-strand break at the targeted site. This delay initiates the cell's own repair processes, which can be used to correct the genome by either non-homologous end joining or homology-directed repair, providing the foundation for many potential applications in medicine and basic research (Ferreira et al., 2020). The applications of the CRISPR-Cas9 system extend far beyond basic genetic research; they encompass profound implications for gene therapy and synthetic biology. Utilizing the precision of CRISPR-Cas9, researchers can create cellular models that are genuine models of human disease and are more accessible to deciphering disease mechanisms and devising potential therapies. Its therapeutic potential is best demonstrated by its ability to correct single mutations for genetic disorders. Bringing it to the clinic, however, is still a daunting prospect, primarily because of concerns about efficiency, specificity, and mechanism of delivery. Solving these problems is crucial for the therapeutic future of CRISPR technologies, as researchers want to optimize the system for safety and efficacy for human therapy (Ferreira et al., 2020). The new discoveries of transposable elements (TEs) integration into genome editing only serve to confirm the revolutionary essence of CRISPR-Cas9. Transposable elements that were previously considered selfish genetic elements are now recognized to possess the potential to contribute positively to genetic variability and adaptation. The CRISPR-Cas9 system has opened novel avenues for exploiting such elements, such as the development of miniature inverted-repeat transposable elements (MITEs) that can serve as functional drivers for genomic editing. This cross-link between TEs and CRISPR technology marks a significant milestone in our understanding of genome dynamics and engineering. As researchers explore these interactions, they are better placed to modulate gene expression and develop sophisticated techniques to capitalize on TEs in genetic engineering technologies (Figure 4.5).

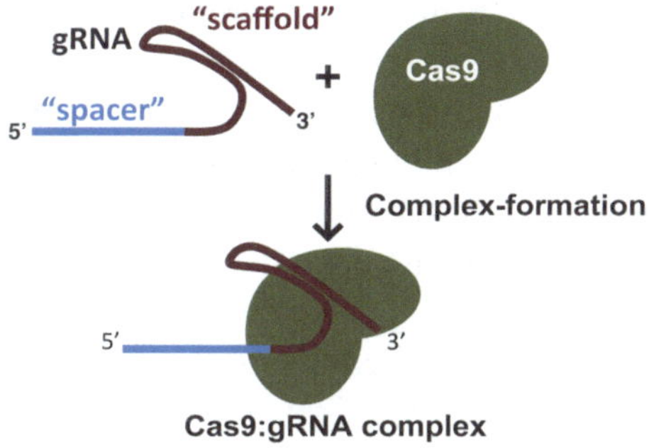

Figure 4.5. The CRISPR gRNA and Cas9 protein.

*Source:*https://blog.addgene.org/components-of-crispr/cas9-our-new-crispr-101-ebook

4.1.2. Applications in Gene Editing

The recent development of gene-editing technologies has revolutionized the landscape of genetic engineering and opened new possibilities for applications in biological research and medicine. Perhaps the most groundbreaking technology is the CRISPR-Cas9 system, which provides exact editing of genomes and enables scientists to investigate gene function and regulation across organisms. This new approach has proven particularly valuable in model systems such as Drosophila, in which CRISPR-Cas9 is utilized to control genetic sequences with unprecedented accuracy and ease. Researchers can therefore gain a better insight into intricate genotype-phenotype relationships, obtain a deeper appreciation of genetic pathways, and eventually pave the way for new therapeutic strategies. The consequences of such research extend beyond basic science; it creates an avenue for subsequent research into the area of genetic therapy for the treatment of genetic disorders and diseases, showing the revolutionary power of gene-editing tools (Bier et al., 2018). Gene editing's use in curing human disease is a profound application of genetic engineering tools like CRISPR-Cas9. This technology has made it possible for scientists to construct reliable cellular and animal models that exhibit human pathological conditions, deriving extremely valuable insights into the mechanisms of disease. Gene editing, for instance, can reverse particular mutations of inherited disease, with therapeutic strategies in the pipeline that were previously unattainable. Aside from primary research, the therapeutic potential of CRISPR technology is vast, particularly in the context of gene therapy, where accurate corrections can have a staggering effect on patient outcomes. But the translation of CRISPR/Cas9 into the clinic is not without various challenges, such as improving delivery means and attaining high specificity of edits, which must be overcome to unlock the full potential of gene editing in medicine (Ferreira et al., 2020). Furthermore, gene editing's versatility is evident in applications in agriculture, where modifications can improve crop resilience, yield, and quality. Gene editing equipment enables the addition of desirable traits, such as resistance to drought or pests, that can aid in food resource security under climate change and growing population scenarios. By reducing the process of developing new crop varieties to a simpler form, gene editing significantly reduces the time and cost needed in the conventional breeding technique. Gene editing's ethical and regulatory environment remains an open question, an issue that requires a balanced approach that considers scientific advancement and societal interests. As studies further elucidate the potential benefits and risks of gene editing in various applications, the present debate will determine the future of genetic engineering and its application to solve global problems, reaffirming the relevance of this technology in the modern world (Bier et al., 2018) (Figure 4.6).

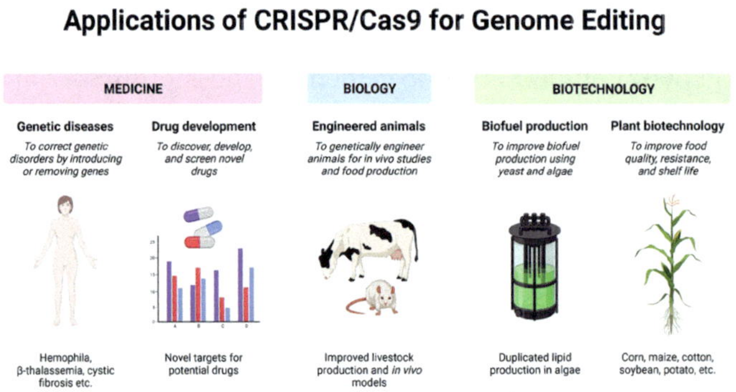

Figure 4.6. Applications of CRISPR/Cas9 for genome editing.

Source: https://cdn.prod.website-files.com/621e95f9ac30687a56e4297e/64a8d7ac64d21b5b0551a280_V2_1671023332817_04d2bd40-2f65-416b-b365-1bad-bd415e24.png.

4.1.3. Ethical Considerations Surrounding CRISPR-Cas9

The rapid evolution of CRISPR-Cas9 technology has initiated high-level ethical controversy among scientists and elsewhere. As a powerful gene-editing tool, CRISPR-Cas9 poses deeply ethical issues, particularly for human germline editing. Designer babies bring about social inequality and the threat of new forms of genetic discrimination. These are not merely hypothetical fears; they provoke thoughtful scrutiny of how new technologies might overstep their intended boundaries, with unforeseeable societal repercussions. Furthermore, the eugenic and genetic manipulation history needs to be recalled so that the lessons of that period can be used to forestall such abuses and not repeat history to emphasize the need for ethical restraints on genomic work and use. The ethical landscape of CRISPR technology is further complicated by the doctrine of informed consent, particularly research into human subjects. Because gene modifications can have implications not just for individuals but also for lineages and populations as well, it is essential to proceed cautiously through such consent processes. The implications of gene modification in embryos, for instance, might make it more complicated for parents to be certain they comprehend the wide-reaching consequences of such an action, one that could go beyond generations. In addition, Fox et al. (2020) identify the need for collaboration between geneticists and the wider community of population genetics. This type of collaboration will enable comprehension of CRISPR-Cas9 application, providing social and ethical frameworks to deal with the potential consequences of genomic interventions in an accountable way. The role of scientists and researchers is also important to balance the uncertainty of modifying complex biological systems. As Freemont et al. (2014) suggest, while synthetic biology holds the promise of building robust biological systems, the inherent randomness of these systems holds the promise of unintended consequences. This randomness warrants strong ethical controls and guidelines that stress transparency and responsibility in scientific practice. Scientists must be trained to detect and limit risks in gene editing, and also to have ethics built into scientific practices. Lastly, the embedding of ethical prudence in the development and application of CRISPR-Cas9 technology is required to establish trust with the people, promote responsible innovation, and address the intricate ethical issues of gene editing technologies like CRISPR.

4.2. RECOMBINANT DNA TECHNOLOGY

Recombinant DNA technology is a foundational subject in the broader context of genetic engineering methods, founded on the ability to manipulate genetic material across various organisms. Recombinant DNA technology encompasses a range of methods applied to the purification, modification, and recombination of DNA fragments that may be expressed within host cells. At the foundation of these methods are recombinant DNA enzymes and cloning methods that can cause the creation of transgenic organisms. UiTM et al. (2017) add that exploration of recombinant DNA technology involves a deep understanding of bacterial transformation techniques, which are basic to successful incorporation of foreign DNA into host genomes. The implications of these approaches extend beyond mere academic interest and are behind some of the advances, for instance, the production of recombinant proteins and transgenic plants, which illustrate the potential and variability of this technology to generalize to agriculture and medicine. Recombinant DNA technology has led to dramatic advances in plant biotechnology, particularly in enhancing secondary metabolism for application in agriculture. Techniques such as Agrobacterium-mediated transformation and particle bombardment have been employed to introduce specific genes into plant genomes to control secondary metabolites. As noted by Yeoman et al. (2001), many techniques of transformation are employed, ranging from the utilization of viral vectors to protoplast fusion, depending on the type of genetic alteration. Notably, researchers have managed to manipulate the expression of developmentally controlled genes and add new genes to enhance desirable traits in crops.

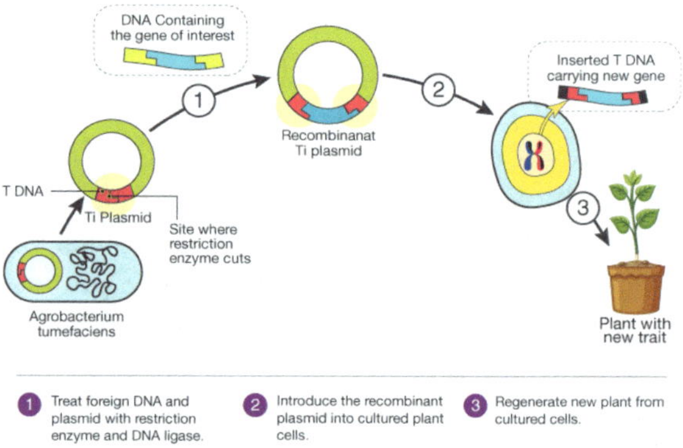

Figure 4.7. Recombinant DNA technology.

Source: https://cdn1.byjus.com/wp-content/uploads/2019/04/Recombinant-DNA-Technology.png.

This has not only ensured higher yield but also allowed for the production of beneficial compounds that can be of significant commercial and therapeutic value. The ongoing research in this field reflects the extensive potential of recombinant DNA technology to transform the landscape of plant science. Applications of recombinant DNA technology extend beyond gene modification to include significant contributions to biomedical research and therapy. For instance, this technology holds the secret to the production of recombinant proteins, therapeutic antibodies, and vaccines, which are critical in the battle against various diseases. Utilization of recombinant DNA techniques in mammalian systems allows scientists to express human proteins in model organisms, providing a cost-effective and scalable tool for the discovery of drugs. As discussed in UiTM et al. (2017), this aspect of recombinant DNA technology gives tremendous advantage as it allows for the rapid creation of medical solutions that can address public health needs. Along with this, the ethical implications of genetic changes in medicine must be appropriately analyzed so that the progress achieved through such technologies is compatible with society's ethics and safety. The discussion in these issues reflects the complexities involved with the use of genetic engineering methods in real contexts (Figure 4.7).

4.2.1. Process of Creating Recombinant DNA

The recombinant DNA production process is a series of well-coordinated processes designed to combine genetic material from two or more origins into one operational unit. Special enzyme reagents, including restriction endonucleases, are employed initially to cut DNA at identified sequences. The cutting allows researchers to isolate genes or segments of interest from the DNA of the donor organism. The DNA in question is then placed into a carrier—a plasmid, for example—that facilitates secondary replication in a host cell. The precision during this initial cleavage and selection of the appropriate vectors is important as it controls the efficiency with which the recombinant DNAs are expressed in the host organism. Once the foreign DNA is inserted into the vector, ligation is an important step in which DNA ligase catalyzes the covalent joining of the DNA fragments. This process not only sequesters the foreign DNA within the vector but also keeps the integrity of the vector and foreign DNA intact, making them compatible with host cell machinery. The selection of vectors is key, with considerations such as replicative origin, markers for selection, and the presence of promoter sequences determining this. Recent advances in the tools of gene-editing, e.g., serine integrases, render these processes much more efficient. These enzymes provide site-specific recombination capability, thus facilitating targeted insertion of the gene, which is capable of maximizing the success rate of recombinant DNA experiments (Stark et al., 2017). Having constructed the recombinant DNA, the second process involves transforming host organisms with the constructed vector. This process implants the recombinant DNA in a suitable host, either bacterial, yeast, or eukaryotic cells, depending on the intended application. The altered cells are then cultured in specific conditions that support replication of the plasmid and expression of inserted genes. Recovery and study of the expressed protein or traits allow scientists to identify the success of recombinant DNA construction. As biotechnology keeps moving forward, methods such as those referenced in computational biology and synthetic biology advance further, offering novel ways to the enhancement of the efficiency and precision of recombinant DNA technology (Elkin et al., 2025) (Figure 4.8).

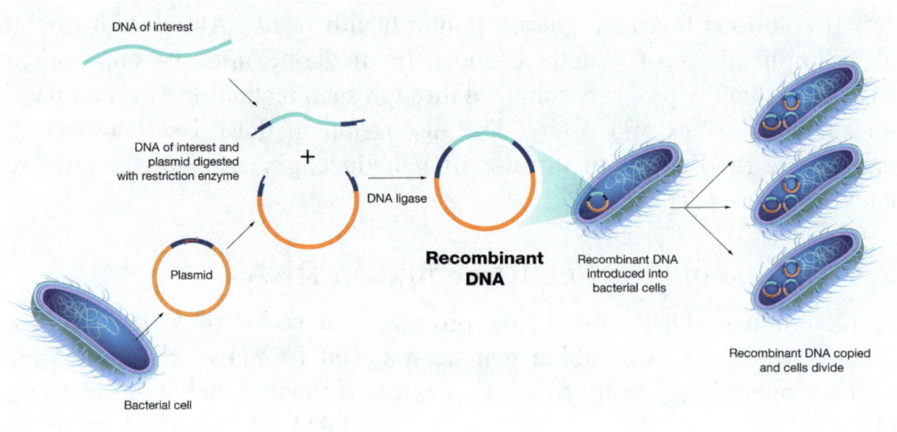

Figure 4.8. The process of recombinant DNA.

Source: https://www.genome.gov/sites/default/files/media/images/2022-04/Recombinant-DNA-technology.jpg

4.2.2. Uses in Medicine and Agriculture

The addition of genetic engineering instruments has had a great impact on medicine and agriculture, enabling unprecedented advancement in them. In medicine, techniques like CRISPR-Cas9 allow gene editing within genomic sequences with ease, and it is simple to create targeted medicines for genetic disorders. These developments hold out the hope that treatment can be made more efficient, since day by day, scientists are figuring out how to design genes responsible for diseases like cystic fibrosis and sickle cell anemia. To be able to correct genetic mutations in their early stages not only has the promise of disease cure but also advances research in personalized medicine, where therapy can be tailored to a personal genetic profile. The projections of these technologies trickle down through health systems, offering hope to many so far incurable conditions. In agriculture, genetic engineering tools provide answers to such key issues as food security and environmental stewardship. Tools like genetically modified organisms (GMOs) enable the production of crops that are pest-, disease-, and environmental stress-resistant, and thus improve yield and reduce reliance on chemical pesticides. This approach not only enhances productivity but also maintains eco-friendly agricultural practices since fewer resource-intensive crops can be cultivated. Furthermore, studies have shown in recent research that the application of genetic algorithms to enhance crop yields has massive innovations in traditional and modern agricultural practices. Noted is the fact that the efficiency of these methods is evidenced by the ability to optimize structure-activity relationships in agricultural commodities, which is evidence of the versatility and efficacy of genetic engineering to meet the demands of an increasing population (Lorentz JÄNTSCHI et al., 2010). The intersection of genetic engineering and evolutionary biology is also critical to advances in medicine and agriculture. Understanding the genetic complexities and diversities of organisms can illuminate how genetic changes might influence traits related to survival and adaptation in varying habitats. The integration of ancient principles of evolution and new genetic methods can enable new approaches to be devised for enhancing crop resilience and producing robust medicinal therapies. With the wealth of genomic research information, as noted in recent reviews, there is a need to have novel theoretical frameworks to suitably utilize this information in furthering our ability to design interventions that can benefit human health as well as agricultural harvests (Arnold et al., 2013). Therefore, applications of genetic engineering techniques to medicine and agriculture are not technological advances per se; they represent a paradigm shift to efficient, integrative solutions to world problems (Figure 4.9).

Figure 4.9. Genetic engineering in agriculture.

Source: https://www.mdpi.com/files/multidisciplinary_topic_graphical_abstract/837/Picture1_1200x630.png.

4.2.3. Safety and Regulatory Issues

The intersection of regulation and genetic engineering is a complex arena controlled by concerns of safety and the inadequacy of present frameworks. The principal regulatory framework governing biotechnology in the United States, the Coordinated Framework for Regulation of Biotechnology, was established decades ago, prior to the looming genetic engineering technology breakthroughs. This framework is therefore not capable of controlling the new risks posed by the new genetic engineering technologies. Most consumer concerns about genetically modified organisms (GMOs) are driven by an impression that regulations in place do not consider the risks to human health and the environment sufficiently. Much of the current controversy surrounds whether such concerns are based on scientific facts or if they represent wider societal fears about technological development. Understanding such duality is critical in determining how the regulatory framework must evolve to restore public confidence and establish better safety measures in genetic engineering (Peck et al., 2017). Further, public discourse around genetic engineering often highlights the need for openness and comprehensive risk assessment procedures. Labelling GMOs campaigns compulsory reveal not just consumers' lack of trust but also an overall appeal for accountability from regulators. This has opened up discussion on whether current systems of oversight are adequate, particularly with regard to previously unconsidered consumer safety concerns. The shortcomings of the Coordinated Framework reveal a mismatch between science discovery and regulatory policy, raising ethical questions about how genetic engineering risks are communicated to the public. Moreover, the advent of new genetic modification technology, such as CRISPR and gene editing, comes along with additional regulatory challenges that existing legislation is not yet geared to deal with, thus necessitating re-examination of safety regulations and administrative capacities (Eicher et al., 2025). The more the genetic engineering technology develops, the more the regulators need to catch up by modifying themselves to meet the new safety challenges while promoting technology innovation. This requires a high-level reassessment of current policies, which may be excessively divorced from the scientific realities of genetic modification. For regulators to be meaningfully effective, they must engage in an open-ended dialogue with stakeholders like scientists, industry, and consumers to better understand the risks and expectations surrounding genetic engineering technologies. Lastly, consideration of adaptive regulatory strategies that allow for flexibility and agility in responding to technology changes is essential. It is only by concerted efforts such as these that one can have a regulatory framework which not only promises safety but also inspires public trust and facilitates continued innovation in genetic engineering (Peck et al., 2017).

4.3. GENE CLONING TECHNIQUES

Techniques of gene cloning are the hallmark of modern genetic engineering, enabling precise manipulation of DNA for research and agricultural uses innumerable. Some of the key methods used in gene cloning include the use of restriction enzymes, which cut the DNA in specific sequences to enable target genes to be inserted into plasmids or other vectors. The process allows researchers to produce recombinant DNA molecules, which can be grown on bacterial cells. The successful integration of foreign genes into host organisms leads to the synthesis of new-function proteins, allowing us to gain further insight into gene expression and control.

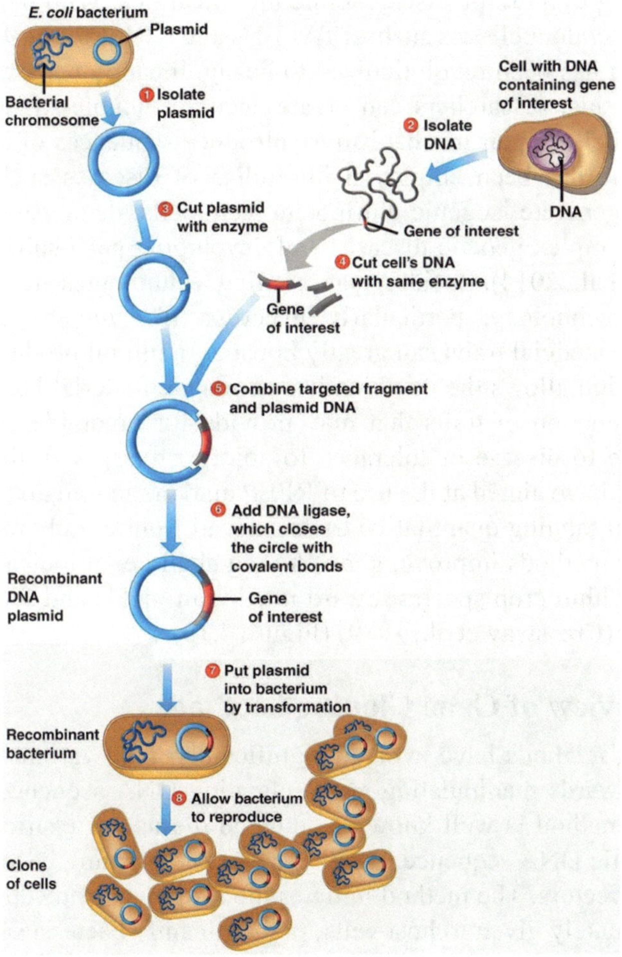

Figure 4.10. Steps in gene cloning.

*Source:*https://microbenotes.com/wp-content/uploads/2018/09/Steps-in-Gene-Cloning.jpg.

Besides, advances in cloning technologies have enabled us to produce transgenic organisms, which have vast potential for agricultural improvement, such as increased pest resistance and nutritional quality. In addition to restriction enzyme cloning, homologous recombination has also become an essential tool in gene cloning, particularly when working with eukaryotic cells. Homologous recombination follows the natural process of recombination in which some portion of the DNA is replaced or repaired by a homologous piece, resulting in more precise genomic modifications. With the discovery of locus-specific endonucleases such as TALENs and CRISPR/Cas9, homologous recombination has been revolutionized to be significantly more efficient. With these instruments, researchers can create accurate double-strand breaks that stimulate the cells' repair mechanism to introduce sequences of interest genes. This technique has been employed in studies of diseases such as Timothy syndrome to generate isogenic pluripotent stem cells, demonstrating its value in studying complex genetic diseases and developing personalized treatments (Geschwind et al., 2014). Besides, gene cloning technologies are also important in plant biotechnology, particularly in crops like tomatoes, where trait isolation for beneficial traits can greatly boost agricultural productivity. Use of protoplast fusion allows the combination of genetic material between species, and the introduction of traits that may provide for favorable characteristics, i.e., resistance to disease or tolerance to abiotic stress, is facilitated. Recent developments have aimed at the use of RFLP markers in tomato breeding, with the promise of tagging quantitative traits derived from closely related species. As molecular methods improve, gene cloning shall be an indicator of genetic innovations within crop species toward maximum yields and food production sustainability (Crossway et al., 1989) (Figure 4.10).

4.3.1. Overview of Gene Cloning Methods

Gene cloning methods have evolved significantly, with various methods that are geared towards manipulating and replicating DNA sequences effectively. One precise method is well known to include the use of restriction enzymes to cut a specific DNA sequence, creating room for targeting genes to be added into plasmid vectors. The method initiates the formation of recombinant DNA, which is ultimately given to host cells, predominantly bacteria such as *E. coli*. The success of the procedure heavily relies upon the compatibility of the ends of DNA fragments, which results in successful ligation and amplification. Another significant technique is the Polymerase Chain Reaction (PCR), which enables one to selectively amplify certain segments of DNA, thus facilitating cloning to become easier. PCRs ability to generate millions of copies of a DNA fragment from a sample so small that it is infinitesimal speaks volumes about its significance in gene cloning and its use in many applications, from medical diagnosis to genetic studies. Advances in gene cloning are also the development of high-throughput techniques, which allow scientists to conduct multiple cloning experiments simultaneously. One of such techniques is the Gibson assembly technique, which streamlines the assembly process of a series of DNA fragments into a single piece using homologous overhangs without restriction enzyme and ligation. It not only streamlines procedures but also conserves time during gene cloning, which is immensely helpful for complex projects that involve assembling large or complex genetic constructions. Whereas the Gibson method has been promising in a variety of applications, there remains a challenge in further optimizing the methodology for certain environments, as mentioned in a study on its application for methanogen transformation (Jennings et al., 2018). Bypassing these through methodology optimizations has the potential to lead to a generally more efficient cloning process. Synthetic biology is a new field that builds on the principles of gene cloning to engineer biological systems for novel applications. By applying engineering principles to biological entities, scientists aim to build intricate gene networks and develop novel organismal functions. The continually more sophisticated methods at their command, such as de novo DNA synthesis and advanced protein engineering, facilitate the engineering of tailored genetic constructs. Techniques in synthetic biology have brought a structured process to gene cloning, promoting standardization and modular design. In applications by scientists, new pathways are being investigated in gene editing and programming of synthetic constructs for the potential of enhancing the manufacture of biological compounds and other biotechnological uses (Heinemann et al., 2006). Finally, the integration of mature gene cloning methods and synthetic biology principles promises unprecedented potential for the resolution of real-world issues and redefining the genetic engineering paradigm.

4.3.2. Importance in Research and Biotechnology

The significance of biotechnology and research in genetic engineering cannot be overemphasized, as they form the foundation for innovative solutions to current issues. The ability to edit genetic material has transformed the understanding of fundamental biological processes as well as brought innovations in medicine, agriculture, and environmental conservation. For instance, researchers currently possess the ability to design crops that yield more and are disease and pest-resistant, thus being able to address food insecurity challenges. Also, genetic engineering technologies such as CRISPR permit precise genome editing, enabling researchers to cure previously deemed incurable genetic diseases. The scope of these technologies is immense, as they provide a platform for further research with the aim of enhancing the quality of life on a global scale, making biotechnology a research field of utmost importance for future progress. Besides agricultural applications, biotechnology research is even more crucial in healthcare applications that have revolutionized the medical industry. The genetic engineering tools enable the tailor-making of drugs specific to the condition and personalized medicine, depending on the individual's own genetic makeup. This targeted intervention not only optimizes the efficiency of treatment but also minimizes side effects normally associated with conventional treatment. For example, genetically engineered biopharmaceuticals, such as insulin, have transformed diabetic management.

Figure 4.11. Biotechnology research.

Source: https://www.scuhs.edu/wp-content/uploads/What-is-biotechnology-research-and-development-.jpg.

Additionally, gene therapy is being investigated as a means of curing genetic diseases at the causative level. Such advances underscore the extensive implications of genetic engineering methods to medicine, thus positioning biotechnology as an integral component of current research efforts aimed at curbing health-related challenges and advancing patient care. In addition, industry-academia collaboration is also needed to propel research in the area of biotechnology, more precisely, marine biotechnology, as highlighted by recent reports. Institutions of higher learning can play a critical role in stimulating innovation by bridging the gap between applied and theoretical studies. Through networking of researchers, these institutions can facilitate interaction resulting in breakthroughs in research findings and the creation of new biotech products. For instance, efforts to map marine ecosystems and explore genetic resources have the potential to uncover beneficial genetic traits for biotechnology applications. Seaman et al. (2001) assert that such cooperation not only benefits academic endeavors but also counsels industries on potential collaborations and economic incentives. Such collaborations ultimately result in a thriving biotechnology sector, ensuring faster translation of research into actual benefits to society as it expands the range of genetic engineering tools (Figure 4.11).

4.3.3. Challenges and Limitations of Gene Cloning

Gene cloning, despite being revolutionary in its potential, is presented with stringent challenges that limit its efficiency and applicability. One of the key challenges lies within the very nature of biological organisms, which are made up of intricate networks of gene interactions that are not well understood. Such a poor understanding leads to unexpected results or aberrant phenotypes in cloned animals. As synthetic biology advances, it strives to overcome such complexities through the use of ideas drawn from different scientific fields and proposing an integrated model of biological system study (Freemont et al., 2014). However, the reliability of gene cloning also rests on the robustness of such frameworks and the reliability of predictive models, which still remain narrow in scope and depth. Further, ethical concerns over gene cloning are of substantial importance that researchers must address. The possibility of cloning technologies' misuse raises significant questions about identity, individuality, and the morality of altering genetic lines, particularly in higher organisms, including human beings. The controversy at hand is mirroring the social impacts and fears of genetic modification. As biotechnology continues to advance, the fear of unexpected ramifications—like the creation of organisms that might upset existing environmental balances—is not unfounded. The Pardee Conference Series addressed these concerns, emphasizing the need for ethical guidelines and regulatory frameworks to govern the application of gene cloning technologies responsibly (Annas et al., 2003). Without proper monitoring, the public acceptance gap and scientific aspiration gap could increase, demotivating progress in genetic engineering. Finally, the technological limitations of gene cloning methods, such as restriction enzyme technology and polymerase chain reaction (PCR), render the work of researchers even more challenging. These tools, though powerful, have low efficiency rates, are capable of inducing mutations, and can result in chimeric genetic end-products. Hence, achieving precise, reproducible cloning outcomes remains a challenge for researchers. Developments in synthetic biology aim to enhance such methods so that more efficient and accurate design cycles may be achieved, which are able to surpass the corresponding challenges of gene cloning. As researchers step forward to develop new instruments and methods to overcome the existing hurdles, the area is on the cusp of potential breakthroughs, which will enable us to gain greater insight and application of genetic engineering instruments (Freemont et al., 2014). However, balancing the technological developments with societal and ethical concerns will be imperative in directing the future of gene cloning onto a responsible and productive track.

4.4. GENE THERAPY

Gene therapy is a recent innovation in the branch of genetic engineering with immense potential for new treatments against a wide range of genetic diseases. Gene therapy functions by the direct manipulation of genes within patients' cells with the goal of repairing genetic defects, alleviating signs of genetic disease, or providing immunity against infectious disease. Pointedly, the CRISPR/Cas9 technology has proven to be a revolutionary instrument in this field, facilitating exact editing of the genome. As noted in current literature, the programmable genome-editing device is capable of facilitating scientists to construct accurate cellular and animal models of human diseases, hence advancing our pathophysiological understanding of these diseases (Ferreira et al., 2020). The therapeutic potential of CRISPR/Cas9 as a gene therapy tool is also underscored by the ability to correct patient-specific mutations, with the potential to cure diseases that were once deemed incurable by conventional therapeutic approaches. The application of gene therapy employs more than corrective measures; it employs new approaches such as synthetic biology and cellular engineering. Synthetic biology involves redesigning cells at the gene level to create new therapies and diagnostic reagents for use in diseases that vary from cancer to diabetes (Gilad and Shapiro, 2017). The new technology not only includes the development of engineered cells with increased specificity for targeting disease but also engages with the integration of molecular imaging modalities with approaches that enable *in vivo* imaging of these treatments. As therapeutic agents, these synthetic cells can be specifically targeted to perform specific activities within the body, having optimal efficacy with negligible off-target activity. Thus, gene therapy, made possible by synthetic biology, is a shining example in biomedical treatment, leading the path towards future developments in personalized medicine. While the bioprotocol of gene therapy is immensely promising, numerous challenges remain to be addressed before attaining complete clinical use. One of the most important points to consider is the improvement of delivery systems of gene-editing technologies, which must be efficient as well as targeted in order to avoid unintended genetic alterations. A better understanding of cellular microenvironments and the development of targeted delivery systems are essential to enhance therapeutic effects (Ferreira et al., 2020). Furthermore, the ethical consequences of genetic alterations require effective regulatory frameworks to ensure patient safety and public acceptance of these technologies. With new research, interdisciplinary discussion among scientists, ethicists, and policymakers will be a part of solving such problems actively, ultimately contributing to the ethical development of gene therapy as an important element of genetic engineering tools (Figure 4.12).

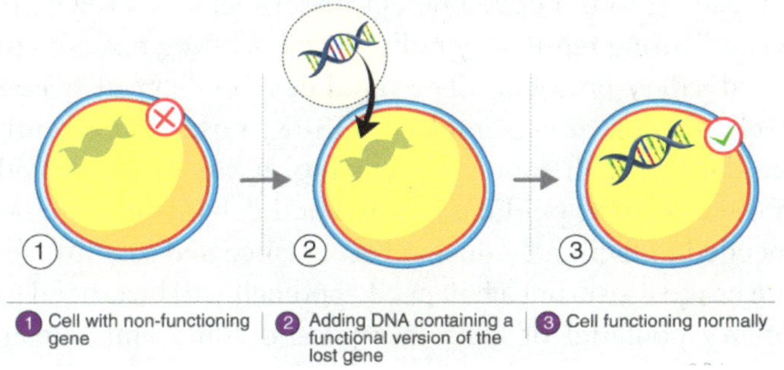

Figure 4.12. Gene therapy.

Source: https://cdn1.byjus.com/wp-content/uploads/2016/07/Gene-Therapy.png.

4.4.1. Definition and Purpose of Gene Therapy

Gene therapy is a new technique of curing genetic diseases by correcting the root cause at the molecular level. The premise of gene therapy is to bring in, change, or remove genes in an individual's cells to repair disease-causing modified gene function or to introduce new functions that have the capability to reverse the signs of the disease. This new approach has gained widespread interest since the evolution of genetic engineering tools such as CRISPR-Cas9 opened up greater precision and efficacy in editing capabilities. Gene therapy is not only concerned with the cure of diseases caused by genetic reasons, like cystic fibrosis or muscular dystrophy, but is also considering using its potential in treating a broad array of more complex diseases, like various types of cancer and viral diseases, and revolutionizing the branch of modern medicine. The objective of gene therapy is not merely to treat; it aspires to enhance the quality of life of patients suffering from chronic and genetic diseases. Through the repair of genetic flaws responsible for these diseases, gene therapy may potentially offer a long-term cure in contrast to other treatments that focus more on symptom treatment than on treating the disease. For instance, therapies like cuprizone are effective in treating inherited retinal dystrophy, restoring vision by delivering functional gene copies. Thus, gene therapy is a milestone shift toward precision medicine, enabling healthcare providers to tailor interventions based on individual genetic profiles. Moreover, the generalizability of gene therapy allows for the exploration of new treatment models, including the potential for cancer immunotherapy, showcasing its ubiquitous applicability to contemporary medical therapy. While the potential of gene therapy is immense, it is weighed with ethical concerns and safety considerations that must be explored in-depth. Historical precedents, such as the problem of early clinical trials leading to the development of serious side effects, remind us of the importance of strong regulatory policies and ongoing research to implement uncompromised safety protocols. The rapid development of gene therapy and associated technologies, for example, CRISPR, raise issues with access and equity in healthcare. With more gene therapies gaining approval and being launched onto the market, public education and debate on these developments are acutely needed to facilitate educated acceptance and to allow for discussion of appropriate usage. Ensuring a balanced approach will be crucial to unleashing the revolutionary potential of gene therapy and addressing societal concerns over genetic engineering methods like CRISPR (Castelyn, 2021) and their implications (Liu et al., 2016) (Figure 4.13).

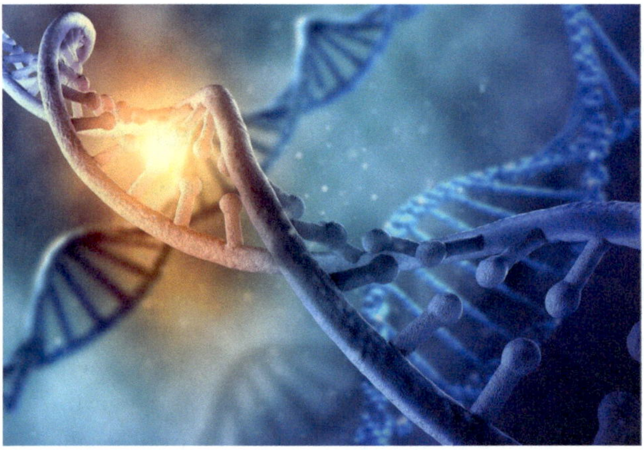

Figure 4.13. Understanding gene therapy.

Source: https://www.qps.com/wp-content/uploads/2023/04/genetic-1536x1075.jpg.

4.4.2. Types of Gene Therapy Approaches

Gene therapy has emerged as a novel approach within genetic engineering, enabling targeted treatment of a variety of genetic diseases. At the center of these strategies is the premise of introducing therapeutic genes into patients' cells to repair or substitute faulty genes. One of the leading forms of gene therapy is based on the use of viral vectors, which are genetically altered viruses that are well-equipped to deliver therapeutic genes to target cells. With the utilization of the inherent capability of viruses to enter cells, researchers have been able to create vectors that transfer targeted genetic material with little harm to healthy tissues. New technologies have emerged, and with them, better-targeted delivery systems have been created that engender improved therapeutic potential of gene therapy, especially in the treatment of genetic disorders and cancer treatment (Ferreira et al., 2020). Yet another infancy-stage gene therapy method is the use of CRISPR/Cas9 technology, which offers exact editing of the genetic code itself. This noble instrument can not only replace faulty genes but also make changes to already existing genetic code to enhance or shut down the activity of particular genes.

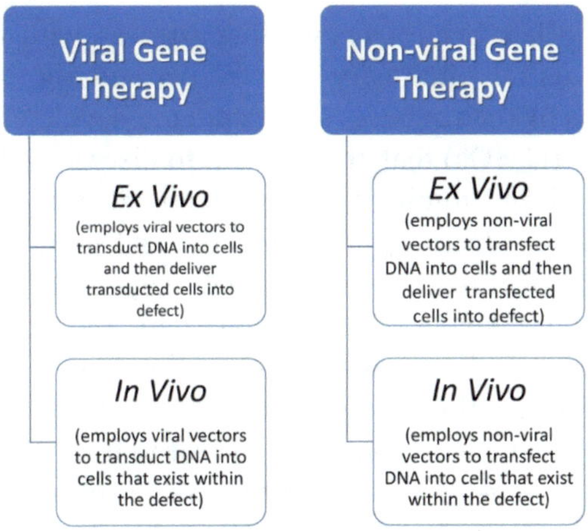

Figure 4.14. Schematic depicting the types of gene therapy approaches.

Source: https://www.researchgate.net/publication/342848883/figure/fig2/AS:1132468744790029@1647012997494/Schematic-depicting-the-types-of-gene-therapy-approaches.jpg.

The use of CRISPR/Cas9 in a range of diseases, including genetic disease, infection, and cancer, is evidence for its use as a tool for disease treatment. Furthermore, the possibility of designing cellular and animal disease models of human diseases with customized specifications extends the disease mechanism scope and aids the development of customized therapy. As researchers continue to further advance the specificity and efficiency of CRISPR/Cas9 delivery systems, the potential for translation of this technology from bench to clinical practice continues to become increasingly possible (Ferreira et al., 2020). Aside from viral vectors and CRISPR/Cas9, other advanced gene therapy technologies include employing non-viral vectors, such as nanoparticles and liposomes, for gene delivery. These vehicles are capable of encapsulating therapeutic genes and delivering them across cell membranes. Non-viral vectors are typically safer and elicit a lower immune response than viral vectors, making them an attractive alternative for certain applications. Recent progress in cancer-specific gene therapy vectors is an excellent case in point; such vectors inhibit cancer cells preferentially without touching normal cells, thereby reducing side effects and maximizing treatment efficacy. This increasing diversity in gene therapy strategies signifies ongoing advances in the field, leading to a holistic approach to confronting genetic diseases and improved patient outcomes (Figure 4.14).

4.4.3. Success Stories and Ongoing Research

Genetic engineering has been a resounding success with numerous success stories foretelling its revolutionary power, particularly in agriculture and medicine. Examples of such success stories are the development of genetically modified organisms (GMOs) that are resistant to disease and pests, which lead to greater crop yield and minimal use of chemical pesticides. For instance, scientists have developed drought-tolerant rice and immensely boosted food security in regions with climate variability. Not only do such innovations prove the ability of genetic engineering technologies to solve pressing agricultural issues, but they also indicate that these technologies can be practically incorporated into traditional farming systems. As such, success stories continue to come forward, which encourage ongoing research activities that are geared toward further enhancing the capacities of genetic resources in a sustainable manner (Hesse et al., 2019). In addition to agricultural advancements, the field of genetic engineering has also witnessed incredible advancements in gene therapy, with very promising outcomes for the treatment of genetic diseases. Such an example is the use of CRISPR-Cas9 technology to repair mutations in sickle cell anemia. Recently, clinical trials have discovered that individuals who were treated using the technology experience phenomenal relief from their symptoms and state, promising a cure that would be permanent. This process serves as an instance of how clean editing instruments can be utilized to treat the causes of diseases rather than their symptoms. Further research in this field is called for to expand such methods and determine the safety and efficacy of them in larger patient populations, thus demonstrating a clear trajectory towards more successful health outcomes through genetic engineering (Hesse et al., 2019). Research into genetic engineering tools goes on into other areas of scientific study, offering a fertile ground for ongoing research projects. Investigations into the creation of gene drives can control infestations of pests and reduce vector-borne illnesses, such as malaria. By genetically modifying the pattern of inheritance in organisms, researchers attempt to eradicate disease-ridden species without disrupting ecological balance. Concurrently, sophisticated computational models are being used to predict the long-term consequences of genetic modification in ecosystems. Such interdisciplinarity is essential in addressing the many-faceted ethical, environmental, and social implications of genetic engineering. As these issues develop, they lay the foundation for future innovations and ensure that genetic engineering continues to be employed responsibly and well (Ahmed et al., 2018).

4.5. SYNTHETIC BIOLOGY

Synthetic biology is rapidly developing, most significantly driven by the advancement of genetic engineering technologies, allowing advanced manipulations of living organisms. Basically, synthetic biology uses techniques in molecular genetics to engineer novel biological systems that can perform intended functions. The science enables scientists to construct new gene networks, plan genomes from scratch, and bioengineer microbes for specific purposes, including biotechnological, pharmaceutical, and biomedical uses. As profiled in today's literature, the relevance of synthetic biology lies in its ability to develop new drugs, treat multidrug-resistant infections, and investigate pathogenic organisms at a fundamental level (Franco et al., 2015). This intricate manipulation of biological material puts synthetic biology on the pedestal of modern genetic engineering, providing space for unprecedented innovation in various scientific disciplines. Furthermore, synthetic biology's ability goes beyond gene editing to encompass a wide variety of biomedical applications. The genetic manipulation of cells is potentially extensive for the development of new diagnostics, therapies, and imaging reagents that are tailored to combat diseases such as cancer and diabetes. Engineered systems can be designed to interact with cellular processes, yielding interactive responses to disease states (Gilad and Shapiro, 2017). In addition, the incorporation of imaging devices into synthetic biology enables scientists to observe engineered cells in real time, which helps to reveal more about how they operate in living organisms. This twofold emphasis on designing genetic constructs and observing their behavior highlights the revolutionary influence of synthetic biology on both fundamental research and practical medical science. With continuing progress in synthetic biology, the ethical implications of its applications become even more relevant. The regulation of genetic material brings subtle challenges in the potential misuse of these technologies, particularly in such uses as bioweapons or genetic modification of humans. Ethical norms must evolve to counter these challenges responsibly while advancing progress in synthetic biology (Franco et al., 2015). Furthermore, the social implications of engineering organisms for specific purposes need to be addressed with an open discussion among scientists, ethicists, and policymakers so that synthetic biology is developed and applied in a way that is safe and beneficial.

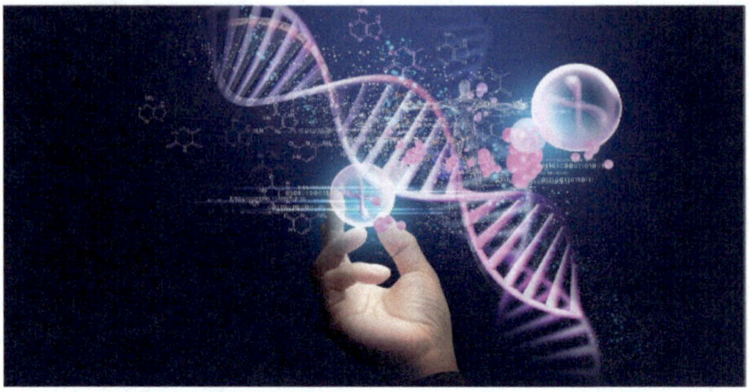

Figure 4.15. Understanding Synthetic Biology

Source: https://www.oecd.org/adobe/dynamicmedia/deliver/dm-aid--a7cef11e-7320-43f4-a2be-fb2a7979594e/synbio.png?quality=80&preferwebp=true&width=2048

4.5.1. Definition and Scope of Synthetic Biology

Synthetic biology is an interdisciplinary field that interfaces biology with engineering principles to design and construct new biological parts, devices, and systems. Synthetic biology differs from traditional genetic engineering in that it emphasizes systematic approaches to modifying living organisms. Through exploiting the fundamental knowledge of molecular biology, synthetic biology leverages techniques such as DNA sequencing, synthesis, and assembly to engineer organisms with new properties. The field of synthetic biology is broad and encompasses various applications like bioremediation, biofuel, and even designing biosensors. For instance, microbes may be used to degrade environmental pollutants or provide renewable energy compounds, and this illustrates how synthetic biology has the ability to directly address the ecological and energy challenges of the modern world. In addition, the domain of synthetic biology is characterized by its high reliance on sophisticated engineering concepts, often in tandem with conventional engineering disciplines. One of the important fields is the standardization of biological components, like the modular devices for electrical engineering, to allow the predictable assembly of biological systems. This standardization allows biological components to be reused across different projects and enhances efficiency and interoperability. Yet, with great advancement, the field is still plagued by the complexities inherent in biological systems such as the variability and unpredictability of living organisms. The challenge to attain greater automation and autonomy in synthetic biology requires more advanced genetic engineering tools to reconcile engineering principles with biological intricacies, leading to more streamlined experimental workflows and applications as illustrated by Beal et al. (2020). The ethical and practical aspects of synthetic biology are central to its significance and scope, evoking reflections on the responsible use of genetic engineering tools. As synthetic biology continues to progress, it raises essential questions in society regarding biosecurity, environmental risk, and the ethical implications of manipulating life. Recognizing that bacteria and other microorganisms usually exist in complex communities, the incorporation of biofilm engineering is a new area of use within synthetic biology. With the capacity to enhance operations within waste treatment and recovery of resources, biofilms identify a rich area to investigate, balancing beneficial uses against potential environmental implications. Understanding such dynamics is crucial to ensuring that advancements in synthetic biology are converted into sustainable benefits with regard to the ethical norms of its application, as detailed later in the ongoing research elaborated in Mukherjee et al. (2020).

4.5.2. Tools and Techniques Used in Synthetic Biology

Synthetic biology advancements have been directly assisted by novel tools and techniques that enable customized genetic modifications. One of the most significant techniques in the field is evolutionary engineering, which allows for improvement of microbial strains through methods that don't necessarily rely on large amounts of biological knowledge. With the emergence of synthetic biology, the method has been advanced with the incorporation of advanced tools for phenotypic perturbation, allowing researchers to edit and screen improved variants more effectively. For instance, integration of high-throughput screening strategies has quickened the identification of desirable traits in engineered organisms, hence speeding up the establishment of strains with enhanced productivity or stress resistance. With such developed mechanisms, evolutionary engineering has emerged as a crucial part of synthetic biology, allowing for the assembly of advanced genetic circuits that are coded to perform specific functions with high efficiency (Jung et al., 2019). Furthermore, the emergence of CRISPR technology has transformed the field of synthetic biology, making it an extremely useful tool for precision genome editing. At first, accepted for its application in model organisms, CRISPR has revealed immense potential in photosynthetic microbes such as cyanobacteria and diatoms, which are now recognized as promising hosts for the generation of renewable chemicals and fuels. Although there has been progress in the application of CRISPR, those that can be utilized to genetically modify these photosynthetic organisms are lagging behind their heterotrophic relatives. The current progress has begun bridging the gap by widening the toolbox for genetic manipulation in these organisms, paving the way for future uses in biotechnology that are in tandem with sustainable methods (Adler-Agnon et al., 2021). The complementarity of genetic engineering tools applied in synthetic biology is not merely an editing capacity; it's a system comprehension of biological building blocks and how they interact within complex genetic circuits. With researchers finding new genetic elements with regulatory powers, the assembly of these components into higher-order logic circuits becomes possible. This technology is central to the design of organisms to perform complex biosynthetic processes, for instance, the production of biofuels or biopharmaceuticals, via pathways previously inaccessible. Lastly, continued tool and method optimization in synthetic biology has the potential to further revolutionize the field of genetic engineering, providing enhanced ability to design and build biological systems that can meet diverse societal needs while being eco-friendly (Jung et al., 2019).

4.5.3. Potential Impacts on Various Industries

Genetic engineering technologies will likely redefine various industries by increasing efficiencies and lowering environmental effects. In the agriculture industry, for example, the application of genetically modified organisms (GMOs) can lead to greater crop yields and lower use of chemical pesticides. This revolution is enormous, particularly when one juxtaposes it with traditional agricultural methods that are often overly reliant on harmful chemicals applied as pesticides. Through methods such as CRISPR technology to genetically engineer some genes responsible for pest resistance or drought tolerance, producers can grow more tolerant plants. Not only does this offer food security, but also minimizes the environmental impact associated with traditional agriculture, thus fulfilling the current sustainability goals and corporate social responsibility roles. The agricultural industry is the greatest proof that genetic engineering can carry with it an enormous leap in more than one parameter, of interest to economics as well as the environment (Azzaro-Pantel et al., 2013). The healthcare industry is also one that is about to be revolutionized by genetic engineering techniques. Precision health, driven by advances in genomics, enables customized treatment protocols based on the genetic profile of a person. This paradigm is important as it promises more effective and focused therapies, particularly in the treatment of genetic diseases and cancer. For instance, gene therapy techniques can fix defective genes that lead to certain ailments, an inferred cure rather than symptom management. These advances create ethical issues regarding access, equity, and impact for genetic privacy. Still, the potential for improved health results and reduced long-term healthcare costs underscores the revolutionary prospects of genetic engineering in medicine. Addressing the root causes of disease, the sector is leading the way in embracing genetic applications, witnessing a distinct shift in health care delivery paradigms (Dufour et al., 2013). Apart from medicine and agriculture, the energy industry is also beginning to consider genetic engineering as a means of developing sustainable solutions. Biotechnology can lead to the production of biofuels from genetically modified microbes that convert organic waste into renewable energy. All these can help greatly reduce the utilization of fossil fuels and eventually contribute to addressing issues of climate change. In addition, genetically engineered algae can produce high volumes of lipids that are well-suited for the production of biofuels, providing an alternative to conventional energy sources. Also, the use of genetic engineering technologies as part of clean energy technologies not only promotes environmental sustainability but also provides potential opportunities for economic development in biomanufacturing and alternative energy technologies. As industries adopt these cutting-edge technologies, they can transition to a more sustainable operating model, boosting environmental and economic resilience through the innovative application of genetic engineering (Azzaro-Pantel et al., 2013).

4.6. BIOINFORMATICS IN GENETIC ENGINEERING

Bioinformatics has evolved into a crucial tool in the application of genetic engineering, allowing for tedious data interpretation and analysis that is needed in understanding genetic material. The integration of genetic engineering and bioinformatics has accelerated the identification and manipulation of genes, rendering genetic changes more precise and effective. For instance, sophisticated bioinformatics platforms make it possible for researchers to screen enormous amounts of genomic data, identifying significant genetic variations such as SNPs. The ability to deconstruct these distinctions is particularly critical in model organisms, like **Bacillus subtilis**, where systems like BacillOndex provide a semantically enriched knowledge base for genetic data mining and phenotypic prediction. Such an integrative approach not only deconstructs the genetic complexities within sequences but also lays the basis for the design of synthetic genetic circuits, thereby enriching the biotechnology and genetic engineering landscape. The use of bioinformatics is also found in synthetic biology where it makes a valuable contribution towards the construction of genetic circuits. Using computational design methods, researchers can design and optimize genetic parts so that they harmonize with biological systems. This is crucial because the successful application of engineered circuits significantly relies on the worth and reliability of participating biological components. The current trend toward bioinformatics-inspired genetic engineering is enabling research into novel functions and enhancing engineered organism robustness. This rational approach not only facilitates technological advances in microbial technology but also offers insight into how genetic changes are tailored to produce desired characteristics more effectively and rapidly. Such findings pave the way for designing adaptive biological systems that respond to targeted technological requirements. In addition, the combination of microengineering and bioinformatics provides unique opportunities for the establishment of genetic engineering. The establishment of microfluidics and microelectronic devices has revolutionized scientists' handling of genetic material entirely and enabled the construction of miniaturized sensing devices in order to enhance the accuracy of experiments. These technologies enable scientists to observe genetic response in real-time, along with making genetic modification easier. The online integration of genetic data and microengineered devices facilitates unprecedented investigation of complex biological systems and an enhanced understanding of gene functions and interactions. Increasingly, the incursion of bioinformatics into microengineering underscores the importance of adequate data analysis in the design of stable genetic circuits. This evolving arena not only enables new biotechnological applications but also illustrates the intermingling of genetic engineering and bioinformatics to achieve greater control over biological outcomes (Figure 4.16).

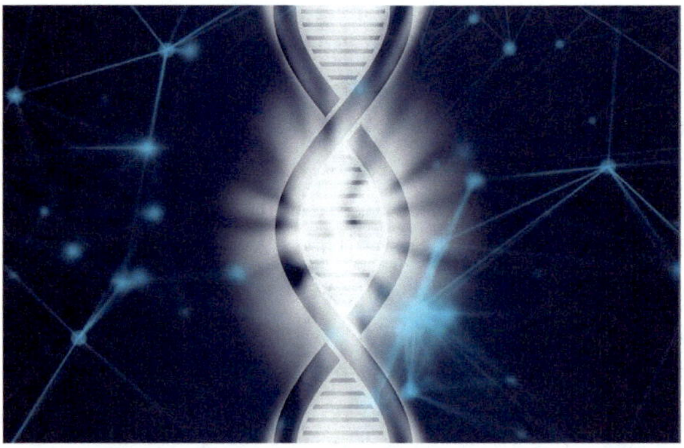

Figure 4.16. Bioinformatics advances genetic engineering applications.

Source: https://blogger.googleusercontent.com/img/b/R29vZ2xl/AV-vXsEhcbKftHQW9fUwza9cJjwOuqAaCWEyhOrAct5f5N4v3pN-bPIgWe4zQLv-hvGCJxP1H54TvNx71-qpa3qsaErUr83BAAykh-9WxkOJ06mSvBfNbk_lNNxDLxTzy-pOr7Y1SYe5FesikusCws/s1600/application-of-bioinformatics-in-genetic-engineering.jpg.

4.6.1. Role of Bioinformatics in Data Analysis

The incorporation of bioinformatics in data analysis has fundamentally transformed how researchers approach genetic engineering. As there is a rise in the amount and intricacy of biological data, traditional analysis methods prove to be insufficient. Bioinformatics provides researchers with highly computational methods by which large datasets could be controlled, visualized, and interpreted, and scientists could perform more precise analyses. For instance, initiatives like BacillOndex show this integration by coming up with an extensive, semantically annotated database for *Bacillus subtilis* where users can mine and analyze diverse datasets with regard to gene interaction as well as SNP analysis. Such tools enable scientists to identify single nucleotide polymorphisms (SNPs) with documented phenotypic effects, thereby elucidating genetic characteristics and their impact on genetic engineering activities, eventually translating into the expansion of the field. The role of computational intelligence (CI) in bioinformatics is also significant, particularly with regard to the analysis of large biological data. When researchers are dealing with complex molecular biology and DNA sequencing data, CI methods become crucial in unveiling meaningful information. Fuzzy logic, Dempster-Shafer theory-based, and Shannon entropy-based tools enable accurate gene sequence analysis such that scientists are able to make accurate assessments within condensed time frames. According to recent studies, judicious selection of appropriate CI tools possesses the ability to significantly enhance data processing rate and accuracy, allowing researchers to respond to pressing biological questions efficiently. The ability to leverage CI in bioinformatics, thus, not only alleviates data analysis but also broadens the scope of investigation in genetic engineering, allowing for more sophisticated genetic tools and technologies to be developed. The interaction between bioinformatics, computational intelligence, and genetic engineering showcases a paradigm shift in biological research. In addition to the design of synthetic genetic circuits, this combination aids in hypothesis generation and system-level studies. According to BacillOndex, tools with integrated capabilities that harmonize various datasets have the capacity to transform discoveries, and thus, these tools are invaluable to researchers engaged in genetic modification and synthetic biology. Second, the ability of large data sets to be compared through the use of established bioinformatics platforms encourages greater understanding of gene function and regulation when subjected to engineered alterations. Ultimately, the ongoing advancement of bioinformatics—along with the creation of computational intelligence—is a critical path to new solutions for genetic engineering, solidifying its status as a quintessential pathway for future biotechnology science.

4.6.2. Software and Tools Used in Genetic Research

The genetic research has been revolutionized by improvements in sophisticated software and tools, which have significantly enhanced the accuracy and efficacy of gene analysis. These tools facilitate a range of applications, from genome sequencing to gene editing. An example of such a development is the use of bioinformatics software, which enables scientists to analyze large sets of data generated by next-generation sequencing technologies. Programs such as BLAST and Galaxy enable scientists to compare genetic sequences and depict genomic data effectively. In addition, CRISPR gene editing-specific software such as CRISPR Design Tool and Benchling make target sequence selection and experiment management easier, thus minimizing potential off-target effects. As research on these technologies evolves, they offer greater understanding of gene function and interactions, underscoring their significance in contemporary genetic research. In addition to bioinformatics and gene editing programs, machine learning and artificial intelligence are entering the field of genetic research in a huge way. The application of AI algorithms allows scientists to analyze and decipher enormous amounts of intricate genetic information that would be beyond the capabilities of humans alone to interpret. Computer software like DeepVariant and GATK use AI to greatly improve variant calling accuracy, greatly increasing genetic diagnostic validity. The ability of machine learning algorithms to identify patterns and predict potential genetic outcomes offers new avenues for personalized medicine and tailored treatments. Referring to the trends in research, the intersection of AI with genetic research is happening at a dizzying rate, and this means that future studies will not only yield more precise tools but also new ways of valuing genetic variation. The convergence of genetic tools and advanced computational techniques is a watershed moment for the life sciences. Furthermore, cloud computing has emerged as an indispensable factor in light of genetic research, with the potential to facilitate collaboration and data sharing across worldwide scientific communities. With the exploitation of cloud-based infrastructure, scientists can leverage a set of bioinformatics tools without incorporating gigantic local computing resources. Cloud platforms, including Amazon Web Services and Google Cloud, offer scalable processing and storage of large genomic data for collaborative projects addressing complex genetic problems in different populations. The cloud platforms also allow for high-throughput analysis of data required for applications from genome-wide association studies (GWAS) to population genomics. Such advancements not only contribute to the accessibility of genetic research tools but also promote open science, whereby data and resources are shared with researchers at no cost. As a result, cloud computing is increasingly indispensable for the development of genetic engineering tools, ushering in groundbreaking discoveries in genetics.

4.6.3. Future Trends in Bioinformatics Applications

As bioinformatics develops, its future application will increasingly allow for the manipulation and understanding of genetic engineering instruments. One big trend is the merging of computational design methodologies with synthetic biology, which will allow the development of novel genetic circuits specially designed for specific purposes. By leveraging powerful algorithms and advanced processing in engineered cells, scientists can even personalize circuits' responses to specific signals, making genetic engineering more effective. Therefore, bioinformatics will serve as a basis for systematic characterization and tuning of biological parts such that synthetic biology builds its desired applications reliably. Hybridization of data-driven approaches with traditional genetic engineering could potentially usher in deep strides in synthetic organisms for specific applications in a broad range of disciplines, such as medicine and environmental science, pointing to the revolutionary nature of these technologies over the near future (Marchal et al., 2013). Ever more accessible and stable de novo DNA synthesis technologies are yet another arena in which bioinformatics is heading. As such synthesis methods advance, they will make it possible for the creation of increasingly sophisticated and functional biological systems, effectively broadening the set of genetic engineering tools to hand. This capability is facilitated by engineering design principles founded on modularity and standardization, as seen in synthetic biology projects. By developing a library of common biological parts that can be recombined as components, researchers can streamline the design process of synthetic biological entities and add more precision and control. Furthermore, computational design will be essential for reprogramming existing proteins and pathways so specific interventions can enhance desirable traits in engineered organisms. Therefore, bioinformatics shall be of critical importance in optimizing and guiding these innovations, eventually culminating in innovative applications in pharmaceuticals and biotechnology (Heinemann et al., 2006). In addition, hybridization between microfluidic innovations and bioinformatics also promises enormous development potential. Coupled with the possibility of marrying synthetic biology and microengineered devices, high-throughput experimentation and analysis become achievable through rapid evaluation of genetic constructs and their activity. This synergy not only improves the investigation of intricate biological systems but also facilitates real-time observation of engineered responses in cellular contexts. Through a synthesis of synthetic circuits and microfluidic platforms, scientists can devise dynamic systems that respond to altering conditions, enabling more effective biotechnological applications. Fine-tuning control and measurement capabilities in these integrated systems assures that bioinformatics will be a major factor in driving innovation and efficiency levels in genetic engineering towards revolutionary solutions for medicine, agriculture, and environmental protection (Marchal et al., 2013).

SUMMARY

The application of genetic engineering techniques is a revolutionary shift in the fields of many sciences, including agriculture, medicine, and environmental science. These technologies, being accurate and able to manipulate genetic material, enable scientists to address intractable challenges such as food security and disease resistance. In agriculture, for example, genetic engineering enables the creation of crop varieties that are resistant to climate change and insects, thereby yielding more and higher-quality nutrition. Moreover, the potential for engineering microorganisms for environmental clean-up illustrates the potential of such technologies. As research continues, regulatory frameworks and ethical considerations will play a critical role in ensuring the responsible use of these technologies while balancing innovation against public safety and environmental responsibility. Further, advances in genetic engineering tools have encouraged interdisciplinary collaboration, fostering a more holistic approach to problem-solving. Researchers from fields such as computational biology, bioinformatics, and environmental science are increasingly collaborating to optimize genetic modifications and take full advantage of them. For instance, actionable analytics, as discussed in Krishna et al. (2017), can help plan phases of genetic modifications, rendering predictive models real-world applications. Besides, the capacity to recognize trends in the market with technologies such as Artificial Neural Networks (ANN), as noted by Chakraborty et al. (2012), demonstrates how these technologies simplify the linking of research and guide the course in the future for genetic engineering. The cooperative arrangement opens the door for innovative solutions to address pressing global issues. Lastly, the destiny of genetic engineering technology hinges on the success in merging technological innovation with ethical and regulatory concerns. While the promise of benefits is great, they need to be weighed against the risks and uncertainties involved with gene alteration. Highlighted by recent research, responsible and actionable decision-making processes are most essential. By establishing comprehensive paradigms that focus on safety, public enlightenment, and judicious use of genetic technologies, society can gain the complete advantage of genetic engineering to promote sustainable development and improve quality of life. Adopting a cautious but imaginative approach will enable these powerful technologies prove useful to the broader community while resolving moral concerns that are certain to arise in their application.

A few of the most significant developments in genetic engineering involve designing and employing a variety of tools by which organisms' genomes come under attack for alteration. Techniques such as CRISPR-Cas9, ZFN (zinc finger nucleases), and TALENs (transcription activator-like effector nucleases) have emerged as powerful tools for gene editing with unmatched precision. CRISPR-

Cas9 has, however, become extremely popular due to its simplicity and efficiency in enabling scientists to design custom RNA sequences that target the Cas9 enzyme to specific genomic regions. This can be employed for both gene knockout and the insertion of a precise mutation, making it simpler to study the function of genes while offering the potential for therapy across many fields, including medicine and agriculture. The impact of these tools on gene studies cannot be underestimated because they have transformed scientific potential and opened biotechnological innovation vistas. In addition to these core tools, the use of genetic transformation procedures, e.g., Agrobacterium-mediated transformation and microinjection, has enabled scientists to integrate foreign genes in plants and animal models with success. Agrobacterium tumefaciens, a gene-transfer-capable bacterium, is now a mainstay of plant genetic modification, enabling the development of genetically altered crops possessing such desirable traits as pest resistance and enhanced nutritional quality. Microinjection, which is standard in animal models like mice, permits direct introduction of genetic material into a fertilized egg nucleus, leading to the creation of transgenic animals. Such strategies, in addition to propelling agricultural yields through improved crop harvests, also uncover significant knowledge regarding gene function and regulation across species, hence influencing fundamental research and applied sciences. The ethical dimensions of genetic engineering tools call for serious consideration as they develop within social contexts. Though proponents identify potential benefits of biotechnology in food safety, promoting health gains, and mitigation of environmental challenges, critics express concerns about its potential unintended consequences, including environmental disturbances and health risks associated with the use of genetically modified organisms. Consumer attitudes vary significantly and are driven by cultural, ethical, and socio-economic factors. Research conducted by the Food Policy Institute indicated that there needs to be ongoing discussion and public awareness of food biotechnology since consumers' acceptance depends on understanding both the benefits and potential risks of genetic engineering. These perceptions will be critical to consider in facilitating responsible advances in genetic engineering methods, so their application will be encouraged through policies grounded in science but with consideration of societal mores and values.

REFERENCES

1. Abdi, H., Alhoniemi, E., Angelov, P., Angelov, I., Angelov, G., Arazo-Bravo, M. J., Atkeson, C. G., et al. (2009). Data-driven soft sensors in the process industry. https://core.ac.uk/download/75803.pdf (accessed on 3rd October 2025).
2. Adamopoulos, K., Afzal, W., Aguilar, J., Al Ba E., Alander, J. T., et al. (2012). Search based software engineering: Trends, techniques and applications. https://core.ac.uk/download/20665886.pdf (accessed on 3rd October 2025).
3. Adler-Agnon, Z., Ahmad, A., Ajjawi, I., Apt, K. E., Armbrust, E. V., Baier, T., et al. (2021). Doing synthetic biology with photosynthetic microorganisms. https://core.ac.uk/download/479169113.pdf (accessed on 3rd October 2025).
4. Ahmed, A., Bruner, K., Gajewski, A., Kiraly, D., et al. (2018). *Triple Helix, Fall 2018*. https://core.ac.uk/download/211337566.pdf (accessed on 3rd October 2025).
5. Aiuti, A., Biffi, A., Winkeler, A., Bucheit, A. D., Tong, A. W., Everts, B., et al. (2014). Anticancer gene transfer for cancer gene therapy. https://core.ac.uk/download/77020742.pdf (accessed on 3rd October 2025).
6. Alansari, Z., Anuar, N. B., Belgaum, M. R., Kamsin, A., et al. (2019). Evaluation of IoT-based computational intelligence tools for DNA sequence analysis in bioinformatics. http://arxiv.org/abs/1904.09268 (accessed on 3rd October 2025).
7. Allenby, N., Hallinan, J. S., James, K., Misirli, G., et al. (2013). BacillOndex: An integrated data resource for systems and synthetic biology. https://core.ac.uk/download/287613592.pdf (accessed on 3rd October 2025).
8. Annas, G., Berwick, R., Campbell, D., Cantor, N., et al. (2003). The future of human nature: A symposium on the promises and challenges of the revolutions in genomics and computer science, April 10, 11, and 12, 2003. https://open.bu.edu/bitstream/2144/22887/1/Conf03-Human-Nature.pdf (accessed on 3rd October 2025).
9. Arnold, S. J., Bejerano, G., Brodie, E. D., Hibbett, D., et al. (2013). Evolutionary biology for the 21st century. https://core.ac.uk/download/12174890.pdf (accessed on 3rd October 2025).
10. Azzaro-Pantel, C., Domenech, S., Ouattara, A., Pibouleau, L., et al. (2013). Economic and environmental impacts of the energy source for the utility production system in the HDA process. https://core.ac.uk/

download/18540149.pdf (accessed on 3rd October 2025).

11. Beal, J., & Rogers, M. (2020). Levels of autonomy in synthetic biology engineering. *Molecular Systems Biology, 16*. https://doi.org/10.15252/msb.202010019 (accessed on 3rd October 2025).
12. Bernardo Franco, F., Padilla-Vaca, F., & Anaya-Velázquez, F. (2015). Synthetic biology: Novel approaches for microbiology. https://core.ac.uk/download/159084117.pdf (accessed on 3rd October 2025).
13. Bier, E., Harrison, M. M., O'Connor-Giles, K. M., Wildonger, J., et al. (2018). Advances in engineering the fly genome with the CRISPR-Cas system. https://core.ac.uk/download/323065301.pdf (accessed on 3rd October 2025).
14. Bolotin, A., Burt, A., Hammond, A., Lorenzetti, A. P., McClintock, B., Castilho, B. A., Feschotte, C., et al. (2018). Modulating signaling networks by CRISPR/Cas9-mediated transposable element insertion. https://core.ac.uk/download/161267635.pdf (accessed on 3rd October 2025).
15. Brunetti, G. L. (2015). Optimization as a design strategy: Considerations based on building simulation-assisted experiments about problem decomposition. http://arxiv.org/abs/1407.5615 (accessed on 3rd October 2025).
16. Castelyn, C. (2021). Shifting perceptions of CRISPR. *Voices in Bioethics*. https://www.semanticscholar.org/paper/2fd31e0657614da5bf2d13bea87dd9d7c191f339 (accessed on 3rd October 2025).
17. Chakraborty, P. S., Chattopadhyay, M., Dan, P. K., Majumdar, A., et al. (2012). Application of artificial neural network in market segmentation: A review on recent trends. http://arxiv.org/abs/1202.2445 (accessed on 3rd October 2025).
18. Chen, Y. C., Liu, F., Sheng, J., Trang, P., et al. (2018). Potential application of the CRISPR/Cas9 system against herpesvirus infections. https://core.ac.uk/download/323080891.pdf (accessed on 3rd October 2025).
19. Coma Company, M., Espinoza Román, H. G., Fruitós Bickham, Ó. A., Lee, C., et al. (2013). Multi-objective design optimization of a 3D-rail stamping process using a robust multi-objective optimization platform (RMOP). https://core.ac.uk/download/41776507.pdf (accessed on 3rd October 2025).
20. Crossway, A., Guri, A., Hoekema, A., Kinshara, A., Morgan, A., Zelcer, A., Paterson, A. H., et al. (1989). Tomato: A crop species amenable to improvement by cellular and molecular methods. https://core.ac.uk/download/pdf/12905733.pdf (accessed on 3rd October 2025).

21. Dufour, D. (2013). Workshop report: Strategies for improving livelihoods through RTB postharvest technologies. http://agritrop.cirad.fr/569117/1/document_569117.pdf (accessed on 3rd October 2025).

22. Eicher, C. K., Maredia, K., & Sithole-Niang, I. (2025). Biotechnology and the African farmer. https://core.ac.uk/download/pdf/7021166.pdf (accessed on 3rd October 2025).

23. Elkin, K. (2025). Artificial in its own right. https://core.ac.uk/download/131214876.pdf (accessed on 3rd October 2025).

24. Ferreira, B. I., Jacinto, F., & Link, W. (2020). CRISPR/Cas9-mediated genome editing: From basic research to translational medicine. https://core.ac.uk/download/288958587.pdf (accessed on 3rd October 2025).

25. Fox, K., Komor, A. C., & Rallapalli, K. L. (2020). Rewriting human history and empowering Indigenous communities with genome editing tools. https://core.ac.uk/download/323064893.pdf (accessed on 3rd October 2025).

26. Freemont, P., Kelwick, R., MacDonald, J. T., Webb, A. J., et al. (2014). Developments in the tools and methodologies of synthetic biology. https://core.ac.uk/download/76995845.pdf (accessed on 3rd October 2025).

27. Gale, G. A. R., Lea-Smith, D., McCormick, A. J., Mills, T., et al. (2019). Emerging species and genome editing tools: Future prospects in cyanobacterial synthetic biology. https://core.ac.uk/download/231838349.pdf (accessed on 3rd October 2025).

28. Geschwind, D. H., Grimley, J. S., Kaykas, A., Krostag, A., et al. (2014). Genome engineering of isogenic human ES cells to model autism disorders. https://core.ac.uk/download/323080298.pdf (accessed on 3rd October 2025).

29. Gilad, A. A., & Shapiro, M. G. (2017). Molecular imaging in synthetic biology, and synthetic biology in molecular imaging. https://core.ac.uk/download/84314121.pdf (accessed on 3rd October 2025).

30. Heinemann, M., & Panke, S. (2006). Synthetic biology—Putting engineering into biology. https://core.ac.uk/download/pdf/12944386.pdf (accessed on 3rd October 2025).

31. Hesse, G., Matthies, C., & Teusner, R. (2019). Beyond surveys: Analyzing software development artifacts to assess teaching efforts. http://arxiv.org/abs/1807.02400 (accessed on 3rd October 2025).

32. Jänschi, L., Diudea, M. V., Bălan, M. C., Sestrăs, R. E., & Bolboacă, S. D. (2010). Results of evolution supervised by genetic algorithms. http://arxiv.org/abs/1009.0915 (accessed on 3rd October 2025).

33. Jennings, M. (2018). Optimizing genetic manipulation of methanogens through faster cloning techniques. https://core.ac.uk/download/158200172.pdf (accessed on 3rd October 2025).

34. Jung, G. Y., Kim, B., Kim, G. Y., Seo, J., et al. (2019). Synthetic biology for evolutionary engineering: From perturbation of genotype to acquisition of desired phenotype. https://core.ac.uk/download/228070113.pdf (accessed on 3rd October 2025).

35. Kobbacy, K. A. H., Rasmy, M. H., & Vadera, S. (2007). AI and OR in management of operations: History and trends. https://core.ac.uk/download/104325.pdf (accessed on 3rd October 2025).

36. Krishna, R. (2017). Learning effective changes for software projects. http://arxiv.org/abs/1708.04589 (accessed on 3rd October 2025).

37. Liu, Y., Wan, L., & Wu, Y. (2016). The analysis of music therapy in children with autism treatment based on computer technology. https://www.semanticscholar.org/paper/1726dbdcb1f777e038c6eeb7813561f54df0c957 (accessed on 3rd October 2025).

38. Marchal, K., Vanderleyden, J., & Venken, L. (2013). Synthetic biology and microdevices: A powerful combination. https://core.ac.uk/download/55705827.pdf (accessed on 3rd October 2025).

39. Mukherjee, M., & Cao, B. (2020). Engineering controllable biofilms for biotechnological applications. *Microbial Biotechnology, 14*, 74–78. https://doi.org/10.1111/1751-7915.13715 (accessed on 3rd October 2025).

40. N/A. (2017). *CBE551: Genetics and Molecular Biology/College of Engineering*. https://core.ac.uk/download/635315902.pdf (accessed on 3rd October 2025).

41. Peck, A. (2017). The failure of federal biotechnology regulation. https://core.ac.uk/download/216810459.pdf (accessed on 3rd October 2025).

42. Seaman, W., & Whitehouse, J. H. (2001). Florida marine biotechnology: Research, development and training capabilities to advance science and commerce. https://core.ac.uk/download/11016354.pdf (accessed on 3rd October 2025).

43. Stark, W. M. (2017). Making serine integrases work for us. https://core.ac.uk/download/84148269.pdf (accessed on 3rd October 2025).

44. Yeoman, M. M., & Zárate, R. (2001). Application of recombinant DNA technology to studies on plant secondary metabolism. https://core.ac.uk/download/56342165.pdf (accessed on 3rd October 2025).

CHAPTER 5

DNA SYNTHESIS AND SEQUENCING

Contents

INTRODUCTION

Synthesis and sequencing of DNA are pillars of modern molecular biology, and their methodology is critical to many areas of science, including genetics, pathology, and biotechnology. The methods used in DNA synthesis and sequencing have evolved as our understanding of DNA structure and function has expanded. The advent of next-generation sequencing (NGS) technology has revolutionized this area, facilitating whole-genome analysis at unparalleled speed and accuracy. Not only has this advancement increased the research pace, but it has facilitated numerous applications for personalized medicine, genetic engineering, and evolutionary biology. Hence, a thorough knowledge of the concepts and techniques involved in DNA synthesis and sequencing is the foundation for framing research and clinical practice of contemporary science. In addition, DNA synthesis and sequencing processes involving complicated mechanisms yield insights into cellular processes inaccessible by other means. The traditional techniques were compromising, particularly when observing complex biological environments in which cellular interaction plays a very significant role in genetic operation. Recent advances, such as plasmonic antennas and zero-mode waveguides, ensure optimal fluorescence detection capacity in crowded cellular media, bypassing the confines of conventional optical means (de Torres et al., 2014). These techniques enable single-molecule observations, which provide improved details on molecular dynamics in real time, particularly during decisive processes like replication and transcription. Therefore, the examination of these technological developments targets the dynamic interplay between DNA synthesis and cellular function, elucidating areas for subsequent research into the vast field of molecular biology. In addition to lab-based techniques, the integration of single-cell analysis is ancillary to the study of DNA synthesis and sequencing, providing a more cohesive view of genetic structure and expression. Traditional biopsies have a tendency to mask cellular heterogeneity, limiting the knowledge to population averages that do not capture the complexities of the individual cellular responses (Bulbul et al., 2018). Single-cell strategies, on the other hand, including advanced tools such as nanopipette technology, allow for precise recovery and analysis of cellular material without cell viability loss. This approach not only raises the level of knowledge of genetic transformation on the individual cell level but also illuminates the molecular basis of pathological changes. By combining DNA synthesis and sequencing and new single-cell analysis, researchers can probe further into the genetic subtleties that control growth, illness, and potential therapeutic moves, opening the doors to visionary advances in medical science.

Definition of DNA Synthesis

The process of DNA synthesis is key to life, as it is the mechanism by which genetic information is copied and fixed. DNA synthesis may be envisioned simply as the construction of a complementary deoxyribonucleic acid (DNA) strand from a template strand. This is achieved primarily through the activity of enzymes known as DNA polymerases, which catalyze the addition of nucleotide triphosphates into the growing DNA chain. While the replication fork moves forward in DNA synthesis, the double helix opens, enabling single-stranded

DNA to act as a template. The accurate addition of nucleotides provides fidelity during DNA replication, which is paramount in ensuring genetic stability from generation to generation of cells. Learning about the complex mechanisms involved in DNA synthesis is crucial to understanding the workings of the genetic code, since it lies at the very heart of heredity and cell function. One of the essential roles of DNA synthesis is its ability to correct errors, which is necessary to ensure genetic stability. Precision during DNA synthesis may be influenced by environmental and replication processes. To prevent errors, cellular processes have designed efficient proofreading functions in DNA polymerases. These proofreading operations allow enzymes to identify and fix erroneous nucleotide pairings upon synthesizing, thus curtailing the mutation rates that could lead to genomic instability or disease. Additionally, recent studies have sought new approaches with regard to DNA storage and synthesis. For instance, research into error-correcting codes for synthetic DNA in the recent past indicates promising directions towards enhancing the reliability of data storage systems where DNA serves as an information encoding material, demonstrating the intersection of biological processes with technology (Lenz et al., 2018). The growing understanding of DNA synthesis also highlights the need for a sophisticated definition of genes that will consider genetic and epigenetic aspects.

Figure 5.1. DNA synthesis.

*Source:*https://sfvideo.blob.core.windows.net/sitefinity/images/default-source/community-blog-images/post-hero-images/dna-blog-2024-_7_.jpg?sfvrsn=50d4f307_2.

Their definition has changed from a fixed, mechanistic concept to a broader one, recognizing their dynamic character. DNA synthesis not only lays down the fundamental gene sequences but also contributes to the epigenetic modifications influencing gene expression and function. Such duality of genes as structural and functional entities renders possible a full understanding of their activities in cell contexts. As continued research reinforces the intricacy of genetic expression and regulation, researchers uphold definitions that include the multihedron nature of genes and interaction with other biological processes (Bolondi et al., 2017). This realization, as it has developed, requires reassessment of the consequences of DNA synthesis within genetics but stresses its applicability in current genetic research and biotechnology (Figure 5.1).

Importance of DNA Sequencing

The significance of DNA sequencing cannot be overstated because it serves as a basis on which the many branches of the biological sciences come to stand, from genetics to virology. By enabling scientists to determine the very nucleotide sequence present in a DNA molecule, sequencing provides seminal information about the functional properties of genomes. This molecular understanding allows for various applications, including the detection of gene disorders, the development of personalized medicine, and exploring evolutionary biology. For instance, sequencing technologies allow scientists to sequence the genomes of various organisms, opening up for the possibility of obtaining medical advances and biotechnology. Furthermore, genomic sequence comparison enhances our understanding of biodiversity and adaptation within species, as testified by the growing body of literature on genomic studies. Apart from helping to extend fundamental biological information, DNA sequencing has enormous practical applications in medicine and agriculture, among others. The ability to sequence DNA with speed and accuracy has revolutionized diagnosis for genetic disease, allowing for tailor-made therapies to improve the outcomes of patients. Additionally, techniques such as NGS can test multiple genes simultaneously, making the identification of mutations that cause disease more efficient. Sequencing can be utilized in agriculture in order to produce genetically modified organisms (GMOs) that have resistance to pests or enhanced nutrition. The intricacies of working with genetic material underscore the importance of familiarity with genomic structure. Consequently, ongoing technological advancements in sequencing remain essential in addressing some of the most pressing challenges facing humanity today. Apart from that, sequencing technology advancements are also facilitating research into complex biological processes, like viral diversity and the interactions between hosts and pathogens. For example, viral sequencing in environmental samples is critical to track viral evolution as well as their transmission patterns. Double-stranded RNA viruses have benefited from improved sequencing techniques, with evidence that the use of additives like DMSO can significantly enhance sequencing yield and quality, especially in challenging samples. The findings describe adding DMSO to reverse transcription, which increases the recovery of sequencing reads by several orders of magnitude, illustrating how innovative techniques are applied to learn more about viral communities and their interactions with hosts (Delwart et al., 2019). In addition, the existence of software packages for compositional data analysis produced by sequencing technologies reflects the significance of tailored analytical approaches to unravel complex datasets, thus addressing particular challenges posed by DNA sequencing (de Pereira et al., 2007) (Figure 5.2).

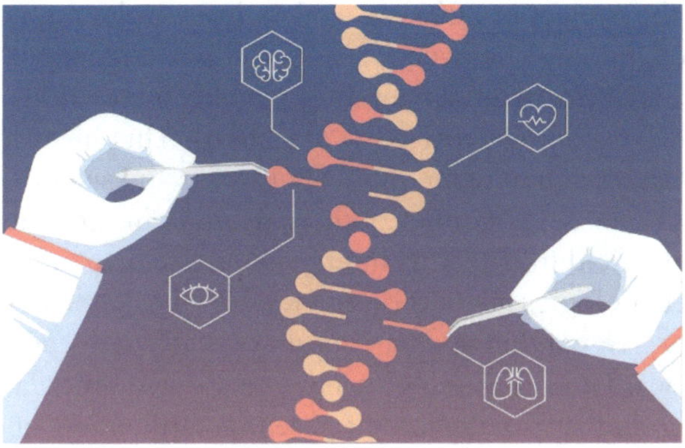

Figure 5.2. Importance of DNA sequencing.

Source: https://praxilabs.com/en/blog/wp-content/uploads/2021/01/41.jpg.

5.1. HISTORICAL BACKGROUND

The origin of DNA synthesis and sequencing is in the 20th century, with its groundbreaking discoveries that changed molecular biology. The discovery of the double helix structure of DNA by James Watson and Francis Crick in 1953 initiated an unprecedented period of research on genetic material. This discovery not only provided the foundation for the understanding of genetic inheritance but also paved the way for subsequent advances in DNA manipulation. As researchers delved into the molecular arrangement of DNA, advances in biochemistry technology picked up steam. Through the late 1970s, methodologies to synthesize DNA sequences were beginning to emerge, allowing researchers to create specialized strands of DNA for purposes of many different applications, thus unlocking the door to genetics, biotechnology, and medicine. The interaction between theoretical biology and technological advances has been the stimulus of the shifting landscape of DNA synthesis and sequencing for decades. With the revolution that had taken place in science, the development of revolutionized DNA sequencing technologies in the 1980s completely revolutionized genetic analysis. The first method, Sanger sequencing, took up the incorporation of dideoxynucleotides, which halted DNA strand elongation at specific bases. The new technique significantly enhanced the speed and precision of DNA sequencing, enabling scientists to sequence larger segments of genetic material and paving the way for the Human Genome Project's ambitious goals. Started in 1990, the international effort aimed to map the entire human genome and provide crucial data on human genetics and illness. The success of the project in 2003, as well as its completion, not only showed the power of collaborative science research but also signaled the necessity of a co-occurrence system in understanding genetic interactions, as referenced in today's healthcare dialogues (Hsu et al., 2016). This historical background highlights the intersection of technological progress and scientific innovation in the production of modern genomics. Other advances in sequencing technology have propelled the field into new frontiers, making sequencing more accessible and cheaper. Technologies like NGS emerged in the latter half of the 2000s, allowing for massive parallel sequencing that greatly reduced the time it took to study genomics. NGS technologies have made genomic data accessible to everyone, promoting an explosion of research utilization and personalized medicine strategies. Personalized medicine emphasizes the use of molecular data generated by sequencing to inform clinical decision-making and is in direct accordance with the European Commission's definition of personalized healthcare (Cabré Juan et al., 2017). This evolutionary framework shows how landmark moments in the past still influence contemporary practices, shaping future directions in deciphering complex diseases and adapting therapeutic interventions. The ever-evolving nature of DNA sequencing not only mirrors technological advancements but also represents the innate human curiosity to understand the complexities of life at the molecular level (Figure 5.3).

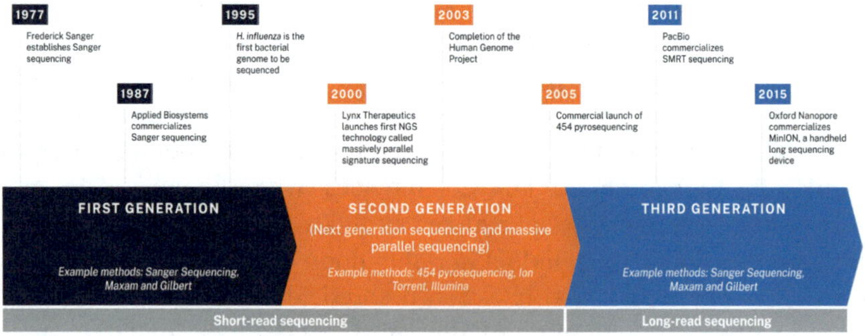

Figure 5.3. The history of DNA sequencing.

Source: https://images.ctfassets.net/ifq1osqu2ba1/4wGTRuwT3HWy3rMPHc3xol/91a39b4e046627da45a1cca69c677f6f/GN-DNA-Sequencing-learn1.png

5.1.1. Discovery of DNA Structure

Identification of the structure of DNAs marked a revolution in molecular biology that changed radically our understanding of genetics. The scientific inquiry that first sought to find out the chemical composition of nucleic acids consequently reached a definitive identification of DNA as genetic material. James Watson and Francis Crick utilized available data, including Rosalind Franklin's X-ray diffraction images of DNA, in the early 1950s. Their hypothesis proposed the double helix model, and how base pairing occurred between adenine and thymine and between guanine and cytosine. This mechanism accounted for the process of genetic replication, where strands unwind and become templates for making new strands. Implications of this finding extended far beyond genetics to become the basis for future breakthroughs in biotechnology, molecular medicine, and forensic science. The description of DNA structure not only created a new paradigm for genetics but also led to the innovation of techniques for DNA synthesis and sequencing. Following Watson and Crick's landmark discovery, scientists sought to find means of manipulating and researching genetic material.

The advent of recombinant DNA technology enabled the combining of DNA from various organisms, allowing for the production of genetically modified organisms (GMOs) and opening the way to critical findings in gene function. Furthermore, as sequencing technology has evolved, it has become clear that building complete genomic libraries is important. From Sanger sequencing to next-generation technologies, the ability to accurately distinguish the nucleotide sequence of DNA has transformed biology and medicine. As recognized by recent studies, the careful design and selection of protocols are crucial to ensure the dependability of such sequencing technologies, thus reaffirming the importance of rigorous methods in genomic analysis (Boone et al., 2018). Improvements in sequencing technology have not only unveiled the zippered topology of DNA but also the vastness of complexity and diversity of genomes, furthering our understanding of biological systems. Recent explorations of microbial genomics are a prime example of future potential for the discovery of new compounds, such as enediynes, through genome sequencing, which releases the hidden biosynthetic potential within microbial environments. As an illustration, studies have identified certain enediyne biosynthetic gene clusters, which contain information that can lead to the discovery of new natural products of significant therapeutic significance (Shen et al., 2016). Besides this, this capacity to rapidly uncover also highlights the importance of sequencing in designing new methodologies for drug making and biological study. With our technology for sequencing DNA continuing to advance, it has the potential to uncover unprecedented genetic information, reaffirming the central role of DNA structure in the pursuit of knowledge in genetics and biotechnology (Figure 5.4).

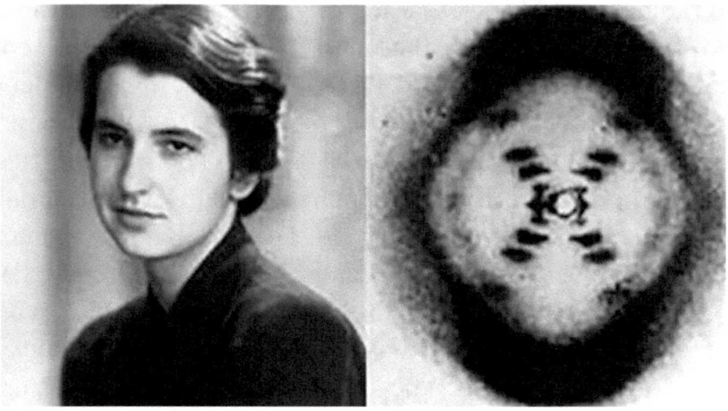

Figure 5.4. Rosalind Franklin and Photo 51.

Source: https://www.kcl.ac.uk/image-library/Press-Photos/Rosalind-Franklin-carousel.xeaf9cd0c.jpg?w=781&h=440&crop=780,440,1,0&f=webp.

5.1.2. Early Methods of DNA Synthesis

The early methods of DNA synthesis were pivotal to the development of molecular biology and were a foreshadowing of modern genetic engineering techniques. Early synthesis of DNA was largely achieved by chemical synthesis. Phosphorylation and protection through chemicals were employed by scientists to synthesize oligonucleotides in a systematic approach. The first significant breakthrough towards automated DNA synthesis was conceived by H. Gobind Khorana and his co-workers in the 1960s. They developed a method known as phosphoramidite chemistry that made nucleotide synthesis more efficient and specific. The foundation of this technique allowed for the synthesis of longer oligonucleotide sequences, which were essential in numerous applications, including cloning vectors and PCR primers. The advances in DNA synthesis introduced a turning point in the field by facilitating enormous leaps in genetic science, therapy, and biotechnology. Later development in DNA synthesis was marked by advancements in the accuracy and rate of synthesis. Among these advancements was the use of automated synthesizers in the 1980s, which enabled high-throughput oligonucleotide synthesis. This discovery greatly enhanced the access of researchers to custom DNA sequences, enabling them to be synthesized for the required DNA in advanced experiments. Machine platforms standardized synthesis processes, minimizing the issue of human error and thereby maximizing reproducibility of results. These machines utilized *Khoranas phosphoramidite* chemistry stepwise, wherein there was greater precision in the addition of nucleotide bases. The accuracy and speed of DNA synthesis opened up extensive applications such as genome sequencing attempts and the new discipline of synthetic biology. As DNA synthesis technology improved, several techniques were devised to deal with the growing demand for heterogeneous DNA sequences. For instance, technologies like polymerase chain reaction (PCR) made use of synthesized primers to amplify single DNA regions, thus revolutionizing molecular genetics. Furthermore, parallel advances in sequencing technologies meant that synthesis methods had to be continuously optimized so that researchers could study more advanced regions of the genome. Over the last few years, second-generation sequencing technologies have necessitated more accurate and efficient synthesis of DNA fragments, further emphasizing the synergy between synthesis and sequencing steps. Eventually, these advances also ushered in the era of synthetic genomics to enable complete genomes to be assembled from synthesized components. This synergy of DNA synthesis technologies and advancements in sequencing technology underscores the original importance of early techniques that served as the foundation for modern research and practice of genetics and biotechnology (Hua et al., 2024; Yang et al., 2025).

5.1.3. Development of Sequencing Techniques

The development of sequencing technologies is a milestone advancement in genetics and molecular biology, providing an expanded understanding of the genetic basis of life. The first milestone was the introduction of Sanger sequencing in the late 1970s, where it was possible to identify nucleotide sequences by employing dideoxynucleotides. The groundbreaking technique paved the path for the Human Genome Project, where mass sequencing technology was used to decipher human DNA. The advancements in NGS technologies came into play as high-throughput and fast sequencing became a requirement that covered millions of DNA pieces at a time. These advances opened unprecedented insight into genomic complexity, with implications for applications in personal medicine, evolutionary biology, and microbiomics. Advances like nanopore and sequencing-by-synthesis methods further increase the efficiency and cost-effectiveness of sequencing, showcasing the phenomenal rate of advances in the field of DNA synthesis and sequencing. The development of sequencing technology has greatly influenced biological research, enabling researchers to investigate genetic differences and their significance in disease and health. Single-cell sequencing technologies have become an especially valuable tool, providing cellular heterogeneity insights that earlier methods were incapable of obtaining. By examining single cells, researchers are able to identify genetic changes and how they affect function, which is needed to better understand diseases such as cancer at a more detailed level. Nanopipette technology has been evolving to extract cellular material for analysis without damaging the single cells, thus being able to measure single-cell behavior more accurately. As researchers increase the focus on how these technological advancements engage with cellular environments, a more refined understanding of gene-environment interactions emerges, eventually driving the fields of genetics and molecular biology, as has been observed in Bulbul et al. (2018). With ongoing advancements in sequencing technologies, the addition of epigenetics to sequencing approaches provides new opportunities for understanding gene control and adaptation. Factors influencing the genetic expression of an organism, such as environmental stimuli and their epigenetic control, remain crucial to producing more effective conservation strategies. This is particularly so in the scenario of the current-day issue of climate change, as outlined in Allen et al. (2018). The ability to monitor epigenetic change alongside genetic sequencing will improve our knowledge of the capability of organisms to adapt to shifting environments. Subsequent research aimed at unscrambling the enigma of epigenetic mechanisms can ultimately reveal future targets for conservation efforts. Closing the progression gap between sequencing technology and epigenetic investigation will enable us to better predict biological responses to ecological change and prepare adaptive strategies for the purpose of preserving biodiversity under fast-paced ecological change (Figure 5.5).

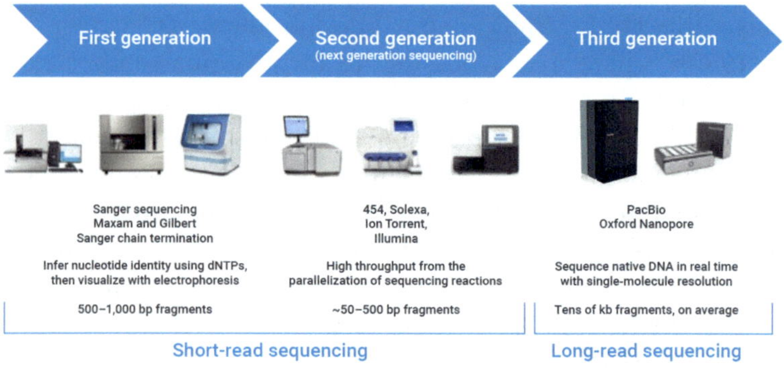

Figure 5.5. Development of sequencing techniques.

Source: https://cdn-ileaamc.nitrocdn.com/saCGjhfsnHihUCILLZSxgfOXwdfXxt-My/assets/images/optimized/rev-7feefbd/www.pacb.com/wp-content/uploads/img_evolution_of_sequencing_technology-2.png.

5.2. DNA SYNTHESIS TECHNIQUES

The evolution of methods of DNA synthesis over the past decades reflects dramatic improvement in speed and precision. Initially, the early sequencing methods, first found in the 1970s, opened the door by providing primitive templates for the analysis of DNA. Methods such as those developed by Sanger, using dideoxynucleotide triphosphates (ddNTPs), introduced a technique for making DNA strands in a controlled environment. But these approaches provided fragments of varying lengths that required extra electrophoresis to be resolved. As the need for better approaches grew, the invention of fluorescently labeled ddNTPs and the emergence of thermally stable DNA polymerases proved to be a great breakthrough. These technologies automated sequencing processes, which resulted in extreme time and cost reductions and increased safety via the elimination of handling radioactive chemicals (Goebel Vázquez et al., 2016). In the transition into the second generation of DNA sequencing, Next Generation Sequencing (NGS) technologies revolutionized DNA synthesis in a profound manner. By employing parallel sequencing techniques, NGS made it possible to analyze numerous DNA fragments at once, increasing throughput significantly while decreasing the need for electrophoresis. Procedures such as Illumina sequencing, involving the utilization of modified nucleotides with reversible terminators, epitomize the complexity of the era. These terminators temporarily inhibit DNA polymerization, allowing for accurate base detection via fluorescent imaging. Moreover, pyrosequencing innovations show a critical shift towards methods founded on the detection of byproducts that are generated as nucleotide is incorporated. Not only were such advancements enhancing the effectiveness of genome sequencing, but they also provided for the synchronization of massive genomic examinations that could not be carried out previously (Goebel Vázquez et al., 2016). With advancements in the field, third-generation sequencing techniques are on their way, pushing the boundaries of what is currently achievable in DNA synthesis and data analysis. Foremost among these is the incorporation of nanopore technology, enabling real-time single-molecule DNA sequencing. This technique uses nanopores consisting of bacterial channels to detect variations in ionic current as nucleotide residues pass through, with direct and efficient sequencing being enabled. Also, methods that utilize ion-torrent technology, where the emission of ions upon base incorporation is followed, are examples of the innovative application of current advances. Beyond traditional DNA technologies, the discovery of the application of synthetic polymers for storing information presents a new and captivating path. Storage of information on synthetic macromolecules may somewhat alleviate the limitations of DNA and pave the way for increased advances in molecular data storage and management methods.

5.2.1. Polymerase Chain Reaction (PCR)

The PCR has become a hallmark technique of molecular biology that can amplify specific segments of DNA. With the use of thermal cycling, PCR greatly increases the number of a given DNA segment so that very small quantities of genetic material can be studied. The process typically consists of three elementary steps: denaturation, annealing, and extension, which altogether yield millions of copies of the target sequence. The precise control of temperature within these steps allows for exponential amplification of target fragments, a feature that is of utility in numerous applications, including cloning, sequencing, and diagnostics. Thus, PCR has become a critical component for use in fields such as genomics and molecular ecology, especially when utilized with library preparation methods enabling the sequencing of complicated genomes and transcripts, paving the way for advances in genetic research and biotechnology. The versatility of PCR is further illustrated by its ability to be modified and adapted for various applications, including quantitative PCR (qPCR) and reverse transcription PCR (RT-PCR). Through these adaptations, researchers are able to quantify DNA and RNA quantitatively, respectively, providing information on gene expression levels and allowing the analysis of transcriptomes. Moreover, the introduction of NGS technologies has enhanced PCR's utility, as it forms the foundation for producing libraries with a wide range of genetic information retained.

Ongoing optimization of PCR protocols, including advances in enzyme specificity and reaction conditions, has significantly improved library generation reliability and efficiency. Such advancements are towards the elimination of bias in the representation of sequences needed for proper genomic analysis. Recent papers suggest library construction through well-informed means made possible through PCR protocols, emphasizing protocol design towards the achievement of reproducible results (Boone et al., 2018). Further, the implications of PCR extend to other multidisciplinary fields, solidifying the significance of interdisciplinarity among molecular biologists and applied scientists. This is evident in research that integrates molecular approaches and environmental and ecological research, in which PCR assists in microbial diversity testing and analysis of biogeochemical cycles. By examining the contribution genetic information makes to ecological processes, researchers are better able to understand systems of life. As found in interdisciplinary research, such as in the Planetary Biology and Molecular Ecology program, the use of molecular genetics techniques outside the field, supplementing field research demonstrates the practical utility of PCR outside traditional laboratory settings (Nealson and Nealson, 2025). The continuous evolution of PCR technology and integration into broader scientific research emphasize the critical role it plays in progressing our understanding of genetic material and its applicability in various fields of biology (Figure 5.6).

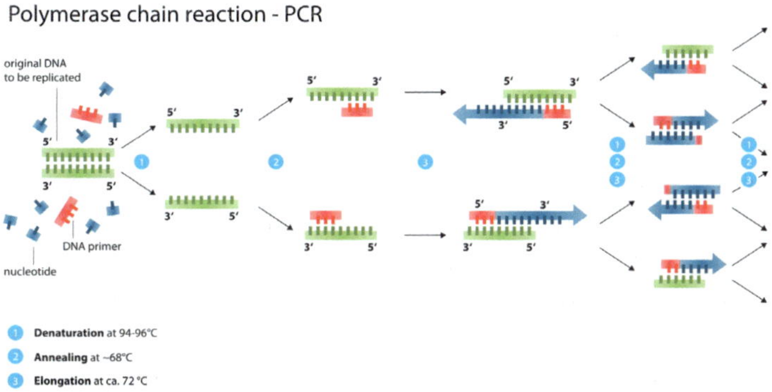

Figure 5.6. Polymerase chain reaction (PCR).

Source: https://i0.wp.com/microbeonline.com/wp-content/uploads/2016/07/Polymerase_chain_reaction.svg_.png?ssl=1.

5.2.2. Solid-Phase Synthesis

The evolution of DNA synthesis techniques has been significantly influenced by solid-phase synthesis methods, which revolutionized the way oligonucleotides are synthesized. The technique uses polymer resins as a solid support on which nucleotides are added sequentially to produce extended chains of DNA. One of the greatest advantages of solid-phase synthesis is its high efficiency, as it prevents contamination and allows the straightforward purification of the synthesized product. In comparison to the classical methods, such as phosphotriester and phosphoramidite approaches, solid-phase synthesis has proved to be stable and more consistent, with the latter playing a central role in the wide use of oligonucleotides in research and clinical purposes. Through efficient synthesis, it has played a part in the steep rise of the oligonucleotide therapeutics market, prompting many FDA approvals over recent years (Musaimi, 2025). Besides its predetermined application in oligonucleotide synthesis, solid-phase synthesis is deeply linked to the development in the production of nucleic acid carriers. With the increased sophistication of therapeutics, the need for precise polymers that are capable of delivering nucleic acids like pDNA and siRNA has arisen. Scientists can synthesize well-defined peptide carriers and multifunctional conjugates that enable enhanced delivery of nucleic acids in biological systems by employing solid-phase supported synthesis. Click chemistry and peptide ligation strategies are employed to site-specifically couple functional moieties, making the carriers more efficient. These advancements highlight the versatility of solid-phase synthesis, which is not only used for oligonucleotide synthesis but also for the assembly of sophisticated delivery systems that amplify therapeutic efficacy (Troiber et al., 2011). The contribution of solid-phase synthesis goes beyond the synthesis of oligonucleotides; it is the basis for modern DNA synthesis software. The solid-phase synthesis method has enabled the technique to exercise unparalleled control over the sequence and composition of synthesized DNA, thereby supporting the discovery of new therapeutic approaches. By providing the capability to introduce modifications in bases, sugars, and backbones in a systematic manner, scientists are able to design oligonucleotides to enhance their properties and therapeutic effects. As the science of DNA synthesis continues evolving, the solid-phase synthesis approach is the pride of it all, as the hub of innovation. Its optimization has a direct implication on the specificity and quality of synthesized strands, paving way to revolutionary breakthroughs in genetic research, targeted therapy, and more (Musaimi, 2025).

5.2.3. In Vitro Transcription and Translation

In vitro transcription and translation are important techniques in molecular biology that enable scientists to synthesize proteins and RNA outside of living organisms. *In vitro* transcription refers to the process of employing purified RNA polymerase along with a DNA template, whose intended gene sequence is present. By adding the nucleotides that are necessary, researchers can generate RNA transcripts for analysis or for further processing. The technique is particularly utilized where gene expression and regulation are being studied because it allows scientists to directly manipulate wanted sequences without the complicating influences of cellular contexts. This effective procedure of acquiring RNA transcripts is important for uses in synthetic biology, functional assays, and the production of recombinant proteins, essentially improving the study of genetic elements and their functional capabilities within organisms. Following the synthesis of the RNA through *in vitro* transcription, the translation of the RNA into proteins is accomplished, in most cases, by the application of *in vitro* translation systems. *In vitro* translation systems typically have ribosomes, transfer RNA (tRNA), amino acids, and other factors of translation, creating a system that mimics ribosomal protein synthesis. Another benefit of *in vitro* translation is its variability because scientists are able to introduce special modifications into translation components, thereby producing proteins with special functional characteristics. The process also facilitates the examination of post-translational modifications and protein folding in an easily controllable environment. The understanding gained from such analyses helps elucidate molecular processes underlying protein function, protein-protein interactions, and protein stability, forming a valuable link between transcription and translation in gene expression. Experimental methods such as *in vitro* transcription and translation become central to advancing our comprehension of biological systems at a molecular level. For instance, studies employing these techniques have illustrated the importance of gene expression modulation to environmental stimuli, as seen in experiments conducted with pathogens. As seen from studies with *Mycobacterium bovis* BCG, transcriptional regulation can give insight into survival strategies of organisms under macrophage interactions, revealing the subtleties of host-pathogen interactions (Wood et al., 2001). Furthermore, analyses of specific mutations, e.g., those in *Neurospora crassa*, can uncover significant aspects of protein function and targeting that are dependent upon successful transcription and translation (Neupert et al., 1987). Therefore, these *in vitro* methods not only inform basic biological research but also have important implications for therapeutic technologies and biotechnological developments (Figure 5.7).

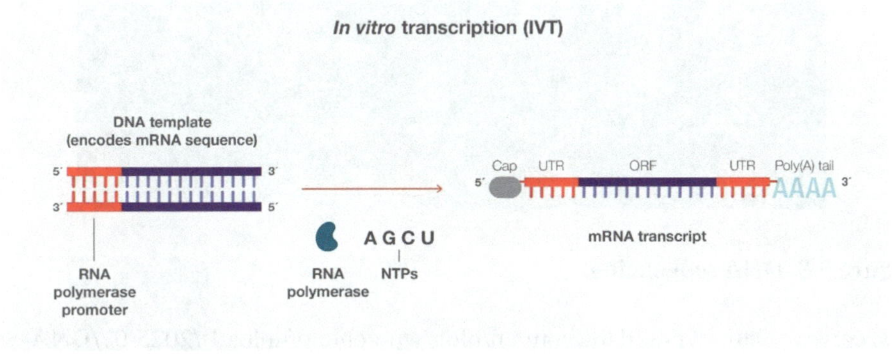

Figure 5.7. *In vitro* transcription.

Source: https://www.thermofisher.com/in/en/home/industrial/pharma-biopharma/nucleic-acid-therapeutic-development-solutions/mrna-research/overview-in-vitro-transcription/jcr:content/MainParsys/image_4952/foregroundimg.img.320.low.jpg/1749570419084.jpg.

5.3. DNA SEQUENCING METHODS

The study of DNA sequencing methods is critical to the advancement of genetic sciences and biotechnology. The most prevalent method is Sanger sequencing, which has been a breakthrough in nucleotide sequence determination. It is a method that relies on the inclusion of dideoxynucleotides, which terminate

DNA strand elongation in replication, forming a set of fragments that may be investigated to establish the template DNA sequence by inference. Sanger sequencing played an integral role in early genomic activities, including the Human Genome Project, and remains employed for low-scale sequencing due to its accuracy and consistency. But it lacks the throughput and cost-effectiveness needed for large-scale use, and thus the focus has moved to NGS technologies. NGS became a revolution in genomics as it allows massively parallel sequencing of millions of fragments in a single experiment. Methods such as Illumina sequencing and Ion Torrent have drastically reduced the time and cost of sequencing to provide large-scale genomic studies at an affordable cost. Most significantly, these methods employ new platforms that facilitate real-time detection of nucleotide incorporation.

Figure 5.8. DNA sequencing.

Source: https://praxilabs.com/en/blog/wp-content/uploads/2025/07/DNA-Sequencing-1-1024x512.webp.

For instance, Illumina technology applies reversible terminator chemistry to sequence millions of fragments within an experiment to produce vast data sets of unprecedented depth. This increase in sequencing efficiency has paved the way for applications in personalized medicine, where information on an individual's genetic code can inform interventions and treatments personalized to them in order to enhance healthcare outcomes. In addition, advances in molecular biology procedures have been added to sequencing protocols to optimize their efficacy and dependability. The combination of systematic evolution of ligands by exponential enrichment (SELEX) and high-throughput sequencing technology is a significant development in aptamer selection methods. By applying droplet digital PCR (ddPCR) to SELEX protocols, researchers have improved the efficiency of aptamer selection, which is critical in therapeutic applications and diagnosis. In addition, the use of such high-throughput methods enables detailed characterization of the initial libraries utilized in selection, making it possible to identify biased sequences that could impact results. Such methodological improvements demonstrate the need for ongoing innovation in DNA sequencing technologies, allowing investigators to negotiate the intricacies of genomic data and expand our knowledge about biological systems (Burnett et al., 2016; ROSELINO et al., 2008) (Figure 5.8).

5.3.1. Sanger Sequencing

Sanger sequencing was a major breakthrough in the history of molecular biology as it offered a viable technique for the identification of the nucleotide sequence of DNA. Frederick Sanger invented this method in the late 1970s. This method makes use of dideoxynucleotides, which are used to end the elongation of DNA strands, thus generating fragments with different lengths that correspond to the template strand. They are separated in a gel electrophoresis process and then can be sized to figure out the sequence by reading the nucleotides sequentially. Sanger sequencing was used in most of the major projects, including the Human Genome Project, because it offered high accuracy over other methods. While scientists endeavored to explore the blueprints of life in various ecosystems, from the activity of microbial communities, Sanger sequencing provided valuable data on the genetic diversity and functional roles of microbes within their ecosystems (Aronson et al., 2014). Though Sanger sequencing led the way in DNA sequencing technologies, recent years have witnessed monumental leaps with the advent of NGS techniques. NGS allows for the simultaneous sequencing of tens of millions of DNA fragments, reducing sequencing project duration and expense. Through this development, researchers have been able to transition from genomic static analysis to dynamic transcriptome analysis and, therefore, capture gene expression patterns in different biological conditions. The scalability of NGS not only facilitates extensive sequencing of intricate genomes but also has the potential to advance in environmental samples, where metagenomic sequencing can clarify the genetic blueprints for complete microbial communities. These developments go towards explaining how improvements in sequencing technologies are revolutionizing our information on genetic material and its ecological influence (Bähler et al., 2008). Even with the plethora of new technologies found today, Sanger sequencing remains a staple in the molecular biology toolbox. Its ability to produce good-quality and consistent sequences renders it a method of choice for applications where precision is necessary, like verification of NGS results or identification of particular gene mutations. Further, the fundamental concepts behind Sanger sequencing serve as a valuable learning ground for students and aspiring researchers confronted by molecular genetics. As the scientific community goes on to recognize the complexities of microbial communities and their interactions, the ease provided by Sanger sequencing remains essential to deconstruct intricate genetic interactions underlying such systems. On this front, Sanger sequencing not only facilitates advanced methodology but also underlines the basic principles of DNA sequencing and synthesis that continue to shape research in modern biology (Aronson et al., 2014) (Figure 5.9).

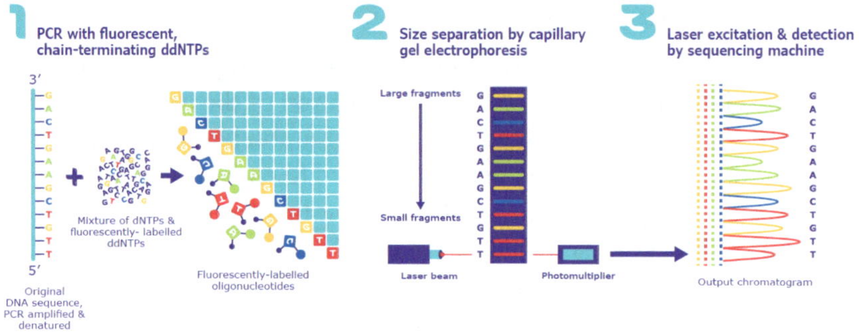

Figure 5.9. Sanger sequencing.

Source: https://praxilabs.com/en/blog/wp-content/uploads/2021/01/sanger-sequencing_steps_process_diagram.png.

5.3.2. Next-Generation Sequencing

The discovery of NGS revolutionized genomics and molecular biology practice by making the sequencing of DNA and RNA fast and affordable. This revolutionary technology offers the highest throughput to sequence millions of fragments simultaneously and thereby speed up the analysis of complicated genomes. In contrast to the traditional methods such as Sanger sequencing that are scale and scope-constrained, NGS can harvest huge genomic variations like single-nucleotide polymorphisms (SNPs) and structural variations. This versatility puts NGS at the forefront of cracking genetic diseases, particularly cancer, by unlocking somatic mutations and gene expression patterns. For this reason, NGS has emerged as a key element not only in research but also in the clinic, paving the way for personalized treatment and targeted therapies, enhancing the efficacy and outcome of patients. In oncology clinics, the use of NGS has revolutionized how clinicians perform cancer genetic testing. With nucleic acid sequencing technologies advancing rapidly, healthcare professionals can analyze a patient's gene pool and identify specific mutations for various cancers. This capability enables the identification of the functional and disease-generating genomic differences and aids in the identification of treatment protocols that are tailored to an individual patient. Conventional assays such as microarrays or qPCR typically fail by assessing only a subset of genes, with large portions of the genome remaining unraveled. NGS overcomes these limitations by enabling broad transcriptome profiling of non-coding RNAs and other essential genomic features. The ability to sequence large volumes of genomic data at high speeds accentuates the pivotal role of NGS in the quest towards precision oncology and the development of targeted therapy in patients. Despite the enormous potential of NGS, it still suffers from limitations that must be addressed to incorporate the technology into routine clinical diagnostics. High-throughput sequencing is also accompanied by intricate technological issues, such as instrument handling, chemistry stability, and complexity in data analysis. For NGS to be widely accepted for molecular diagnostics, it is essential to have harmonized quality standards and protocols that would ensure accurate results. Large NGS technology companies have recently introduced an attempt at bringing bench-top instruments with lower throughput and shorter run times, which would facilitate broader clinical applications. To realize the ambitious goals of reducing diagnostic expenses and raising the detection of new gene targets, collaborative efforts in further optimizing NGS protocols and analytic processes are essential. Only with established quality criteria can NGS be successfully implemented into daily routine procedure, bringing a solid foundation for molecular genetic diagnostics advancement (Adey et al., 2012; Fortina et al., 2016) (Figure 5.10).

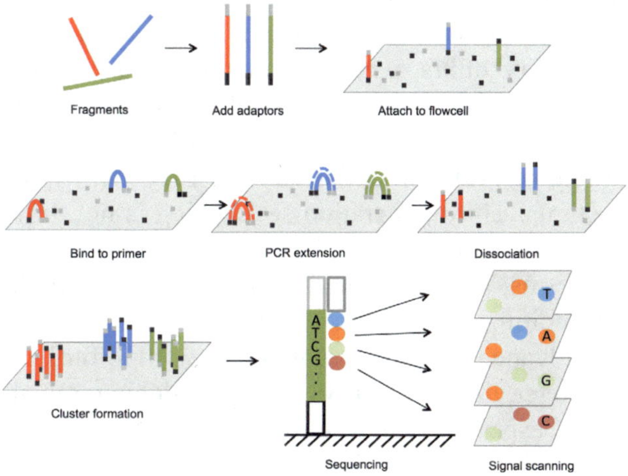

Figure 5.10. Next-generation sequencing.

Source: https://praxilabs.com/en/blog/wp-content/uploads/2021/01/NGS-1024x784.png.

5.3.3. Third-Generation Sequencing (TGS)

The TGS is a significant enhancement in genomic analysis that is marked by the ability to directly sequence single DNA molecules in real time. Unlike its predecessors, which utilized amplification strategies that had the potential of including biases or errors, the TGS reads the nucleotide sequence directly without the need for pre-amplification. Such ability has been achieved through advanced strategies such as nanopore sequencing and single-molecule real-time (SMRT) sequencing platforms. These approaches are capable of producing long-read sequences, often more than tens of kilobases in length, thereby enhancing the resolution of difficult genomic regions that are difficult to investigate with short reads. By doing this, the density of information derived from TGS not only reconstructs genomes but also facilitates the identification of structural variations and epigenetic marks, thus creating valuable insights into genomic structure and function. Beyond its scientific advantages, third-generation sequencing has unprecedented implications for various fields, including microbiology, personalized medicine, and ecology. As an illustration, the ability to sequence complex microbial consortia rapidly allows researchers to explore the vast diversity of microbes and their role in ecosystems (Aronson et al., 2014).

This is greatly applicable given that microorganisms play a central role in biogeochemical cycles and ecosystem interactions. Aside from that, TGS enhances the understanding of microbial dynamics, an area of emphasis for assessing reaction to environmental change. For personalized medicine, the long-read capability offers a more in-depth characterization of disease-associated genetic variations, making it possible for more effective therapeutic interventions. With the rapid advancement of TGS technology, then, there is the possibility to transform our understanding of genetics and larger-scale ecological interactions. Third-generation sequencing is not, however, without limitations. One of the most fundamental issues is the high rate of errors inherent in some TGS platforms, which can make data interpretation challenging and necessitate strong bioinformatics tools. Although base-calling algorithms have been enhanced by advances, further research is needed to further develop these technologies. Moreover, the cost of TGS, though reducing, remains prohibitive for mass adoption, especially in resource-poor environments. Though the pressure to enhance the cost-effectiveness and precision of third-generation sequencing is still on the rise as it is increasingly recognized as a vital tool for widespread genomic exploration. With the evolution of science, it's critical to address these issues in a bid to maximize the opportunities of TGS in scientific and applied biological research (Goebel Vázquez et al., 2016) (Figure 5.11).

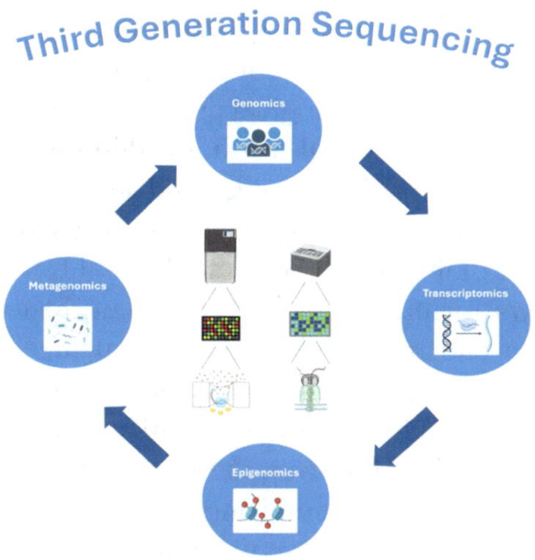

Figure 5.11. Third-generation sequencing (TGS).

Source: https://pub.mdpi-res.com/biomolecules/biomolecules-14-00568/article_deploy/html/images/biomolecules-14-00568-ag.png?1715592257.

5.4. APPLICATIONS OF DNA SYNTHESIS

Applications of DNA synthesis have spurred monumental advancements in various fields of research and industry. In medicine, novel therapies such as gene editing and the development of personalized medicine rely most on synthesized DNA. Through the creation of individualized genetic constructs, researchers can find and modify genes for certain diseases, paving the way for therapies that are individually tailored to a genetic profile. This not only enhances the efficacy of treatment but also minimizes side effects because therapeutic interventions can be made to surpass the individual genetic makeup of a patient. Moreover, the manufacture of DNA constructs makes it possible to significantly enhance learning about gene function and reveal complex biological pathways, which is critical to designing new drugs and therapies to combat genetic diseases and cancer. Another potentially valuable application is in the field of synthetic biology, in which DNA synthesis forms the basis of creating engineered biological systems. Scientists manufacture genes and pathways from scratch to introduce new functions into microorganisms, and this is being utilized to create biofuels, drugs, and other value-added biochemicals. The versatility of synthesized DNA enables scientists to design organisms to convert raw materials into high-value products efficiently. In addition to that, the engineered beings can be designed to serve sustainable needs, fixing some of the most important problems of today's world such as energy deficiency and climate change. By the use of genetic engineering of the very fabric of life, DNA synthesis not only answers life's biological puzzles but also provides real-world solutions with significant commercial worth. In addition to that, DNA synthesis is revolutionizing the field of data storage and information systems. Recent advances have shown that synthetic polymers can be used to store data such as DNA, addressing some of the limitations of traditional DNA-based storage. The novel approach allows for more monomer alphabet, hence supporting higher data storage densities and more effective retrieval mechanisms. As pointed out in recent literature, research on such synthetic alternatives can change information management technologies and significantly enhance our ways of storing vast data sustainably. Moreover, the use of DNA synthesis technology on data storage systems provides promising prospects for the development of next-generation computing architectures that utilize the advantages of biological substrates, which can lead to innovations in processing and energy efficiency (Aksakal et al., 2019).

5.4.1. Genetic Engineering

Genetic engineering is now a key tool of modern biotechnology that allows for the precise modification of an organism's genetic code. With advanced techniques such as CRISPR-Cas9, scientists can effectively target and edit individual genes, enhancing desired traits or introducing new functions. This precision is the basis for many applications, ranging from agriculture in which plants can be designed to yield more or be disease-resistant, to medicine in which gene therapy holds out the hope of cures for genetic diseases. Advances in DNA synthesis and sequencing technologies have helped take these applications far; with increasingly more accurate and efficient sequencing methods, they make it ever simpler to identify genetic targets for modification, thus making genetic engineering less complicated and allowing researchers to explore nuanced genetic interactions with higher resolution than before (Johnston et al., 2025). The implications of genetic engineering extend beyond commerce and health. Ethics have become more pronounced with the development of the potential for genetic alteration. Biodiversity concerns, dangers of unforeseen consequences, and the socio-economic impact of biotechnology have generated heated controversy. For instance, farm use of genetically modified organisms (GMOs) challenges ecological balance and food safety. As DNA synthesis and sequencing technologies improve, enabling precise genetic alterations, the regulatory landscape must catch up to enable responsible use while fostering innovation. Equilibrating these ethical issues is essential to win public trust and realize the full potential of genetic engineering toward more sustainably solving global issues like food security and medicine (Gronenberg et al., 2013). Furthermore, the cross-talk between high-throughput sequencing technology and genetic engineering has accelerated genomic research as well as functional genomics. Emerging technologies allow for in-depth probing of multifactorial traits due to complex genetic interactions. For example, pathway-based sequence analysis of phenotypes significantly increases the detection of genetic etiologies related to specific phenotypes, revealing complex pathways involved in metabolic processes. These innovations not only expand our understanding of genetic underpinnings but also reflect the confluence of sequencing technology and genetic engineering in practical settings. As researchers push further the boundaries of what is achievable with these tools, the scope of genetic engineering holds promise for spectacular progress in numerous arenas, including biofuel development and medical therapies (Gronenberg et al., 2013) (Figure 5.12).

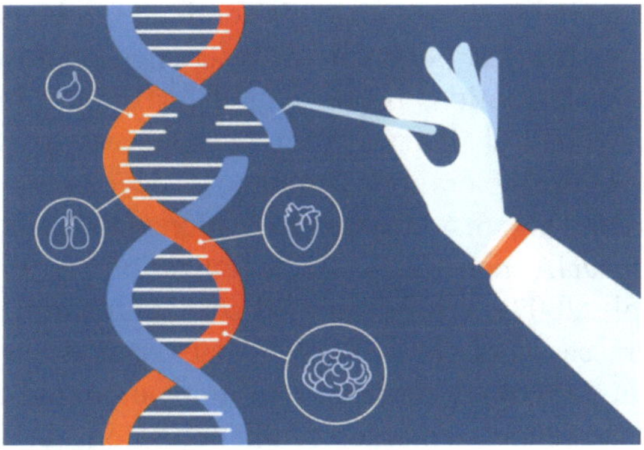

Figure 5.12. Understanding genetic engineering.

Source: https://t4.ftcdn.net/jpg/02/47/30/89/360_F_247308963_CT-F84BsVoXOOa6B0I6PTsRDNQ5FMrK9t.jpg.

5.4.2. Synthetic Biology

Synthetic biology is an innovative extension of traditional biomanufacturing techniques, facilitating the design and construction of new biological parts, devices, and systems. In short, synthetic biology revolves around DNA strand synthesis that can be employed in the production of genetic circuits and engineered organisms for particular applications. As researchers strive to advance towards more efficiency and versatility in the synthesis of DNA, innovations such as microfluidic devices are turning out to be game changers. These devices facilitate inexpensive and efficient synthesis of designer DNA constructs, which is critical to advancing the field of synthetic biology (Carr et al., 2007). This growth in ability not only accelerates research but also expands the possible applications across a great number of fields like medicine, agriculture, and environmental science, ultimately expanding our understanding and mastery over biological systems. The role of DNA sequencing in synthetic biology cannot be overstated, as it makes the important knowledge acquisition possible to verify the integrity and functionality of synthesized constructs. By implementing complex sequencing technologies, researchers can confirm that the produced DNA sequences are identical to the intended design. Such evidence is required because it ensures precision and enables the identification of potential errors which could be acquired during synthesis (Densmore et al., 2018). By combining precise DNA sequencing and advanced synthesis technologies, biotechnologists feel assured to make breakthroughs as they can better comprehend intricate biological pathways and interactions. As synthesis and sequencing tools and methods continue to advance, the discipline of synthetic biology has the potential to experience explosive growth with novel biotechnological solutions being crafted for the resolution of some of the world's most serious challenges. Moreover, access to synthetic biology technologies is one of the factors contributing to its development and utilization. With microfluidics and automated systems developed to facilitate the synthesis and screening of DNA constructs, democratization of this field is also possible (Carr et al., 2007). With decreased costs and simpler gene synthesis and verification, researchers from different backgrounds in institutions can contribute to synthetic biology projects, and innovation and collaboration are driven. Moreover, as schools embrace synthetic biology in the curriculum, a new generation of scientists will be ready with the capabilities to exploit these technologies. This window not only expands the scope of synthetic biology research but also demands interdisciplinary approaches, which allow innovative solutions to issues such as environmental sustainability, healthcare, and others to be developed. Hence, the common ground areas of DNA synthesis and sequencing are critical to the advanced march of the synthetic biology discipline and fulfillment of its vast potential.

5.4.3. Therapeutic Applications

Therapeutic applications of DNA sequencing and synthesis have revolutionized modern medicine, particularly through genetic engineering and personalized medicine. Such technologies allow researchers and physicians to tinker with genetic material, paving the way for new treatments of genetic disorders. Simple techniques, such as gene therapy, allow for the introduction of correctable genes to replace defective ones, with optimism regarding illnesses such as cystic fibrosis and hemophilia. Synthesizing specific DNA sequences specifically engineered to address diseases allows therapies to be more focused, effective, and less encroaching. With advancements in the biotechnology industry, its coupling with DNA synthesis technologies foretells a paradigm revolution in therapy paradigms that promises to bring hope to diseases for which there were no good treatments before (Johnston et al., 2025). Moreover, the uncovering of high-throughput sequencing has revolutionized therapeutic research through the facilitation of wide genomic profiling of patients. This approach allows one to identify genetic mutations that cause various diseases, and subsequently, one is able to create targeted medicines tailored to an individual's unique gene profiles. By sequencing cancer genomes, researchers can discover mutations that may render tumors susceptible to specific drugs, thereby enhancing the efficacy of treatment while reducing off-target side effects. As a result, personalized medicine not only enhances patient outcomes but also advances the understanding of disease pathogenesis at the molecular level. Such capability of high-throughput sequencing not only enhances drug development but also supports treatment selection, in effect linking genetic information with therapeutic interventions (Burnett et al., 2016). In addition to direct therapy, uses of DNA synthesis and sequencing involve the production of biologics, such as monoclonal antibodies and vaccines. Plasmid manufacturing at high speed for antibody production has been instrumental in developing therapy for diseases ranging from autoimmune disease to many forms of cancer. Moreover, mRNA vaccine manufacturing, most forcefully highlighted through the COVID-19 pandemic, reveals the significant impact that DNA synthesis technologies have had on public health worldwide. This new application of sequencing also allows monitoring of vaccine reactions at the genetic level, enabling vaccines to be rapidly adapted as new strains become available. The growing weight of evidence for the efficacy of the therapies reasserts the central role played by DNA synthesis and sequencing in informing therapeutic applications and responding to global health emergencies (Johnston et al., 2025).

5.5. APPLICATIONS OF DNA SEQUENCING

Applications of DNA sequencing are vast and have revolutionized many areas, including medicine, biodiversity conservation, and environmental science. In medicine, DNA sequencing plays a crucial role in personalizing treatments based on one's genetic makeup to enable targeted therapy, particularly in oncology. Medical practitioners are able to make personalized interventions that are more effective and less toxic using the study of genomic variations. DNA sequencing also enables disease-causing mutations to be detected, resulting in earlier detection of disease as well as genetic disorder management.

With the technology improving continuously, its use in routine clinical practice has the potential to transform preventative medicine and overall outcomes for patients, underlining the critical value of sequencing in healthcare. Beyond the clinical sector, DNA sequencing plays a pivotal role in biodiversity conservation efforts. It assists scientists in gaining insight into the evolutionary past and genetic diversity of aquatic animals that are typically under stress for a variety of environmental stressors. Using NGS tools, scientists are gaining vital information about the molecular evolution of aquatic animals as well as their functions in ecosystems. This genetic information not only emphasizes the need for the conservation of genetically diverse species but also helps predict the conservation measures needed to address the issues of these ecosystems. Bioinformatics tool development enables scientists to efficiently process genomic datasets of enormous proportions, thereby enhancing the understanding of aquatic biodiversity and supporting conservation efforts for such vital resources (Afiqah-Aleng et al., 2019). Environmental science has uses for DNA sequencing in the study of microorganisms that lay the basis for ecosystem processes. These microorganisms, which typically occur in vast numbers, play an important role in biogeochemical cycling and ecological balance. The explosive growth of molecular techniques has considerably increased our ability to identify and examine microbial communities in a wide range of habitats, from terrestrial to aquatic habitats. Microbial dynamics are crucial to comprehend not just for the understanding of their contribution towards the health of the ecosystem but also for climate change and other anthropogenic impacts (Aronson et al., 2014). By making use of DNA sequencing, researchers are able to decipher the complex dynamics of such communities, bringing them one step closer to a better strategy in controlling and conserving ecosystems. The growing convergence of DNA sequencing and environmental research is revealing the intrinsic importance of this technology to the resolution of contemporary ecological problems.

5.5.1. Genomic Research

Genomic research has transformed our understanding of biological systems profoundly, particularly through applications of DNA synthesis and sequencing technologies. With the discovery of NGS, researchers have become capable of generating massive sets of genetic data that enable the study of molecular evolution and biodiversity. This technology revolution has introduced an era of paradigm shift in ecological conservation research, particularly in marine ecosystems, which are precedence hotspots of biodiversity affected by extreme environmental risks. By using bioinformatics tools to decode the genomic data produced by NGS, it is possible to provide answers to important questions regarding aquatic life's evolutionary history and establish better conservation strategies (Afiqah-Aleng et al., 2019). Incorporation of genomic data thus delivers the complete picture as well as an effective approach toward understanding and conserving these vital ecosystems. The relevance of genomic studies goes beyond conservation to a broad array of disciplines like medicine, agriculture, and biotechnology. In medicine, for instance, human genome sequencing has enabled personalized medicine, providing tailored treatment based on specific genetic profiles. Further, the identification of genetic variants associated with diseases has enhanced our capacity for diagnosing and treating illnesses.

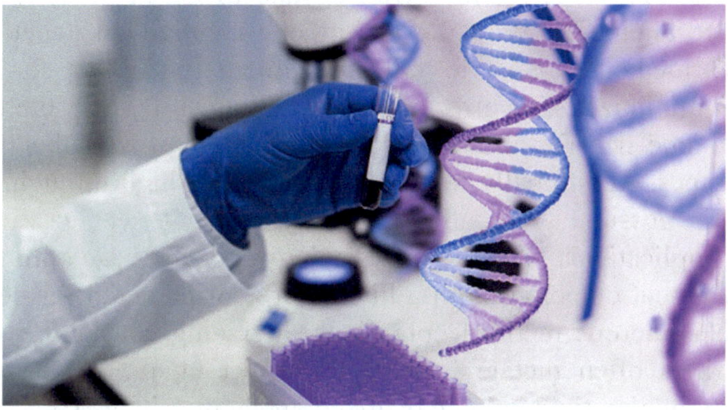

Figure 5.13. Exploring genomic research.

Source: https://www.fomatmedical.com/wp-content/uploads/2023/10/dna-doctor-does-lab-tests-1100x619.jpg.

In agricultural biotechnology, genomics technologies are employed to enhance the characteristics of crops, increase production, and provide safety in case of global warming. The enhancements underscore the importance of DNA sequencing and synthesis in enhancing not only our scientific appreciation but also addressing the world's most important issues. Thus, cross-disciplinary integration of genomics can yield creative solutions that leverage genetic knowledge for tangible benefits to society. The ongoing breakthroughs in genomic research place emphasis on thorough knowledge of the supporting genomic architectures and their functional implications. The case in point is the research on mutant species such as *Penicillium paxilli,* which indicates the roles of genomic differences on metabolite formation and their implications on ecological processes and biotechnological applications (Young et al., 1998). By examining genetic rearrangements and mutations, researchers can identify major pathways and mechanisms that regulate organism behavior and adaptation. Such discoveries provide insight into the molecular basis of traits and serve as groundwork for further research in genetic engineering and synthetic biology. While genomics research continues, the use of sequencing techniques with computational techniques will undoubtedly increase our ability to manipulate DNA for new applications, ultimately deciding the destiny of future biological investigations (Figure 5.13).

5.5.2. Personalized Medicine

Personalized medicine has emerged as a revolutionary approach in the healthcare industry, closely linked with advancements in DNA synthesis and sequencing technologies. With the use of one's individual genetic makeup, personalized medicine seeks to tailor treatment modalities for optimized efficacy and minimized side effects. This paradigm shift alters the focus from a traditional one-size-fits-all modus operandi to improved therapies based on genetic types. Additionally, with the inclusion of high-throughput genetic analysis, it is possible to identify clinically relevant genetic variants with therapeutic implications. With these technologies, doctors can optimize patient care and treatment outcomes with a high degree of potential for personalized medicine in numerous medical conditions, including cancer therapy, where genetic markers often dictate therapy (Schrijver et al., 2011). With rapid advancement in the field of personalized medicine, it also poses pressing challenges and ethical issues. The rapid rate of genomic technologies also has a tendency to outrun the ability to attribute clinical utility to found genetic variants. While it aids in the provision of high-throughput mutation screening and extensive pharmacogenomic screening, there is a pressing need for robust guidelines and frameworks to enable reliable interpretation and application of genetic information to the clinic. Such intricacies are faced when health systems incorporate additional biomarkers into routine testing and management algorithms. Addressing these paradigms necessitates collaboration between clinicians, geneticists, and bioethicists to facilitate the ethical application of genetic testing and navigate potential ramifications on patient confidentiality and consent protocols (Schrijver et al., 2011). The Human Microbiome Project likewise embodies the emerging paradigm of personalized medicine by demonstrating the convergence of advanced human genomic data with microbial populations. The project emphasizes the significance of viewing the human body as an ecosystem and, therefore, the symbiotic relationship of the human genome with the microbiome. With increasing knowledge in this field, the relevance of personalized medicine transcends individual genetic effects to encompass the role of microbiota in disease and health. It challenges us to redefine what we define as health and the inherent, intricate properties of biological systems. By integrating insights from both genomic and microbiomic research, personalized medicine promises to transform medical practice, pushing the boundaries of our current medical models into a new paradigm of more holistic and individualized therapeutic models (Figure 5.14).

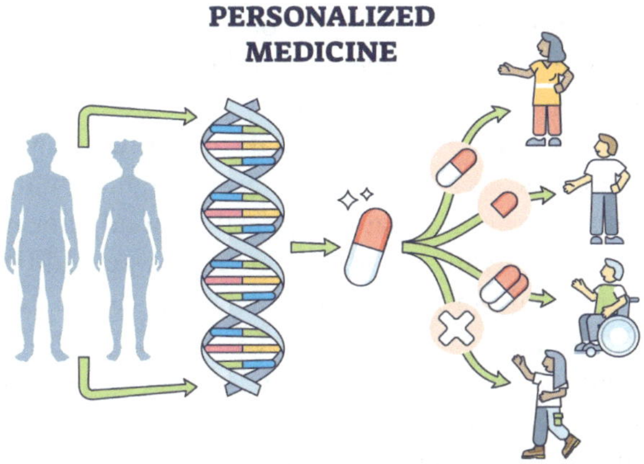

Figure 5.14. Personalized medicine.

*Source:*https://www.medznat.ru/uploads/images/post_content/7182/image_3/1690458980image_3.webp.

5.5.3. Forensic Science

Forensic science has emerged as a cornerstone of criminal investigations, particularly with the advances in DNA synthesis and sequencing technologies. These techniques enable the analysis of genetic material recovered from crime scenes to determine suspects and victims. The reliability and predictability of these techniques facilitate the comparison of DNA profiles, which have the potential to incriminate or exonerate people involved in criminal enterprise. Moreover, forensic DNA analysis can also provide invaluable information on the timing of biological events. Research analyzing post-mortem tissue samples indicates that gene expression can vary with the post-mortem interval (PMI), thereby providing invaluable information on the time of death. Forensic experts can establish a more comprehensive scenario of events preceding a crime by being aware of such nuances, ultimately enhancing the investigative process. The impact of DNA sequencing and synthesis is not only identification, but also ethical matters that are integral to forensic science. The existence of multiple databases of DNA, including those composed of DNA fingerprints and known gene sequences, raises vital ethical and legal problems. Forensic experts must devise means of addressing such problems in order to safeguard individual rights while also seeking the common good in the administration of justice. According to the literature, the ethical issues related to the storage and management of DNA data necessitate close examination of privacy rights and abuse potential (Stewart et al., 2006). Balancing prospects related to DNA technology with social responsibilities becomes important as forensic science plays a prominent role in law enforcement and the legal process. Furthermore, advances in DNA sequencing and synthesis technology have revolutionized forensic science with newer techniques that are more accurate and faster in analysis. Some examples are NGS, which can thoroughly analyze genetic data from small or damaged samples that are typically found in crime scenes. Such capability is particularly invaluable in cold cases, where traditional approaches have failed. NGS not only provides a broader view of the genome, allowing for better identification, but also helps in the study of genetic variation among populations. The ability to screen multiple DNA loci at once provides a better picture of human population associations and evolutionary histories. Last but not least, these advances in DNA technology provide much firmer bases for forensic science and its applications in contemporary society (Figure 5.15).

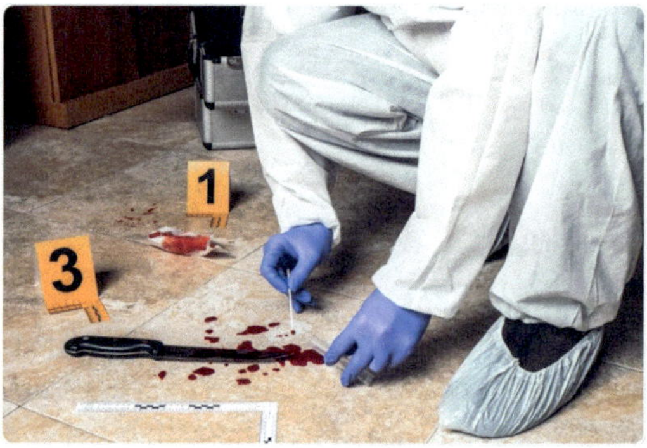

Figure 5.15. Forensic science.

Source: https://www.azolifesciences.com/image-handler/picture/2021/7/shutter-stock_1366514525.jpg.

5.6. CHALLENGES AND LIMITATIONS

DNA synthesis and sequencing have various challenges inherent in their domain of activity, particularly in describing the genetic elements such as the short tandem repeats (STRs). The traditional sequencing technologies, i.e., the short-read sequencing, possess far-reaching limitations in resolving the STRs due to the read length limitation, which is not inclined to capture the full context of these complex genomic regions. These restrictions lead to higher mapping errors and lower sensitivity for identifying expansions and interruptions in these repetitive sequences. Since STRs have been shown to significantly contribute to genetic variability and are responsible for several neurological and muscular disorders, the failure of standard methods to fully characterize them causes significant hindrances in genetic studies and clinical diagnostics. Technologies for long-read sequencing, such as PacBio and Oxford Nanopore, provide a theoretical solution, but their implementation in routine genetic analysis poses further logistic and computational challenges worthy of ongoing investigation (Chaushevska et al., 2025). Another major shortcoming of DNA synthesis and existing sequencing technologies is the resource intensity and cost-effectiveness of these processes. As DNA synthesis moves towards its transition from research-oriented applications to common consumer applications, such as data storage, error susceptibility, and latency of storage gain prominence. DNA-based storage processes, for example, in the revolutionary DNA-DISK platform, remain prone to slow retrieval processes and complexity in data management. Although these newer technologies have now enabled data to be stored in DNA, production and sequencing of the DNA itself are still time- and resource-intensive to the point that this technology is now hampered by a fundamental barrier to its mass deployment when compared to the conventional digital storage technologies, which are as efficient without the added biological complexity (Li et al., 2024). In addition, ethical and security issues introduce additional complications in terms of DNA sequencing and synthesis. As the potential to utilize DNA as a storage medium for information grows, concerns of data privacy and genomic data security become more real. The intrinsic singularity of genetic data raises severe ethical issues relating to ownership, access, and misuse of private genetic information. Securing data in DNA-based systems demands sophisticated encryption and protection procedures, thus complicating the incorporation of such technologies into actual programs. For instance, despite technology like DNA-DISK being optimistic, its use has to withstand the intersection of technological progress and moral responsibility. While researchers and practitioners address the technical deficiencies, a balanced response to the ethical concerns is critical to build trust and ensure that society benefits from advances in DNA synthesis and sequencing (Chaushevska et al., 2025; Li et al., 2024).

5.6.1. Accuracy and Error Rates

Accuracy and error rates in DNA synthesis and sequencing are important for producing consistent results in genetic studies and biotech applications. These steps' accuracy is regulated by the inherent fidelity of DNA polymerases, the enzymes responsible for constructing new DNA strands. Alterations in polymerase accuracy result in varied error rates that have a major impact on subsequent analysis. For instance, studies have found that certain polymerases contain a minimum of 0.08% mutation frequencies, while others tend to introduce errors at higher frequencies, and hence the need to select the appropriate enzyme for specific applications. Quality versus speed is also an issue, whereby high-throughput sequencing platforms will opt to compromise on data quality at the expense of speed, which can lead to errors and compromise genomic data quality (Delwart et al., 2019). Hence, understanding such dynamics is essential to achieve correct sequences in DNA analysis. The incorporation of additives, such as dimethyl sulfoxide (DMSO), has proven promising in maximizing sequencing yields, particularly for challenging templates such as double-stranded RNA (dsRNA). Usage of DMSO during reverse-transcription has shown to enhance sequencing yield more than two orders of magnitude, especially when RNA and complementary DNA (cDNA) levels are minimal (Delwart et al., 2019). Its enhancement is useful in viral genome analysis as well as the immune response of eukaryotes. Besides, DMSO treatment has been shown to speed up sequencing without compromising single-stranded RNA (ssRNA) recovery, suggesting its versatility. By maximizing sequencing conditions, researchers can improve accuracy and minimize the potential for misinterpretation of data, which maximizes the validity of genetic analysis in research and clinical environments. Error rates for DNA synthesis and sequencing are not only a function of the type of polymerase and additives applied but also of the overall design of the experiment. For instance, gapped plasmid constructs like those derived from pUC18 provide a stable system for the determination of polymerase fidelity *in vitro* replication and are useful for providing information on error rates for different enzymes (Bebenek et al., 2009). Systematic examination of such constructs has reiterated their suitability for the standardized protocols and provided additional insight into the factors leading to nucleotide misincorporation. This ability to adapt is important in the development of reliable sequencing technologies that are reproducible in various applications, from clinical diagnostics to synthetic biology. Therefore, constant improvements in methods and error estimation approaches will have an important role to play in pushing the DNA synthesis and sequencing field forward, such that the evolving needs of the scientific community are suitably catered to.

5.6.2. Cost and Accessibility

Advancements in DNA sequencing and synthesis technologies have greatly affected cost and accessibility of genomics. Previously, due to high costs, their use was limited to specialized labs and affluent research institutions. Advances in NGS technologies, though, have reduced the genomic analysis time and cost significantly. According to recent research, the implementation of NGS by clinical laboratories has made it financially feasible for oncologists to analyze somatic mutations and gene expression alterations in cancer patients. The democratization of genomic tools has unshackled opportunities for personal medicine, providing more accurate and effective modalities of treatment to more patient populations for longer durations, which was previously unavailable due to economic limitations (Fortina et al., 2016). Moreover, as the costs of DNA synthesis and sequencing decrease, it becomes more accessible to smaller laboratories and research groups, further broadening the reach of genomic analysis. The reduction in costs has enabled scientists to explore a greater range of genetic variations and their implications for human health. For instance, the use of high-throughput sequencing platforms such as the massively parallel reporter assay (MPRA) allows comprehensive screening of therapeutic sequences without submitting them to gigantic economic burdens (Aleshe et al., 2020). It represents a shift towards a larger research setting in genomics, where smaller institutions can contribute meaningfully to genetic engineering and therapy studies, advancing scientific knowledge as well as clinical applications. Despite the huge strides made in lowering costs and broadening access, issues are rampant in the promotion of equity in the spread of genomic technologies. Access is not only determined by the economic price tag in DNA synthesis and sequencing but also by geographical and institutional variations in the scientific infrastructure. The majority of regions, particularly developing countries, still have to contend with limited resources, and hence, it proves to be an uphill task for them to capitalize on such advances in genomics. Thus, while the future of increased access to DNA sequencing technology holds promise, there is a need for the science fraternity to strive to address such imbalances to ensure all populations benefit from the advances in genetic research and personalized medicine (Fortina et al., 2016).

5.6.3. Ethical Considerations

The recent advances in DNA sequencing and synthesis technologies have opened up a vast array of medical and research applications. But these applications come with huge ethical implications that need to be addressed. One of them is the potential for abuse of genetic data, resulting in employment, insurance, and other forms of discrimination based on one's genetic profiles. With pharmacogenomics and personalized medicine becoming more and more feasible, ethics in consent, privacy, and potential exploitation form themselves. Their resolution is crucial to ensure public trust and adoption of genetic technologies, as highlighted in the significance of evidence-based guidelines in pharmacogenetics (Paepe et al., 2013). A full ethical system can be used to balance innovation with societal values and ensure that the benefits of DNA synthesis and sequencing will be equitably shared. A second key ethical issue beneath DNA synthesis and sequencing is the impact of gene editing technologies, such as CRISPR. While these technologies have unparalleled potential to treat genetic diseases, they raise basic questions of how far humans should go to change the natural genome. The likelihood of designer babies does not just pose technical issues but also ethical and social concerns in terms of fairness, eugenics, and what is considered normal in health. Furthermore, the environmental implications of synthetic biology cannot be overlooked, as designed organisms can perturb existing ecosystems if accidentally or intentionally released. The controversy surrounding these issues highlights the importance of collaborative dialogue among scientists, ethicists, policymakers, and the community to navigate the complex ethical issues of DNA sequencing and synthesis (Batiuk et al., 2017). The development of an ethical framework of principles and policy could help to mitigate risk while supporting responsible scientific progress. Finally, the commercially-driven aspects of DNA sequencing and synthesis raise yet more ethical concerns, most significantly in terms of accessibility and equity in health care. In reducing the cost of sequencing technologies, inequalities in access to the technologies risk exacerbating existing health disparities. Wealthier societies would thus reap the rewards of targeted medicine, with poorer groups still deprived of necessary genetic analysis and therapy. This is intended to highlight an immediate ethical responsibility to ensure that progress benefits all sections of society and not just privileged sections. As these technologies become increasingly part of routine clinical practice, considerable effort must be devoted to lobbying for policy arrangements that ensure equitable access to genetic materials and treatments. It is through bringing these ethical issues into alignment that the best potential of DNA synthesis and sequencing can be realized for the greater good.

SUMMARY

In summarizing the discussion surrounding DNA synthesis and sequencing, one should note the sophisticated developments and methodologies that have characterized this field. Evolution from traditional Sanger sequencing to modern NGS has transformed the landscape of genomic research. This revolution enables scientists to screen vast reservoirs of genetic information rapidly and efficiently, with applications ranging from evolutionary biology to medicine. Furthermore, advances in library preparation protocols and enzyme specificity have enhanced sequencing library quality and complexity, enabling breakthroughs and precision in genomic research. As highlighted, well-designed sound protocols can significantly minimize biases, thus providing stable representations of heterogeneous sequences in libraries (Boone et al., 2018). Added is the accommodation of some of the issues presented by various forms of RNA, particularly double-stranded RNA (dsRNA), in sequencing. Failure to define dsRNA sequences has hindered previous studies regarding immune function and viral diversity. New methods, such as the application of dimethyl sulfoxide (DMSO) during cDNA synthesis, have achieved impressive gains in sequencing yields for dsRNA viruses and indicate the potential of dramatic breakthroughs in viral genetic insight. Not only are such methodologies useful for increasing recovery rates when RNA levels are minimal but also indicate the necessity of optimizing methodologies to reach unobtainable genomic sequences (Delwart et al., 2019). The creation of such approaches is nothing but a testament to the ongoing desire to overcome barriers in molecular biology. Ultimately, the future of DNA synthesis and sequencing is all contingent on ongoing innovation and determination when confronting new scientific challenges. Researchers must be not only capable of enhancing existing technologies but also of creating novel approaches that incorporate interdisciplinary thought from genomics, bioinformatics, and computational biology. As the pursuit of increased genetic knowledge continues, it is important to keep an eye on the ethical implications of such technology, especially in areas with potential applications affecting human health and biodiversity. The balance between technological effectiveness and ethical accountability will mark the next wave of research in DNA synthesis and sequencing, so that breakthroughs are developed for the benefit of scientific discovery rather than imposing a successful but unconscious response to the intricacies of genetic data.

The technology of DNA synthesis and sequencing has gone far in recent years, with immense relevance to biotechnology and medicine. The major developments include the creation of improved and higher-speed sequencing technologies, which allow us to understand genetic material and their roles better. Especially significant is what has been learned about the molecular basis of drug resistance in disease-producing pathogens. Studies on the protozoan pathogen *Leishmania* (L.) donovani illustrate how genetic variation contributes to drug resistance. Such findings highlight the importance of taking account of genetic backgrounds in efforts to identify resistance markers, because the very diversity of human populations renders it impossible to establish universally valid tools of surveillance for drug resistance. These perceptions are vital, as they guide therapeutic measures and epidemiological strategies, which are essential in the fight against global health issues related to resistant pathogens (Debrabant et al., 2012). In addition, the incorporation of metagenomic and metatranscriptomic analysis within DNA sequencing studies has provided novel avenues for elucidating microbial ecosystems, especially in human health environments. By examining extensive metadatasets of human fecal samples, researchers have begun to delineate a core metatranscriptome, characterized by functions transcribed reproducibly across various individuals. This foundational research highlights the importance of specific microorganisms and their ecological communities in shaping human health outcomes. The ability to monitor and reconstruct ecological interaction networks allows scientists to appreciate both the resilient and variable elements of gut microbiome function over time. Such an integrated vision of microbial dynamics has explicit consequences for DNA synthesis technologies, which hold the key to profiling and engineering these microbial communities for improved health interventions (Abu-Ali et al., 2018). Eventually, these advances in DNA synthesis and sequencing not only sharpen our view of genetic diversity but also challenge applications from precision medicine to personal therapy. The interplay of different genetic backgrounds and the molecular basis of traits, such as drug resistance, renders the genetic research landscape complicated. This is further compounded in the clinic, where treatments must be tailored to consider individual patients' unique genomic profiles. There is a necessity for persistent studies and technological innovations to offset such hurdles so that DNA synthesis and sequencing methods can keep pace with the intricacy of population genetic interactions. Eventually, persistent efforts at integrating the findings from drug resistance research and microbial ecology will consolidate the shared interest of improving health outcomes and advancig biotechnological innovations in molecular genetics (Abu-Ali et al., 2018; Debrabant et al., 2012).

REFERENCES

1. Abu-Ali, G. S., Branck, T., Chan, A. T., Drew, D., et al. (2018). Meta transcriptome of human fecal microbial communities in a cohort of adult men. https://core.ac.uk/download/286364483.pdf (accessed on 3rd October 2025).

2. Adey, A., Gehring, A., Benet-Pagès, A., Bainbridge, M., Bentley, D., Bergmann, C., Clark, D., et al. (2012). Diagnostic applications of next generation sequencing: Working towards quality standards. https://core.ac.uk/download/19418347.pdf (accessed on 3rd October 2025).

3. Afiqah-Aleng, N., Danish-Daniel, M., Mohd Nor, S. A., Razali, R., et al. (2019). Applications of next-generation sequencing technologies and computational tools in molecular evolution and aquatic animals' conservation studies: A short review. https://core.ac.uk/download/286549011.pdf (accessed on 3rd October 2025).

4. Aksakal, R., Du Prez, F., & Steinkönig, J. (2019). Molecular access to multi-dimensionally encoded information. https://core.ac.uk/download/322828930.pdf (accessed on 3rd October 2025).

5. Al Musaimi, O. (2025). Oligonucleotides: Evolution and innovation. https://core.ac.uk/download/634262413.pdf (accessed on 3rd October 2025).

6. Aleshe, B., Ayoub, P. G., Brown, D., Hollis, M., et al. (2020). Creating new β-globin-expressing lentiviral vectors by high-resolution mapping of locus control region enhancer sequences. https://core.ac.uk/download/323311026.pdf (accessed on 3rd October 2025).

7. Allen, C., Allendorf, F., Alley, R., Andrewartha, H., Anway, M., Araújo, M. B., Ardura, A., et al. (2018). Synthesizing the role of epigenetics in the response and adaptation of species to climate change in freshwater ecosystems. https://core.ac.uk/download/158347914.pdf (accessed on 3rd October 2025).

8. Aronson, E., Izard, J., Klatt, C., Zhou, Y., et al. (2014). The unseen world: Environmental microbial sequencing and identification methods for ecologists. https://core.ac.uk/download/286364389.pdf (accessed on 3rd October 2025).

9. Bähler, J., Marguerat, S., & Wilhelm, B. T. (2008). Next-generation sequencing: Applications beyond genomes. https://core.ac.uk/download/77000177.pdf (accessed on 3rd October 2025).

10. Batiuk, E., Kiyohara, C., & Nguyen, T. (2017). Engineering synthetic antibody by expanded genetic code. https://core.ac.uk/download/84118620.pdf (accessed on 3rd October 2025).

11. Bebenek, K., Connolly, B. A., Biles, B., Cline, J., Eckert, K., Evans, T., Firbank, S., et al. (2009). Plasmid-based lacZα assay for DNA polymerase fidelity: Application to archaeal family-B DNA polymerase. https://core.ac.uk/download/2721638.pdf (accessed on 3rd October 2025).

12. Bolondi, A., Caldarelli, F., Camilloni, G., Di Felice, L., et al. (2017). What is a gene? A two-sided view. https://core.ac.uk/download/80309530.pdf (accessed on 3rd October 2025).

13. Boone, M., Callewaert, N., & De Koker, A. (2018). Capturing the 'ome': The expanding molecular toolbox for RNA and DNA library construction. https://core.ac.uk/download/158346023.pdf (accessed on 3rd October 2025).

14. Bulbul, G., Chaves, G., Olivier, J., Joseph, A., Ozel, R. E., et al. (2018). Nanopipettes as monitoring probes for the single living cell: State of the art and future directions in molecular biology. https://core.ac.uk/download/323063966.pdf (accessed on 3rd October 2025).

15. Burnett, J. C., Chomchan, P., Ho, M., Rossi, J., et al. (2016). High throughput sequencing analysis of RNA libraries reveals the influences of initial library and PCR methods on SELEX efficiency. https://core.ac.uk/download/323063058.pdf (accessed on 3rd October 2025).

16. Cabré, J. A. (2017). Personalized medicine of cystic fibrosis. https://core.ac.uk/download/83830434.pdf (accessed on 3rd October 2025).

17. Carr, P. A., Chen, L., Jacobson, J., et al. (2007). Parallel gene synthesis in a microfluidic device. https://core.ac.uk/download/4435153.pdf (accessed on 3rd October 2025).

18. Chaushevska, M., Alapont-Celaya, K., Schack, A. K., Krych, Ł., Garrido Navas, M. C., Krithara, A., Madjarov, G. (2025). Get ready for short tandem repeats analysis using long reads—the challenges and the state of the art. *Frontiers in Genetics, 16*. https://www.semanticscholar.org/paper/3bca970ab6c190594afe0092773766d77d84cc3b (accessed on 3rd October 2025).

19. de B Pereira, C. A., Dougherty, E., Stolovitzky, G., Thygesen, H., Brentani, H., Braslavsky, I., Shmulevich, I., et al. (2007). Sim cluster: Clustering enumeration gene expression data on the simplex space. https://core.ac.uk/download/pdf/287399.pdf (accessed on 3rd October 2025).

20. De Paepe, A., & Vanakker, O. (2013). Pharmacogenomics in children: Advantages and challenges of next generation sequencing applications. https://core.ac.uk/download/55848729.pdf (accessed on 3rd October 2025).

21. de Torres, J., Ghenuche, P., Grigoriev, V., Moparthi, A., et al. (2014).

Plasmonic antennas and zero mode waveguides to enhance single molecule fluorescence detection and fluorescence correlation spectroscopy towards physiological concentrations. http://arxiv.org/abs/1403.1390 (accessed on 3rd October 2025).

22. Debrabant, A., Dube, A., Mehta, A., Mukherjee, A., Sidhu, A. B., Coelho, A. C., Fairlamb, A. H., et al. (2012). Molecular mechanisms of drug resistance in natural *Leishmania* populations vary with genetic background. https://core.ac.uk/download/6176183.pdf (accessed on 3rd October 2025).

23. Delwart, E., Díaz-Muñoz, S. L., & Wilcox, A. H. (2019). Next-generation sequencing of dsRNA is greatly improved by treatment with the inexpensive denaturing reagent DMSO. https://core.ac.uk/download/323068531.pdf (accessed on 3rd October 2025).

24. Densmore, D., Lashkaripour, A., & Ortiz, L. (2018). Automating functional enzyme screening & characterization. https://core.ac.uk/download/215936573.pdf (accessed on 3rd October 2025).

25. Fortina, P., Londin, E., & Paolillo, C. (2016). Next generation sequencing in cancer: Opportunities and challenges for precision cancer medicine. https://core.ac.uk/download/98337676.pdf (accessed on 3rd October 2025).

26. Goebel Vázquez, C., & Henao Mejía, L. M. (2016). Evolution of genome sequencing techniques. https://core.ac.uk/download/62908349.pdf (accessed on 3rd October 2025).

27. Gronenberg, L., Harper, M., Lee, C., Liao, J., et al. (2013). Comprehensive detection of genes causing a phenotype using phenotype sequencing and pathway analysis. http://arxiv.org/abs/1303.0455 (accessed on 3rd October 2025).

28. Hsu, M. C. Y. (2016). A co-occurrence framework conceptualized for bridging the gap between basic science, clinical research and clinical practices. https://core.ac.uk/download/142068225.pdf (accessed on 3rd October 2025).

29. Hua, L., Jiang, Y., Lu, D., Wang, B., Han, B., Hou, J., & Sun, Z. (2024). Advancements in semen sample analysis: Evaluating the efficacy and progress in human DNA removal techniques. *Journal of Men's Health*. https://www.semanticscholar.org/paper/7080c0ba305e745e4e7c9665fda7edf9b274b89a (accessed on 3rd October 2025).

30. Johnston, R. S., Norton, D. J., & Baldwin, T. (2025). Space and biotechnology: An industry profile. https://core.ac.uk/download/pdf/42821064.pdf (accessed on 3rd October 2025).

31. Lenz, A., Siegel, P. H., Wachter-Zeh, A., Yaakobi, E., et al. (2018). Coding over sets for DNA storage. http://arxiv.org/abs/1801.04882 (accessed on 3rd October 2025).

32. Li, K., Lu, X., Liao, J., Chen, H., Lin, W., Zhao, Y., Tang, D., et al. (2024). DNA-DISK: Automated end-to-end data storage via enzymatic single-nucleotide DNA synthesis and sequencing on digital microfluidics. *Proceedings of the National Academy of Sciences of the United States of America, 121.* https://www.semanticscholar.org/paper/7108a75a94cab93454a680b5bfc2336a5143e6f6 (accessed on 3rd October 2025).

33. Monahan, I. M., Mangan, J. A., Everett, M. J., Li, M. S., Butcher, P. D., Waddell, S. J., et al. (2001). cDNA-RNA subtractive hybridization reveals increased expression of mycocerosic acid synthase in intracellular *Mycobacterium bovis* BCG. https://core.ac.uk/download/2710689.pdf (accessed on 3rd October 2025).

34. Nealson, K. H., & Nealson, M. S. (2025). Planetary biology and microbial ecology: Molecular ecology and the global nitrogen cycle. https://core.ac.uk/download/pdf/42807174.pdf (accessed on 3rd October 2025).

35. Neupert, W., Stuart, R. A., & Tropschug, M. (1987). Deficiency in mRNA splicing in a cytochrome c mutant of *Neurospora crassa.* https://core.ac.uk/download/12168243.pdf (accessed on 3rd October 2025).

36. Roselino, A. M. (2008). Molecular biology in tropical dermatoses. https://core.ac.uk/download/pdf/37451202.pdf (accessed on 3rd October 2025).

37. Schrijver, I., & Wiita, A. P. (2011). Clinical application of high throughput molecular screening techniques for pharmacogenomics. https://core.ac.uk/download/323063319.pdf (accessed on 3rd October 2025).

38. Shen, B., Rader, C., Yang, D., Van Nieuwerburgh, F., Hindra, Ge, H., Crnovčić, I., et al. (2016). Strain prioritization and genome mining for enediyne natural products. https://core.ac.uk/download/84042091.pdf (accessed on 3rd October 2025).

39. Stewart, J. A. (2006). DNA, data and ethics. https://core.ac.uk/download/55367.pdf (accessed on 3rd October 2025).

40. Troiber, C., & Wagner, E. (2011). Nucleic acid carriers based on precise polymer conjugates. https://core.ac.uk/download/12174828.pdf (accessed on 3rd October 2025).

41. Yang, Y., Wen, X., & Wang, L. (2025). Advancements in DNA methylation technologies and their application in cancer diagnosis. *Epigenetics, 20.* https://www.semanticscholar.org/paper/ecc2bff17d09231bf27c4fb389c540837b1a009a (accessed on 3rd October 2025).

42. Young, C. A. (1998). *Paxilline Negative Mutants of Penicillium paxilli Generated by Heterologous and Homologous Plasmid Integration: A Thesis Presented in Partial Fulfilment of The Requirements for The Degree of Master of Science in Molecular Genetics at Massey University, Palmerston North, New Zealand*. https://core.ac.uk/download/157568893.pdf (accessed on 3rd October 2025).

CHAPTER 6

SYNTHETIC GENE CIRCUITS

Contents

INTRODUCTION

Advances in synthetic biology have enabled the development of synthetic gene circuits, which are man-made systems engineered to control cellular functions and behaviors in a predictable manner. By mimicking the natural regulatory mechanisms of cells, such circuits are invaluable for understanding complex biological processes. One of the most convincing qualities of synthetic gene circuits is that they have the ability to act like programmable devices, such that genes can be turned on or off based on certain environmental signals. Not only does this deepen understanding of cellular functioning, but it also lays the foundation for the potential to develop biotechnology innovations. The use of computational methods in the design and analysis of these circuits has been a pivotal strategy, making it both efficient and accurate to apply, hence opening doors to developments in the future. Efficient synthetic gene circuits, though, are the biggest challenge, particularly as the components are usually optimized for their indigenous environments and not synthetic. Due to the inherent biological system complexity, combining various elements into a smooth circuit can result in unreliable interactions and behavior of the circuit. To eliminate these issues, researchers have begun applying mathematical modeling with directed evolution—a method that maximizes circuit function by systematically modifying components through empirical alteration.

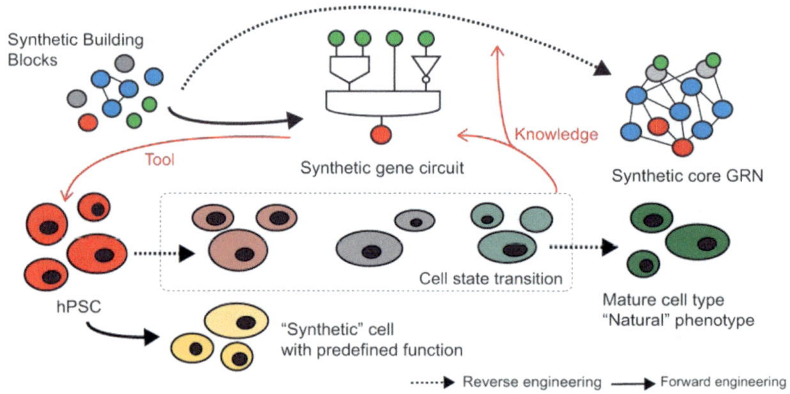

Figure 6.1. Synthetic gene circuits and cellular decision-making in human pluripotent stem cells.

Source: https://ars.els-cdn.com/content/image/1-s2.0-S2452310017301336-fx1_lrg.jpg.

The combination allows directed alteration, improving circuit reliability without the need for full biophysical knowledge. By offering glimpses into potential targets of mutations and reducing the evolutionary search space, such methods are a noteworthy contribution to synthetic biology research, enhancing the stability of synthetic gene circuits and enabling broader applications in biotechnological research (Arnold and Haseltine, 2007). Furthermore, the interfacing of synthetic biology with microengineering and electronics has drastically altered the scene for the construction of gene circuits. As evidenced by current advancements, the focus is increasingly being placed on enhancing not just abiotic elements but also on strengthening the processing capabilities of designed biological systems. This merging allows new miniaturized devices and sensors to be developed that can rapidly respond to modifying environmental cues, owing to genetic circuits that have been designed to be specific about their functions. Fundamental to the advancement is the use of computational design approaches, which facilitate circuit response optimization and tuning. A rigorous definition of biological elements is an indispensable requirement for effective circuit design. Closing the gap between the biological and mechanical fields allows researchers to access the enormous potential of synthetic gene circuits, flipping the fundamentals of biological systems on their head, in addition to creating beneficial applications (Marchal et al., 2013) (Figure 6.1).

Definition of Synthetic Gene Circuits

Synthetic gene circuits are constructed networks of genetic components that can be precisely engineered to control cellular processes in a specified manner. Operating on the same principle as electrical circuits, genetic designs are able to accumulate various inputs to produce desired outputs, enabling organisms to perform specific functions. Using modular design principles, synthetic biologists can integrate different genetic elements, such as promoters, inhibitable RNA, and coding sequences, into modules that regulate levels of gene expression in response to environmental stimuli or intrinsic cellular signals. Through modularity, the development of highly varied applications, from biosensors to detect toxins to therapeutic systems directly against disease in biological systems, is made possible. Through the controlled manipulation of these genetic circuits, researchers are opening the door to new solutions for complex biological problems and advancing fields such as medicine, agriculture, and environmental science, thereby unleashing the revolutionary power of synthetic biology in contemporary research and applications. The concept of synthetic gene circuits falls under the wider platform of synthetic biology, which aims to design and construct new biological components, devices, and systems. A major characteristic of such circuits is that they are capable of reaching certain functionalities by interfacing with the native cellular pathways without substantially disrupting the host's endogenous regulation networks. As technologies in synthetic biology advance, the focus has progressively turned towards enhancing these circuits for enhanced stability, efficiency, and specificity across diverse host organisms. For instance, successful use usually means the careful selection of chassis, or host cells, with minimal background activity in their endogenous gene regulatory circuits, allowing for more distinct expression of the engineered circuits. This optimization not only improves the performance of synthetic circuits but also encourages safety and containment measures critical to the safe application of synthetic biology (Kallergi et al.,

2020). Thus, synthetic gene circuit development is a quintessential example of innovative convergence of engineering principles with biological systems. Moreover, the design and use of synthetic gene circuits are intrinsically constrained by biosafety and societal challenges. Since these circuits are designed for applications as diverse as environmental bioengineering and medical therapeutics, careful consideration of their potential harms and ethical implications becomes necessary. Employing genetic precautions in design, such as "safelocked" organisms that will not survive outside certain laboratory settings, is a proactive approach to preventing unintended consequences and containment (Mukherjee et al., 2020). By involving stakeholders in a dialogue between scientists, ethicists, and the public, researchers are able to solve these challenges while stimulating innovation in synthetic biology. By having such cooperation, not only does it make synthetic gene circuits safer and more appealing, but it also aligns their development with what society expects and values, ultimately leading to a responsible integration of these technologies into applications (Figure 6.2).

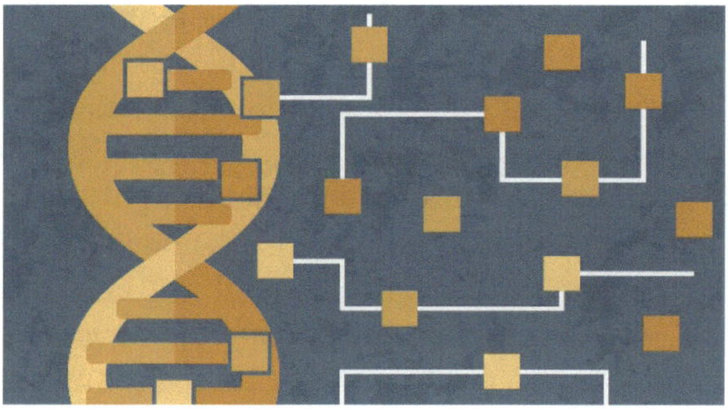

Figure 6.2. Synthetic gene circuits.

*Source:*https://engineering.asu.edu/wp-content/uploads/sites/36/2020/04/shutterstock_1302691477w-scaled-synthetic-gene-circuits-16x9-hero.jpg.

Importance in Biotechnology and Synthetic Biology

Synthetic gene circuits' contribution to the fields of synthetic biology and biotechnology cannot be overstated. They are a technique for engineering biological systems, and they enable the researcher to manipulate cellular behavior with significant precision. Synthetic gene circuits can carry out complex functions such as signal processing, decision-making, and metabolic pathway automatic control through the combination of genetic components. The engineering principles embedded in synthetic biology—standardization, abstraction, and uncoupling of design from construction—make such circuits possible to construct (Heinemann et al., 2006). Such designed intent makes it possible to achieve more controlled biological responses, as well as optimize individual potential for applications ranging from biomanufacturing to therapeutics, and highlights the revolutionary impact of such advancements in today's science. In addition, synthetic biology aims to address some of the most basic biotechnology challenges, for example, developing sustainable production processes for drugs. Synthetic gene circuits are likely to produce beneficial compounds by reprogramming microbial metabolic pathways. The fact that the circuits are modular means that various genetic elements can be assembled together, ensuring flexibility and versatility in design (Heinemann et al., 2006). As a result, the scientists are able to make the most out of bioactive compound production, and therefore, there are improved yields and reduced costs. This new approach not only optimizes the effectiveness of bioproduction processes but also offers potential for the development of novel bioproducts, which are of prime importance to the reduction of global health and environmental problems. The study of artificial cells is yet another field of synthetic biology that emphasizes the importance of synthetic gene circuits. These membrane-bound cell-like structures can be engineered to act similarly to cells but with specified parameters of operation, such as size and biomolecular composition. Microfluidic technologies have enabled the development of these artificial cells in high-throughput formats that make it possible to examine synthetic gene circuits at larger scales. This union of artificial cell construction and synthetic gene engineering allows for the reconstitution of biological function and complex behavior in the laboratory. Ultimately, the successful fabrication of these artificial cells can revolutionize biotechnology by providing vehicles for drug delivery, biosensing, and even for the creation of life-like systems, illustrating the vast potential of synthetic biology for application in the future.

Overview of Applications and Implications

The applications of synthetic gene circuits are already shifting rapidly, with immense implications in various fields, notably biotechnology and medicine. Synthetic gene circuits are designed biological pathways that can respond autonomously to certain stimuli, which can have the potential to revolutionize drug development as well as therapeutic interventions. For instance, synthetic gene circuits can be designed to regulate metabolic pathways so as to increase the production of pharmaceuticals, making it simpler to create new drugs with better efficiency. Their dynamic capacity to control gene expression allows one to design complex therapeutic methods, such as targeted anticancer treatments, causing cell death in cancer cells but not in normal tissue. As emphasized by the recent synthetic biology breakthroughs, these constructed circuits not only increase drug output but also enable one to synthesize compounds with the required properties, which accentuates their revolutionary potential in modern medicine (Heinemann et al., 2006). Moreover, synthetic gene circuits hold much environmental sustainability value, particularly in microorganism engineering for bioremediation. By inserting circuits that provide them with certain metabolic capabilities into bacterial strains, researchers can create such organisms to degrade poison or convert waste into viable biofuels. For example, *Bacillus subtilis* engineered strains can be designed to speed up the breakdown of harmful chemicals in contaminated environments, thus providing a valuable solution to pollution issues. Furthermore, the development of tools such as BacillOndex enables the integration of vast datasets and allows scientists to derive valuable information on *Bacillus subtilis* genetic interactions and metabolic pathways. This capability depicts the relevance of synthetic gene circuits in meeting environmental issues, demonstrating an aggressive approach to harnessing biological systems for ecological balance and sustainability (Allenby et al., 2013). The potential of synthetic gene circuits extends beyond the fields of environment and medicine into agriculture, where they offer solutions for enhancing the resilience and productivity of crops. By constructing gene circuits controlling responses to environmental stresses, such as drought or pathogen attack, plants can be made to acquire protective characteristics that increase yield and reduce reliance on chemical pesticides or fertilizers. Such development not only increases food security but also lessens the ecological footprint of conventional agriculture practices. The promise of synthetic gene circuits lies in their potential to orchestrate complex plant responses and, ultimately, create crops that can thrive under changing climates. As researchers continue to find new frontiers to break in synthetic biology, the range of uses and applications of synthetic gene circuits attests to their central role in shaping the future of biotechnology and sustainability across various industries (Heinemann et al., 2006).

6.1. HISTORICAL BACKGROUND

Historically, the trajectory of artificial gene circuits reflects the intersection of both systems biology and genetic engineering, integrating overall cross-pollination of scientific disciplines over time. Early advances in recombinant DNA technology provided the platform for the manipulation of genetic material, enabling scientists to express foreign genes and gain insight into the roles of genetics. Genetic constructs were mostly dependent on the utilization of plasmids as vehicles, enabling the introduction and expression of new genes in host organisms. This period, typically characterized by its focus on deciphering genetic codes and molecular-level interactions, laid the groundwork for even more sophisticated designs that incorporate synthesizing synthetic gene circuits. These initial advances are known to be essential, as they show the gradual expansion of experimental knowledge and technical expertise that would serve as the foundation for later synthetic biology activities, ultimately leading to the design, build, test, and learn approach paradigms of synthetic gene circuit research (Laubichler et al., 2008). In recognizing the historical subtleties propelling synthetic gene circuits, it is also necessary to appreciate the specific technological innovations that have characterized the field. The construction of the BioBrick standard during the early 2000s is a notable example of how collaborative research initiatives propelled the advancement of modularity within genetic components towards the consistent assembly into useful circuits. This trailblazing infrastructure not only enabled the exchange of genetic parts but also streamlined the engineering process, creating an open culture of science. Subsequent advances in high-throughput sequencing technologies further revolutionized the landscape with the capability to speedily characterize biological components. These developments lie at the heart of addressing the problems posed by the low number of transcriptional regulators available for use in circuit construction, highlighting the continued necessity of ingenuity in synthetic biology (Vecchione et al., 2019). Further, historical studies of signal transduction pathways have played a critical role in ongoing synthetic gene circuit improvement. Extracytoplasmic function (ECF) σ factors, recently discovered to be strong building blocks for the construction of orthogonal synthetic circuits, demonstrate the potential of synthetic biology to leverage natural biological systems for technology. ECFs, along with anti-σ factors, highlight how knowledge regarding these regulators has progressed, allowing for new avenues to address issues such as cross-reactivity and circuit function in diverse cellular settings. As the field continues to harness these biological discoveries, the balance between basic and applied studies reflects a fertile past. It is crucial for understanding the challenges and breakthroughs of today and tomorrow in the field of synthetic gene circuits (Vecchione et al., 2019).

6.1.1. Early Developments in Genetic Engineering

The early history of genetic engineering began in the mid-20th century through fundamental breakthroughs in molecular biology. The age was marked by discoveries when Watson and Crick elucidated DNA's double helical structure, providing the starting point for understanding genetic material for subsequent manipulation. When biologists understood the intricacies of gene expression and function, they started making tools to intentionally alter these genetic sequences. Early methods employed recombinant DNA, which employed enzymes such as restriction endonucleases to cut DNA at specific spots so that desired genes from other species could be spliced in. This groundbreaking research established the basis for synthetic biology innovations, which enabled scientists to examine gene function and create genetically modified organisms (GMOs) engineered to perform specific functions in agriculture, medicine, and industry. The creation of techniques for transforming organisms helped in further developing the science of genetic engineering, like the creation of transgenic plants and animals. An example is that the first genetically modified plant was a tobacco plant, which was produced by using *Agrobacterium tumefaciens* to insert foreign genes into the plant's genome. This was a breakthrough, demonstrating that it was possible to create organisms with new traits, such as resistance to herbicides or improved nutritional content. Moreover, the identification of transposons, or jumping genes, as vectors for genetic engineering provided a mechanism for producing more advanced genetic changes. These strategies eventually resulted in engineered strains that carried stable properties through generations, thereby providing a foundation for synthetic gene circuits engineered with specificity to respond to environmental cues and regulate biological processes as dictated by the transposon-based toolkit established in fungal systems (Baek et al., 2025). As far as molecular advances are concerned, the advent of CRISPR-Cas9 technology was a leap into a new dimension for genetic engineering by facilitating accurate modification of genetic material in a wide variety of species. By hijacking components of the bacterial immune system, researchers have found ways to edit genomes with unprecedented accuracy and ease. The versatility of CRISPR-Cas9 has enabled revolutionary applications, from gene therapy for genetic disorders to increasing plant and crop yields. The system not only reduced the complexity of designing synthetic gene circuits but also increased their applications to RNA editing and diagnostics (AYESA et al., 2025). Even as scientists further develop CRISPR technology and its broad applications, they are faced with challenges such as off-target effects and delivery mechanisms. Overcoming such challenges remains essential to achieving the full potential of genetic engineering for synthetic biology and future applications.

6.1.2. Milestones in Synthetic Biology

Synthetic biology has been marked by a series of significant landmarks that have served each in its turn to pave the way for the development of synthetic gene circuits capable of controlling biological systems. The ability to engineer and construct novel biological parts first emerged in the early 2000s with the development of initial tools, such as standardized biological parts, or so-called BioBricks. The modular DNA pieces enabled scientists to construct complex gene circuits piece by piece. Most importantly, the ability to screen these circuits within living organisms paved the way for genetic engineering milestones. This design modularity has enabled the construction of programmable systems of gene expression so that the temporal pattern and amplitude of gene expression in response to environmental stimuli can be regulated. The breakthrough in synthetic gene circuits that followed has led to applications ranging from agriculture to medicine, revealing the extent of the impact of synthetic biology on many domains of application. With improvements in research, the cooperation between synthetic biology and computational modeling came to represent a turning point that was quite pivotal in the field. Computational tools enabled modeling synthetic gene circuits, predicting their behavior, and designing them for optimal function before physical placement. This was an answer to some of the inherent complexities of biological systems, where interactions among numerous components would yield unpredictable outcomes. Synthetic gene circuits were then engineered more accurately and predictively. Notably, this computational innovation synchronized with experimental approaches, as researchers began to integrate high-throughput strategies with microfluidics to facilitate the rapid testing of circuit building blocks. This synergistic dialogue between computation and experimentation was a paradigm shift of the highest importance, accelerating how rapidly scientists were able to create and optimize synthetic gene circuits and thus driving understanding of both foundational biology and applied biotechnology (Leal-Alves et al., 2024). Synthetic gene circuits grew increasingly crucial in addressing global issues like food security and medicine. For instance, synthetic biology in plants has shown the value of such circuits to confer desirable characteristics such as enhanced drought tolerance and nutrient use efficiency (Khan et al., 2025). Achieving advanced synthetic circuits allows for the accurate control of gene expression as a reaction to environmental change, thus transforming agricultural production. Furthermore, synthetic gene circuits are also being applied in medicine, where they are being employed to produce new therapeutic devices. Through the design of microbial networks capable of generating drugs or targeted treatments, the implications for drug synthesis and patient therapy are staggering. As these milestones collectively demonstrate, ongoing progress in synthetic gene circuits continues to redefine our understanding of biological systems and yields revolutionary solutions to some of society's most pressing issues.

6.1.3. Key Figures and Their Contributions

The development of synthetic gene circuits has been developed through key players who have opened new frontiers utilizing pioneering methodology and technology for synthetic biology. George Church is one of the leading contributors whose genomic engineering research has paved the way for the field to be explored with complex genetic manipulation. Church's research involves high exposure to CRISPR technology that makes exact genetic sequence alterations possible. His research not only brings the relevance of synthetic circuits in fundamental research to the limelight but also their possible applications in medicine and biotechnology. By promoting the integration of several biological tools, Church's vision enhances how synthetic circuits interact with various cellular environments, ultimately contributing to the design of stable gene circuits. In addition to Church, there is Drawal Piramal, who has become an expert in the engineering of viral vectors for use in the delivery of genes. Piramal's research indicates how challenging it is to build effective delivery systems for synthetic circuits, as the specificity of the delivery mechanism itself is what typically guarantees therapeutic effect. His research is particularly valuable for its emphasis on controlled release and targeted action, how much more effectively genetic circuits can be applied under particular cellular conditions.

Figure 6.3. George Church.

Source: https://wyss-prod.imgix.net/app/uploads/2016/08/05095301/George_Church_headshot_1500x1000.jpg?w=800&h=1067&crop=faces&auto=format&q=90&fit=crop.

This study interfaces with novel methods such as applying pseudotyping and designed proteins to enhance the specificity of gene targeting. Piramal's work has critical implications in that it induces more innovation of therapeutic application on the basis of synthetic circuits for improved outcomes. Moreover, the advancement in synthetic gene circuits is enhanced by research teams that collaborate with diverse strategies, including the development of degradation tags and protease-based systems for temporal control. As identified in the concerted work of a number of partners, the focus is to develop circuits that are very specific and have controlled responses. In this concerted research, what is particularly new is the utilization of proteins that only activate in a specific intracellular environment, thus preventing unintended impacts on non-target cells. Such a multilayered strategy is critical for the therapeutic future since it addresses the two-pronged challenge of choosing a particular cell type and ensuring a robust safety mechanism against off-target activity. The works of these trailblazers collectively boost the capacity of synthetic gene circuits in addressing complex biological problems, advancing the general field towards novel therapeutics (Cao et al., 2008; Chong et al., 2020) (Figure 6.3).

6.2. COMPONENTS OF SYNTHETIC GENE CIRCUITS

The operation and engineering of synthetic gene circuits are based on certain significant building blocks that may be understood based on their biochemical activity and their interaction. Promoters are at the center of these circuits, controlling the transcription of genes and thereby deciding the gene expression dynamics. Promoters can also be engineered to react to specific signals, thereby allowing gene circuits to be able to regulate cellular action in line with environmental states. Furthermore, ribosomal binding sites (RBS) play an equally important role through the facilitation of translation to ensure proteins are produced efficiently. These elements together form the layers of construction of synthetic circuits, allowing the fine-tuning of cellular activities. With engineers designing increasingly sophisticated and delicate circuits, the fine-tuning of such devices is necessary, especially with their intrinsic complexities when used in novel contexts (Arnold and Haseltine, 2007). Besides promoters and RBS, the regulatory elements like transcription factors and RNA thermometers also play crucial roles in the fine-tuning of the performance of synthetic gene circuits. Transcription factors, on the other hand, can turn on or turn off gene expression based on specific stimuli, providing fine control over circuit behavior. RNA thermometers, in turn, can modulate gene expression in response to temperature fluctuations, demonstrating the versatility of synthetic adaptation. These two are regulatory components that enable a range of strategies of genetic engineering since they can be employed together to create circuits that can execute complex decision-making, such as that in natural cellular operations. Besides, the application of mathematical modeling in circuit design allows for the prediction of circuit behavior under various conditions, demonstrating the principles of synthetic biology (Arnold and Haseltine, 2007). All these factors combined eventually lead to the production of artificial cells, which are nothing but simplified forms of cellular systems in nature. Artificial cells are equipped with biological membranes, transcription-translation machinery, and genetic building blocks that cooperate in concert to reproduce essential cellular processes. Such a structure allows researchers to interrogate cellular dynamics with minimal perturbation from intrinsic cellular complexity. Artificial systems serve not only educational purposes but also have immense potential for biotechnological applications such as drug delivery and environmental sensing. By bridging the divide between natural and artificial cells, scientists will be able to examine fundamental questions on the minimal requirements for cellular function, replication, and evolution, enabling us to gain better insight into synthetic biology and its implications.

6.2.1. Promoters and Their Role in Gene Expression

The role of promoters in gene expression lies at the heart of the mechanism of synthetic gene circuits since the regulatory sequences that control the initiation of transcription. Promoters get attached to RNA polymerase and transcription factors, which enable the recruitment of the proteins to specific sites of a gene and ensure correct transcription of genetic information. Synthetic biology enables scientists to specifically regulate gene expression through the design and integration of synthetic promoters. By employing particular promoter sequences, researchers can control the timing and quantity of gene expression, ultimately enabling them to create complex genetic circuits that may respond to environmental signals or cellular cues. As the current understanding of promoter function evolves, novel designs such as those based on bacteriophage λ cI derivatives become feasible, allowing for the design of two-function promoters that further enhance the flexibility and functionality of synthetic circuits, as evidenced by recent work (Broedel et al., 2016). The complexity and diversity of promoters in a synthetic gene circuit can actually influence the performance and robustness of designed systems. As scientists push synthetic biology frontiers, there is a greater need for orthogonal expression systems. Orthogonal systems provide the possibility of several promoters being used independently without cross-interacting with one another, critical in the reduction of undesirable interactions within circuits that require stringent control. New libraries of promoters and regulatory elements, such as the use of heterologous sigma factors from *Bacillus subtilis*, show a significant advancement of the toolset available to build gene circuits. Such platforms allow for the titration of the transcriptional response to individual sensor stimuli in a quantifiable manner, providing a framework for constructing sophisticated gene circuits without interference among them (Bervoets et al., 2018). In the synthetic gene circuit context, continued promoter design and utilization are necessary to achieve intended synthetic biology results. With scientists trying to create more sophisticated circuits based on the complexity of natural biological circuits, it will be critical to completely understand how promoters work. The integration of synthetic promoters with newly designed transcription factors, built for orthogonality, allows for increasingly precise genetic circuits. This ongoing research not only enhances the efficiency of gene expression but also promotes greater scalability and flexibility in synthetic biological systems. With the advancement of the field, the innovative utilization of promoters in synthetic circuits illustrates the potential of synthetic biology to create tailored solutions for biotechnological issues, hence bridging the gap between applied work and practical application (Bervoets et al., 2018; Broedel et al., 2016) (Figure 6.4).

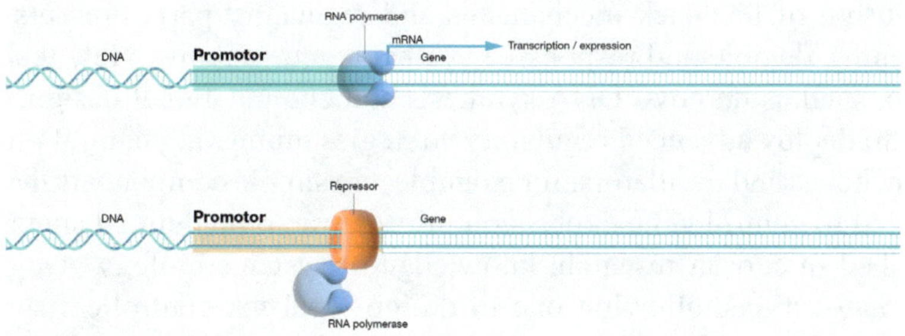

Figure 6.4. Promoter.

Source: https://www.genome.gov/sites/default/files/media/images/tg/Promotor.jpg.

6.2.2. Regulatory Elements and Feedback Mechanisms

Regulatory elements and feedback mechanisms, in the synthetic gene circuits' complicated world, play a key role in managing gene expression and ensuring system robustness. Regulatory elements like promoters, ribosome binding sites, and terminators are sites of control responsible for directing the flow of genetic information along the engineered circuits. By optimizing these components, scientists can manipulate the expression of individual genes in response to outside inputs or internal cellular states. This degree of control is required for effective implementation of synthetic biology applications, for which repeatability and predictability of behavior are desirable. In addition, awareness of these regulatory elements enables investigators to design circuits that rapidly adapt to changes in environmental factors, thereby significantly enhancing the function and utility of synthetic biological systems. Feedback regulation plays an extremely crucial role in maintaining the stability and robustness of synthetic gene circuits. Feedback regulation is either positive or negative feedback loops, with each having distinct functions in gene regulation. Positive feedback amplifies the response to a stimulus, creating an exaggerated output, while negative feedback damps responses to prevent overproduction and maintain homeostasis. A balance between these two forms of feedback is necessary; that is, a circuit synthesizing a therapeutic protein must precisely regulate output levels to avoid toxicity. Researchers are now employing sophisticated modeling techniques to predict feedback dynamics in engineered networks and how the interactions can be utilized to optimize performance. This information not only informs the design of more effective circuits but also underlies the principles of control theory applied in synthetic biology. The inclusion of feedback mechanisms and regulatory parts presents a route for designing flexible and responsive synthetic gene circuits. New technology advances, such as de novo DNA synthesis and computational design, make it possible to deploy advanced regulatory strategies mimicking natural processes. Toggle switches and oscillators, for example, are simple components that can be constructed to control gene expression in response to predefined signals. And, as described in current research, knowledge of systems biology is significant in these innovations, allowing one to design feedback controllers stabilizing synthetic circuits under stochastic fluctuations because of genetic noise and environmental variability. Thus, the precise interaction between the regulatory elements and feedback loops not only enhances the predictability of synthetic circuits but also emphasizes the engineering-centered approach of synthetic biology, in which such elements are specifically designed to serve a particular functional purpose (Du et al., 2010; Heinemann et al., 2006).

6.2.3. Genetic Parts and Standardization Efforts

Synthetic biology has brought about a paradigm shift in our understanding of genetic engineering and its use. This change centers around the concept of genetic parts—discrete modules of genetic content that can be combined into artificial gene circuits. Genetic parts, just like the engineer's modular components, enable complex biological systems with known functions to be designed. By enabling assembly of standardized, exchangeable modules, researchers can simplify gene circuit design and optimization of gene circuits tailored for specific functions, with far-reaching implications for biomanufacturing, therapeutic medicine, and general biological research (Chen et al., 2012). As the field advances, the need to standardize these genetic building blocks is rendered even more apparent through stringent characterization and quality control protocols that can ensure reliable performance across various applications. Standardization efforts in the area of genetic parts play significant roles in various applications in synthetic biology. To start with, consistency of genetic components allows reproducibility in experimental processes, which is necessary when making comparisons across studies. Secondly, having a standardized system of genetic parts allows collaborative research and information exchange since researchers can easily build on one another's work. The process of building repositories of characterized genetic elements is an imperative step toward a common understanding of how to use these elements in different applications. By fostering a culture in which genetic elements are properly characterized and made available, researchers can efficiently design systems that respond predictably to inputs and reduce the inherent complexity of working with biology (Marchal et al., 2013). Lastly, such standardization efforts complement the overall advancement of synthetic gene circuits to facilitate innovation in academia and industry. Aside from facilitating standardization, integration of computational design methods lies at the core of genetic circuit assembly. Such methods enable scientists to simulate and predict the behavior of synthetic circuits before they are constructed, hence enhancing the productivity and efficiency of genetic engineering processes. Computational systems are able to explore complex interactions of artificial gene circuits, wherein sequences of gene expression are optimized to respond to environmental signals and operational demand. This synergy of computational modeling and laboratory experimentation is an indication of the extent to which well-defined genetic parts must be utilized, as success in such models hinges greatly on the quality of the data the models are based on (Chen et al., 2012; Marchal et al., 2013). Thus, continuous investment in standardization and characterization efforts is a key path to the realization of the full potential of synthetic gene circuits, ultimately contributing to innovative solutions for today's health, agricultural, and environmental challenges.

6.3. DESIGN AND CONSTRUCTION

Design and construction of synthetic gene circuits require diligent planning and precision, a signature of the complexity inherent in biological systems. Standard methods of DNA sequence assembly typically pose challenges, notably because repetitive elements disrupt the efficiency of homologous recombination. As seen, conventional synthetic circuits typically consist of standard promoters and terminators that can undermine the process, necessitating new solutions to achieve maximum assembly efficiency. This issue motivated the creation of unique nucleotide sequences (UNSes) for efficient ordered assembly. The use of these computer-synthesized UNSes gives the option to mix repeating units of promoter-gene-terminator without affecting circuit function. Not only is this more efficient in the construction process, but it also promises modular construction for synthetic biology, with the potential for researchers to manipulate genetic configurations easily and precisely (Boehm et al., 2014). As synthetic biology gains momentum and artificial cells gain prominence, it is crucial to understand gene circuit design and assembly. Artificial cells, with their minimalist architectures, may provide insight into designing adaptive gene circuits with predictable behavior. These minimalist systems comprise biological membranes, translation and transcription machinery, and genetic modules governing cell dynamics. By designing gene circuits directly for use in artificial cells, researchers can explore fundamental questions about cellular function with far less complexity. Able to engineer stable and manipulable artificial cells, synthetic biology will not only be advanced but also bridge the gap between artificial constructions and natural cellular processes. This project raises essential questions about information encoding, needed elements for robust function, and communication efficacy in engineered systems (Ding et al., 2014). With advancements in the field, applications of these synthetic gene circuits extend beyond mere design and assembly; they hold the promise to revolutionize biotechnology and medicine. For instance, optimized circuits can be utilized in metabolic engineering to enhance the production of valuable compounds, e.g., biofuels or pharmaceuticals, in microbial hosts. Again, the ability to include more complex logic functions in gene circuits allows room for the creation of cells that can sense and respond to environmental signals, again replicating natural systems. The methods used to construct such circuits, for example, UNS-guided assembly, have opened new research frontiers. This continued research not only improves our understanding of genetic architecture but also offers new directions to the design of autonomous, self-tuning artificial systems, suggesting a future where biological engineering is being applied to transform various domains (Boehm et al., 2014).

6.3.1. Methods for Designing Synthetic Circuits

Synthetic circuit design within synthetic biology consists of an array of advanced methodologies that take advantage of computational strategies and experimental systems. One of the central methods is using computational modeling to make predictions on circuit functionality before constructing it physically. This allows researchers to make forward-looking predictions on interactions between genetic elements within a cellular environment and optimize their designs in a cycle-by-cycle manner. With the use of software platforms, biosystems are able to be simulated and, accordingly, optimal designs for output functions can be determined. Modeling is a central component of rational synthetic circuit design, as it dictates the process of component and arrangement selection, ultimately leading to engineered systems being more predictable and efficient. Standardization of biological parts is central to moving synthetic circuits forward, enabling them to be incorporated into complex designs. By ensuring that the building blocks of genetic circuits are well characterized, the robustness and utility of the circuits can be improved. Efforts to enumerate and define stock biological parts have been supported by many initiatives, which entail developing libraries of reusable genetic components between multiple projects. This not only reduces the resources and time needed for design but also ensures consistency in parts, encouraging modularity and scalability of circuit design. Such approaches have been emphasized as requirements for effective application in basic research and technological innovation in synthetic biology (Marchal et al., 2013). Furthermore, integrating synthetic biological circuits with microfabricated devices presents new opportunities for improving the function of these systems. The convergence of synthetic biology and microfluidics enables precise control of the microenvironments and manipulation of biological constituents on a microscale. The convergence leads to sophisticated platforms where cellular behaviors can be monitored and quantified in real time. The employment of microengineered devices offers scientists another analytical avenue for studying the dynamics of synthetic circuits in living systems, finally leading to the development of even more advanced biotechnological functions. As the research continues to explore the synergies between synthetic circuits and microelectronic technologies, the potential for innovations in biomanufacturing and disease treatment becomes increasingly possible (Chen et al., 2012).

6.3.2. Tools and Software Used in Circuit Construction

Synthesis of synthetic gene circuits involves the use of various advanced tools and software that enable designing, modeling, and analysis of genetic parts. A crucial component of the process is the use of computational design strategies that enable scientists to predict how the genetic circuits will behave before they are actually built. This predictive capacity is essential because it allows researchers to iteratively refine their designs, minimizing the amount of time and resources spent in trial experiments. In the last few years, the incorporation of software packages that simulate genetic interactions has been decisive, offering platforms for in silico visualization of circuit behavior. This shift toward computational modeling enhances not only the efficiency of circuit design but also develops further understanding of the intricate dynamics that govern biological systems, as would be expected for the general goals of synthetic biology defined in the literature (Marchal et al., 2013). An additional key aspect of modern circuit assembly is the integration of standardized biological parts or BioBricks, which facilitate modular construction in synthetic gene circuits. These pre-characterized components streamline engineering through the application of reliable building blocks that have been thoroughly tested for discrete functions. Computer platforms that contain large libraries of such components enable researchers to select the correct components based on desired circuit specifications. In addition, the incorporation of design platforms such as CAD for biological systems allows intuitive graphical modeling of genetic circuits and makes it easier to design complex assemblies for engineers. The harmonious relationship of standardization and advanced software not only makes it possible to rationally design biological systems but also produces enhanced robustness and functionality for synthetic circuits, which transcends the inherent aspect of complexities in biological engineering (Freemont et al., 2014). Furthermore, analysis and synthesis of synthetic circuits are greatly supported by extremely sophisticated microfluidics and high-throughput automation technologies. These offer precise control over experimental conditions such that numerous variants of circuits can be rapidly tested in parallel. High-throughput sequencing and analysis software play an important role in determining the functionality and performance of constructed circuits at the genomic level. The union of microengineering and synthetic biology has allowed the development of minuscule devices that integrate electronic components with biological processes, leading to novel opportunities for biosensing and diagnostics applications. As a result, such technologies not only enhance the capability of circuit building but also make ways to novel approaches in biotechnology, demonstrating tremendous synergy of technological advancement and biological engineering as purported in research studies currently (Marchal et al., 2013).

6.3.3. Challenges in Circuit Assembly and Optimization

The complexity of processes involved in synthetic gene circuits provides significant challenges in the assembly and optimization phases of design. The biological components in these man-made systems usually remain uncharacterized, resulting in inconsistency in their response when trying to generate target functions. This unpredictability is due to intricate interactions between biological pathways, where they may be modulated by environmental and genetic backgrounds. The assembly process also has possible sources of error, including the incorrect integration of parts or the formation of undesired interactions among parts. Overcoming these issues entails the wholehearted application of advancements in both synthetic biology and associated engineering principles. An important aspect in overcoming such issues is the calibration of assembly methods with precision and high-reliability error-checking mechanisms to ensure functional integrity in the product circuit, as mentioned in Freemont et al. (2014). The scalability challenge is another key hindrance in the optimization of synthetic gene circuits. When circuits are being developed to be used in real-world applications, say, in biomanufacturing or treatments, it is crucial that they work as intended at a greater scale. Nevertheless, numerous techniques developed are tried at a small scale, which may not be a complete picture of how they will work in more complicated scenarios. When circuits are constructed and put into operation, their performance may be different from what is expected due to emergent properties that occur once they are scaled up. It is therefore important to construct models that can accurately predict performance at various scales. This requires strong empirical testing and iterative model adjustment with empirical data, emphasizing the need for a constant design cycle enabling feedback in each stage, as emphasized by Chen et al. (2012). Lastly, bringing disparate biological elements into integrated gene circuits is an important challenge in synthetic biology that influences assembly as well as optimization procedures. The variation in signaling pathways and levels of genetic expression across organisms complicates the standardization of components. Consequently, synthetic biologists must ensure compatibility among components from diverse sources, which in most cases requires cumbersome characterization and validation. Techniques like synthetic parts libraries and modular design principles are being developed to facilitate this integration, necessitating a more formalized approach towards circuit design. With the emergence of novel tools and methods, it is now more necessary to assess their effectiveness in overcoming the assembly and optimization problems, ultimately resulting in more reproducible and efficient synthetic gene circuits, as emphasized by recent studies Freemont et al. (2014).

6.4. APPLICATIONS IN MEDICINE

Synthetic gene circuit applications in medicine represent a revolutionary front that integrates engineering sciences with biological networks to address advanced medical challenges. By genetic engineering, synthetic biology enables the development of novel therapeutic goods, diagnostic equipment, and therapeutic approaches that could not be previously achieved. With controlled patterns of gene expression, researchers can engineer cells to react dynamically to stimuli in their environment, making them even more desirable for medical therapies. With this designer response, the outcomes of treatments and side effects are minimized through targeted drug delivery systems reacting specifically to exact disease conditions. The ability to recapitulate gene expression systems *in vitro* through the application of cell-free synthetic biology platforms also displays such versatility by enabling scientists to study genetic circuits without the limitations of live cellular systems (Agarwal et al., 2019). Bacterial therapy and immunotherapy are two prime examples of how synthetic gene circuits can be used effectively to treat disease. Engineers have created bacteria that are capable of detecting tumor markers and delivering therapeutic agents directly within the tumor microenvironment without inducing significant toxicity to the surrounding healthy tissues. This approach is extremely promising for conditions such as cancer, wherein traditional therapies always present immense problems in terms of efficacy and side effects. Moreover, the evolution of synthetic gene circuits enables the design of regenerative medicine methods that facilitate the engineered repair or replacement of impaired tissues and organs. These novel application milestones represent a paradigm shift in treatment, positioning synthetic biology as a vital tool for the treatment of many medical conditions (Gilad et al., 2017). The integration of synthetic gene circuits with high-performance imaging further boosted their potential in the clinical setting. Molecular imaging methods are essential in tracking the function and location of engineered cells within the body. In addition to using reporter genes in synthetic circuits, researchers can employ noninvasive imaging technologies, such as bioluminescence and magnetic resonance imaging, to monitor therapy in real time. This merging of synthetic biology and imaging enhances the precision of treatments, delivering optimized patient care through real-time monitoring. With synthetic gene circuits being continually developed and evaluated in clinical trials, their capacity to solve complex biomedical problems will rise, revolutionizing the future of medicine through personalized and adaptive therapeutic interventions (Gilad et al., 2017) (Figure 6.5).

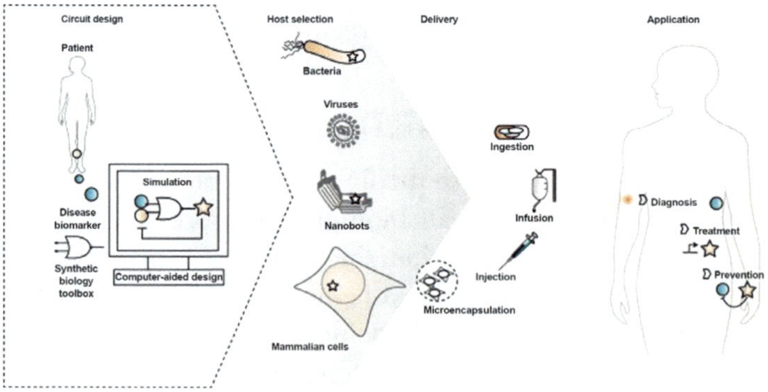

Figure 6.5. Development process for gene circuits addressing medical conditions.

Source: https://media.springernature.com/m685/springer-static/image/art%3A10.1038%2Fs41551-018-0215-0/MediaObjects/41551_2018_215_Fig1_HTML.jpg.

6.4.1. Gene Therapy and Disease Treatment

The cutting-edge field of gene therapy is on the forefront of new disease treatment, particularly in the case of genetic disease and cancer. With synthetic gene circuits, researchers can engineer responses that heal specific pathophysiological states at the molecular level. This groundbreaking approach has the potential to offer the selective modulation of gene expression in such a way as to enable the development of specific therapies that may potentially correct genetic mistakes or boost immune assault against cancer. Given the widespread influence of cancers such as this, which is one of the leading causes of mortality globally, the necessity to develop precise therapeutics cannot be overemphasized. Advances in cancer pathway insights have optimized patient outcomes with tailored approaches. However, there are drawbacks since resistance to drugs usually overshadows the efficacy of therapy, so it is essential that techniques not only effectively transfer therapeutic genes but also overcome resistant pathways (Bivona et al., 2019).

Synthetic biology deployment also illustrates the promise of bacterially delivered therapies as a new paradigm for cancer treatment, which can be effectively integrated into gene therapies. Synthetic gene networks can be employed to engineer bacteria selectively to target cancer cells to facilitate localized therapy with decreased systemic toxicity. Bacteria can be designed to deliver therapeutics locally within the tumor microenvironment, circumventing constraints from traditional drug delivery systems. Synthetic gene circuits' real-time modulation capacity supports modulating therapeutic outputs according to the tumor's dynamic landscape. Phasing over pharmacokinetic hurdles and the complexity of the tumor microenvironment, new strategies offer possibilities for better treatment outcomes. Such system integration is a possible transition from the conventional therapies to more effective, personalized approaches to cancer treatment (Bivona et al., 2019). However, efficient application of gene therapy depends on the efficacy of delivery systems, which must possess precise targeting and regulation of gene expression. Synthetic molecular circuits developed in recent times have presented hope in creating intelligent viral delivery systems with conditional targeting of target cell types according to surface antigens. Such systems implement sophisticated technologies, such as pseudotyping and designed proteins, to enhance specificity and safety.

By selectively expressing and targeting the release of therapeutic factors, such breakthroughs in synthetic biology could be a game-changer in the science of gene therapy. Having control over these delivery systems is the ongoing challenge that must be overcome, as maximum therapeutic specificity will be critical in reducing off-targeting effects and increasing therapeutic success. The capacity to synthesize different engineering strategies in synthetic gene circuits is the hallmark of the potential to revolutionize gene therapy and disease treatment (Figure 6.6).

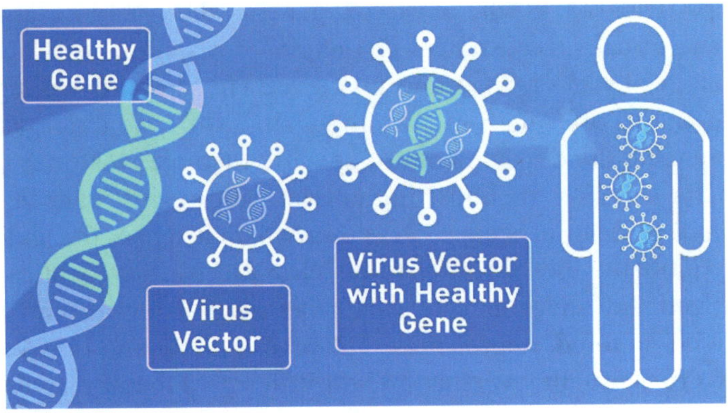

Figure 6.6. How gene therapy can cure or treat diseases.

Source: https://www.fda.gov/files/How%20Gene%20Therapy%20Can%20Cure%20or%20Treat%20Diseases_1600x900.png.

6.4.2. Synthetic Circuits in Vaccine Development

Synthetic gene circuits have revolutionized the process of vaccine development, offering new ways to enhance immune responses. Such engineered circuits can be programmed to control antigen or immunostimulatory molecule expression, regulating the immune response against a given pathogen. Synthetic circuits controlling the timing and quantity of antigen presentation are capable of regulating the duration and magnitude of the immune response, thereby improving vaccine effectiveness. This extreme accuracy engineering allows for the construction of vaccines that have the capacity to design their adaptability to varying host conditions and disease states, thereby they are more efficient and versatile in application against infectious diseases. Additionally, the versatility encapsulated in synthetic biology platforms provides researchers with the ability to respond rapidly to emerging pathogens, an imperative need in the current global health emergency. In addition to optimizing immune responses, synthetic circuits enable the design of vaccines that can regulate their release and delivery autonomously. For instance, such circuits can be incorporated into microbial systems, which can be engineered to release the vaccine to infected tissues selectively and circumvent off-target effects. This novel use draws upon pioneering work in synthetic biology, where bacteria have been engineered to target therapeutic molecules into tumor microenvironments and their prospects for fighting complex health disorders. Applying similar methodologies to vaccines, researchers can take advantage of the engineered bacterium's ability to target tumors to site vaccines specifically within the body. With the ongoing development in the field, this intersection of synthetic biology and immunology promises a new dawn in the development of highly potent vaccines that take advantage of cutting-edge technology platforms. In addition, the incorporation of synthetic circuits into vaccine designs has the potential to greatly minimize the occurrence of drug resistance in pathogenic microorganisms, especially for the case of viral and bacterial infections. With circuit-based approaches, it is now feasible to optimize the interactions of vaccines with the immune system to further facilitate the body's ability to recognize and defend against pathogens. Such an approach is particularly pertinent in the fight against high-mutation rate diseases such as influenza and HIV. Advances in synthetic circuits can be guided by prior work focused on their application in therapeutic interventions, such as bacterially mediated cancer therapy, which underscores their potential for precise intervention in complex biological systems. As researchers explore further into such frontiers, the application of synthetic circuits to the design of vaccines promises not only more potent immune responses but healthier, more resilient health interventions.

6.4.3. Personalized Medicine and Tailored Therapies

The incorporation of personalized medicine into current treatment paradigms is an important milestone in medical treatment. The method centers on patient individualized profiles, integrating genetic, environmental, and lifestyle factors to create a tailor-made intervention. Leveraging advances in synthetic gene circuits, scientists can now design therapies that evolve based on the unique pathophysiological characteristics of a patient's disease. A good example is the treatment of glioblastoma, where heterogeneity of cancer and its complexity demand personalized strategies. As noted, systems medicine provides a comprehensive view by employing a number of biological and computational techniques to design customized treatment strategies considering both the specific characteristics of the tumor and the overall health status of the patients. This customized process is intended to enhance treatment efficacy as it considers the various reactions that have been noted with patients on existing therapies, thus achieving better clinical outcomes and reducing side effects (Zeng et al., 2023). Outside solid tumor cancers, personalized therapy is also used in the management of autoimmune diseases, where traditional treatment means usually have unreliable outcomes across patient populations. New advances in synthetic biology have produced new opportunities for the production of controlled therapeutic systems. For example, genetically modified cells that react autonomically to disease biomarkers and adjust the delivery of cytokineblocking proteins are a very sophisticated, tailored treatment mechanism. These advances allow real-time and dose-dependent therapeutic response that is particularly significant for the treatment of rheumatoid arthritis. This specificity not only minimizes the potential for side effects but also maximizes treatment effectiveness, offering an exciting frontier for personalized medicine. With the application of synthetic circuits and immune modulation, researchers are able to create interventions that evolve with the patient's illness and environmental variables, thereby optimizing therapeutic adherence and patient quality of life (Hell, 2018). As the field of personalized medicine progresses, the need to integrate robust synthetic gene circuits becomes ever more important. These circuits would enable strict control over gene expression in response to a stimulus and facilitate the development of responsive therapies tailored to the needs of individual patients. With the heterogeneity encountered in genetic expression and treatment responses, the application of synthetic biology in the design of adaptive gene circuits optimizes the possibility for individualized results. For instance, the incorporation of synthetic circuits that sense and react to inflammatory cytokines can redefine treatment strategies in chronic illness. The ability to engineer therapies that dynamically respond to changing biological contexts encourages a paradigm shift towards more effective management schemes and tailored solutions. The proactive method not only highlights the synergy between personalized medicine and synthetic biology but also opens up prospects for novel breakthroughs in customized therapeutic formulations (Hell, 2018; Zeng et al., 2023) (Figure 6.7).

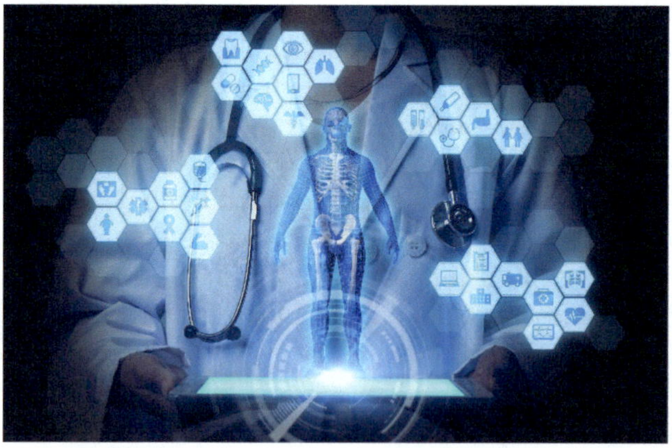

Figure 6.7. Precision Medicine

Source: https://events.marketsandmarkets.com/blog/wp-content/uploads/2022/06/precision_medicine_body.jpg

6.5. ENVIRONMENTAL APPLICATIONS

The environmental application of synthetic gene circuits has excellent potential for mitigating contemporary ecological challenges. By designing microbial populations that express sophisticated communication and division of labor, researchers have managed to reduce personal metabolic burdens while enhancing overall adaptability to various environmental stresses. This new approach leverages the principles of synthetic biology to construct well-defined microbial communities, which are more effective in responding to stressors, such as toxins or climate change. When the designed systems become mainstream, they advance the boundaries of conventional microbial applications, and novel solutions in waste treatment, bioremediation, and other environmental management challenges are accessible. Such artificial systems' development thus constitutes an innovative avenue to developing high-performance biological organisms that can be harnessed for specific ecological tasks and challenges and can ultimately give rise to sustainability and ecological health (Zhang et al., 2023). In addition, employing synthetic gene circuits significantly enhances the effectiveness of bioremediation. Recent advances involving CRISPR-Cas systems have demonstrated their use in constructing digital circuits to detect and respond to environmental pollutants.

By creating programmable gene circuits in microbes, researchers can create bacteria that not only detect pollutants but also respond with targeted remediation approaches, such as degrading harmful chemicals or immobilizing heavy metals. For instance, the implementation of Boolean logic gates in microorganisms allows for sophisticated decision-making in response to environmental signals. This innovation accentuates the importance of translating synthetic biology design ideas into productive applications that directly benefit environmental monitoring and remediation (Yu et al., 2022). The continued evolution of such systems can improve the capability of regulating and minimizing human activity on the environment. Finally, the programmability of synthetic gene circuits welcomes their application in the production of bioindicators for environmental monitoring. By designing microbes using genetic circuits that respond to specific environmental conditions or pollutants, researchers can produce living sensors providing real-time data on the health of an ecosystem. This application of synthetic biology not only improves our capacity for surveillance but may also facilitate the early identification of environmental stressors, such that timely interventions can be made. The capability to construct such bioindicators in modular elements can have the effect of tailor-made solutions geared towards environments that are aquatic, terrestrial, or urban. With the increasing synthesis of synthetic gene circuits, their application in environmental frameworks illustrates an innovative juncture of technology and ecology that formulates stronger ecological systems amidst human-induced change and stress (Zhang et al., 2023).

6.5.1. Bioremediation Using Synthetic Circuits

Application of synthetic circuits in bioremediation presents a promising pathway for remediating environmental contamination with new microbial engineering. By creating and deploying artificial gene circuits, scientists can design microorganisms that can degrade toxic substances in contaminated environments. Genetically engineered microbes can be designed to react adaptively to contaminants, which will make them more efficient at detoxifying soil, water, and industrial waste. Microbial actions being controlled by synthetic circuits allow for the design of sophisticated responses to varying levels of contamination. This adaptability not only makes the bioremediation processes more efficient but also minimizes the environmental footprint of the interventions, in keeping with sustainable environmental strategies. This coupling of synthetic biology and environmental remediation, therefore, unlocks a new paradigm for managing pollution and rebuilding ecosystems. Recent progress in synthetic biology has enabled the engineering of microbial communities that can perform sophisticated functions, significantly enhancing bioremediation strategies. Bioremediation has traditionally used indigenous microorganisms, but with synthetic gene circuits, it is possible to design clonal populations with specific features like differentiation and communication. Through the facilitation of interactions between the engineered strains, scientists can mimic cooperative action that naturally occurs with microbial consortia, thus having faster degradation rates of contaminants. For instance, such populations can possess common metabolic pathways and coordinated reactions to environmental cues and therefore improve remediation processes. Design potential for synthetic microbial interactions offers novel solutions to industrial processes, particularly in the remediation of complex contaminants like heavy metals and organic solvents. This marks a significant shift toward utilizing the complexity of microbial ecology using the synthetic approach, highlighting the future of environmental biotechnology. The uses of bioremediation based on synthetic circuits extend beyond straightforward pollutant degradation; they represent a more general comprehension of microbial interactions and ecosystem function. As the literature highlights, synthetic biologists are now targeting the simulation of sophisticated behaviors observed in natural microbial consortia, which are typically more multifunctional and robust than clonal populations. By integrating cell signal transduction mechanisms into built organisms, researchers are able to create systems that react with memory and adaptive behavior, leading to more effective bioremediation. Research into such built systems not only enlightens microbial ecology but also brings knowledge regarding innovative environmental remediation strategies. As the field continues to advance, the integration of synthetic biology with ecological rules holds the promise of creating practical, sustainable solutions for solving environmental issues, thus ushering a new era in the war against pollution and the enhancement of ecosystem resilience (Brune et al., 2012; Chen et al., 2012) (Figure 6.8).

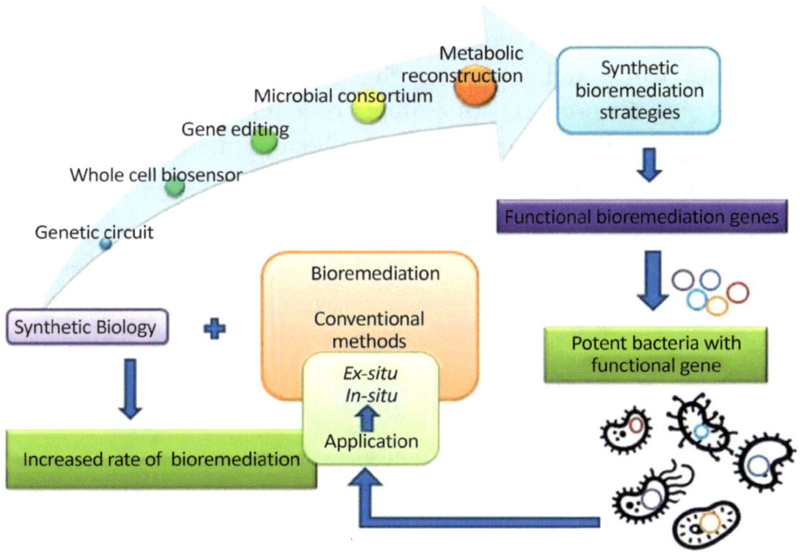

Figure 6.8. The strategies of synthetic biology applicable for bioremediation.

Source: https://www.researchgate.net/publication/341496419/figure/fig1/AS:11431281258716512@1720134684214/The-strategies-of-synthetic-biology-applicable-for-bioremediation.tif.

6.5.2. Engineering Microbes for Biofuel Production

Microbial engineering for the production of biofuels is a significant move towards the quest for sustainable energy. Synthetic gene circuits can be employed to design microbial metabolism in such a manner that the biomass is converted into biofuels efficiently. This not only enhances the yield but also allows for the utilization of a wide range of feedstocks, including agricultural waste and non-food biomass. The application of synthetic biology as a design approach enables one to explore various metabolic pathways, optimizing the process whereby microbes are utilized in transforming these substrates into valuable products of energy. Notably, combining Bayesian model selection with this design process is able to augment the models' ability to predict and explain, and attain a worthy balance between function and simplicity in microbial engineering applications, as the experimental efforts that leverage the modelling potential given by Barnes et al. (2011) stipulate. One of the greatest arguments in favor of engineering microbes for biofuel production is related to microbial communities. Rather than employing clonal populations, synthetic biology allows the design of microbial consortia capable of performing complex functions in concert. Such communities, designed, can enhance biomass conversion efficiency and adjust to changes in environmental conditions, thereby enhancing the prospects of biofuel production systems. The interaction between microbial cell communication and signaling networks among various microbial populations can be engineered to develop novel functions important for industrial applications, such as the synthesis of biofuels.

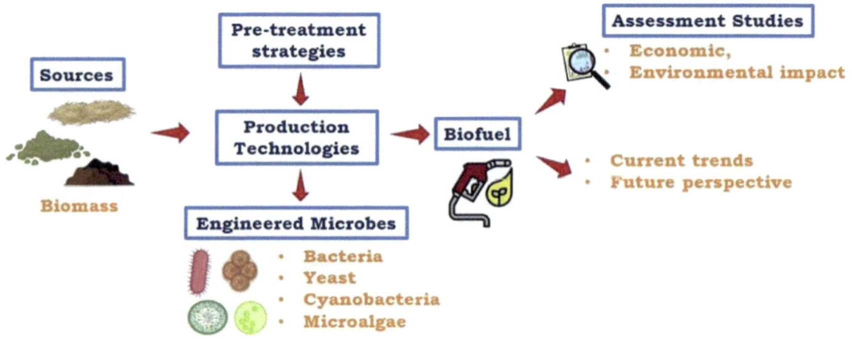

Figure 6.9. Advances in engineered microbes for sustainable biofuel production.

Source: https://ars.els-cdn.com/content/image/1-s2.0-S0926669024019654-ga1_lrg.jpg.

Current advancement in the same line has been promising towards increased efficiency and durability in bioenergy production, basing its success on lessons acquired via natural microbial consortia that are functionally effective under various conditions, as discussed when considering microbial interaction and its potential application in biofuel consumption according to Brune et al. (2012). Furthermore, synthetic gene circuit regulation of gene expression allows for the fine-tuning of microbial metabolic pathways. Such a level of control is essential for the development of efficient and regulatable systems for biofuel production. Through the adaptation and switch-like characteristics in genetically engineered microbes, the performance of biofuel-producing strains can be fine-tuned under varying operation conditions. The inclusion of such artificial circuits enables a platform for calibration of cellular responses, optimizing the microbes to efficiently transform substrates into target products. This new approach enables in-silico prototyping, in which experimental design is used to decipher the interactions among different genetic elements and their overall effects on metabolic output. Ultimately, these innovations mark a crucial step in the manufacture of biofuels, answering to the universal call for sustainable and renewable energy systems, as envisaged through the bioengineering paradigm of Barnes et al. (2011) (Figure 6.9).

6.5.3. Impact on Sustainable Agriculture

The use of synthetic gene circuits within agriculture holds a revolutionary promise towards achieving sustainable agriculture. The circuits can be designed to render the crops more resilient to environmental stresses such as drought or salt, making it increasingly relevant with the onset of climate change. By enabling plants to operate at higher efficiency in poor environmental conditions, synthetic gene circuits can enhance food security without compromising the application of chemical inputs such as fertilizers and pesticides. This accords with the notion that sustainable agriculture must prioritize ecological integrity and resource-use efficiency. As commodity crops evolve, their evolution induced by biotechnological innovations will impinge heavily on the lives of millions globally, hence good guidance of such developments is crucial and imperative (Habib et al., 2007). The journey toward the implementation of synthetic gene circuits in agriculture demands an analytical understanding of the underlying biophysical interactions across various scales. Semiconductor-based material knowledge can be used to inform the construction of effective biointerfaces, enhancing the scope for monitoring and controlling biological response in crops (Anikeeva et al., 2018). This might lead to the development of novel signaling pathways to enhance plant productivity and resistance, thereby promoting sustainable agricultural systems. By combining both synthetic biology and materials science information, scientists and farmers can create pathways for increased efficiency and sustainability in commodity crop farming. This combined approach is key to making certain that agriculture not only feeds the world today, but also does so in a way that preserves the environment for future generations. In addition, the financial impacts of gene circuit construction in agriculture cannot be overstated because they hold the power to reengineer supply chains and market dynamics. Because they can reap more yields using fewer ecological footprints, farmers are capable of practicing more sustainable practices, which translates into greater profitability and reduced environmental degradation. But this transformation should also be executed via a collective effort of agricultural industry stakeholders, including policymakers, researchers, and farmers, to successfully handle disruptions and innovations. The use of foresight exercises, as suggested by current research, is essential in predicting emerging trends, enabling stakeholders to make informed decisions that will drive sustainable agriculture into the future (Habib et al., 2007).

6.6. ETHICAL CONSIDERATIONS

The development of synthetic gene circuits introduces a myriad of ethical issues that must be scrutinized due to the potential consequences of gene manipulation. Behind this discussion is the concern of biosafety, more precisely, the unintended effects engineered organisms will inevitably impose on the environment. With the evolution of synthetic biology, organism design with novel characteristics has raised questions regarding their potential environmental impact and the stability of established ecosystems. The evaluation process and regulating bodies need to consider long-term ecological stability in light of the unpredictable nature of such changes. Failure to properly address these biosafety concerns could have irreversible consequences on biodiversity, leading to more general consideration of the moral responsibility of scientists and engineers who engage in synthetic gene circuit design and synthesis, particularly focusing on the necessity of an ethically conscious approach to research and development. Aside from environmental safety, ethical concerns regarding synthetic gene circuits also center on their effects on society, specifically access and equity. With advances in genetic engineering becoming cheaper and more sophisticated, an imbalance in access could occur that benefits specific populations or nations compared to others. This then raises questions about who should be able to access such technologies and for what purposes. As synthetic biology grows, it is important that adequate governance systems are established to offer equally equitable access to genetic innovation but consider the heterogeneity of global society. Without doing so, the risk of exacerbating existing disparities is an imminent concern that threatens the ethical bases on which scientific advancement needs to be made. Thus, an inclusive discussion involving stakeholders is required to solve these ethical concerns seamlessly. Finally, ethical concerns in synthetic gene circuits also cover the philosophical controversy regarding playing God when it comes to genetic engineering. This idea raises underlying questions of the amount of intervention in natural processes that would be advisable for human beings to undertake and the ethical aspect of creating life within a laboratory setting. The potential for synthetic biology to alter the underlying processes of life raises questions of the motives and implications behind doing so. Arguments on the basis of whether certain genetic alterations are of benefit to humanity's wider ethical agendas may result in mutually opposing opinions among different religious and cultural groups, thus necessitating the need to discuss the implications of such technologies openly. Exploration of such intricate ethicological realms provokes a careful approach to synthetic gene circuits as it encourages society to harmonize innovation with ethical limitation, particularly as Freemont et al. (2014) and Hall and Howe (2012) suggest that society requires an engineering design cycle responding to such complexities.

6.6.1. Potential Risks and Safety Concerns

As synthetic gene circuits continue to pick up in some of the most diverse fields possible, from medicine to environmental science, the potential risks involved in their application and construction have been all the more acute. Among the most compelling of these concerns is that of dual use in synthetic biology, where the same technologies that are supposed to benefit human well-being can also be used in twisted ways for ill, including bioterrorism. The decreasing price of synthesizing DNA and access to genomic sequences enable anyone with ill intentions to utilize such capabilities free from checks that otherwise prevail in the professional setup. The emergence of DIY communities for biology is a demonstration of this phenomenon, triggering apprehension about the level of competency and ethical framework that controls such experiments. As is highlighted, focus becomes placed on access rather than the multifaceted human practices that are rooted in institutional knowledge, thereby complicating the story of biosecurity in relation to synthetic gene circuits (Jefferson et al., 2014). Besides security, safety concerns related to synthetic gene circuits relate to unintended consequences resulting from their use in living organisms. Engineering organisms to generate favorable characteristics can have unforeseen ecological implications, such as the potential to disturb existing ecosystems or generate novel pathogens. Release of GMOs into the environment, with the possible result of synthetic biology use, raises the concern of upsetting the ecological balance. Regulatory frameworks are typically lagging behind technology, and there are additional risks if these are not accompanied by proper evaluation and monitoring mechanisms. Therefore, biosafety concerns, particularly long-term environmental effects, require stringent evaluation frameworks to forecast and mitigate risks as synthetic gene circuits are increasingly adopted for applications (Chaturvedi et al., 2015). Secondly, ethical views matter in dealing with the possible risks and safety concerns of synthetic gene circuits. As researchers push the boundaries of genetic modification, discussions about the ethics of tampering with living organisms have gained more traction. The possibility to engineer pathogens or designer organisms that are capable of harmful activities terrifies not just scientists but also the general population. With the global demand for science and technology coupled with responsive governance, it is necessary to imbue ethical aspects in policy-making structures governing synthetic biology. The integration of ethics into research and development entails an understanding of varying perceptions of risks and benefits across various cultural and geographical spaces. Through the promotion of public engagement and the application of multifaceted ethical viewpoints, stakeholders can work together toward an improved system of governance that addresses the polyfaceted risks of synthetic gene circuits (Chaturvedi et al., 2015).

6.6.2. Ethical Implications of Genetic Modifications

Synthetic gene circuits placed genetic modification at the center of debates about the ethical implications of changing living organisms. With recent advances in molecular genetics, researchers can now design sophisticated networks of genes that significantly enhance the biological functions of microorganisms and higher organisms. Such a capability holds fantastic potential, particularly in medical biotechnology, where GMOs would be used to produce medicines or vaccines. But ethical consequences cannot be overlooked. Misgivings are regarding the possibility of unforeseen consequences, including ecological perturbation or the creation of harmful pathogens, and forcing the scientific community to be cautious. Ethical standards must be formulated to guide research and ensure that innovations never outstrip ethics. Recent literature suggests that wrongful application of genetic technologies poses serious threats, emphasizing the importance of cautious regulation in synthetic biology studies (Franco et al., 2015). A further ethical concern for genetic modifications is possible impacts on social justice and equality. With gene editing and synthetic biology technologies spreading, disparities in access to those technologies will arise. More proportionate benefits to wealthier countries and individuals from advancements in genetic modifications may come, with marginalized groups having little or no access. This creates an immediate moral imperative of ensuring these technologies are allocated equitably, especially since they are being developed to address severe health issues. Consequences go beyond the individual's health and may have effects on populations and communities at large. Moral discussion on this issue has to revolve around fostering equity and preventing further social stratification through biotechnological means. By placing emphasis on equal access to genetic technology, society is able to work towards the elimination of inequalities that undermine the basic values of justice and equality, thereby making it essential to address such ethical concerns in the framework of synthetic gene circuits (Marder et al., 2008). Further, the potential of GMOs creates challenges to the value of nature per se and the limits of human intervention in terms of ethics. This moral question questions the philosophical debate regarding the rights of GMOs and scientists' responsibilities as custodians of nature. As much as the payoffs of synthetic biology can be colossal, they raise issues concerning the natural order and how much mankind ought to dominate it. They must take into account not only the scientific significance of gene changes but also their effect on the environment and their philosophical concerns about altering life itself. These attempts must meet the moral responsibility to safeguard biodiversity and maintain ecological integrity, knowing that artificial gene circuits are part of a complex ecosystem. Therefore, the ethical debate on gene modification entails an interdisciplinary approach grounded in scientific progress and moral and ecological imperatives (Franco et al., 2015).

6.6.3. Regulatory Frameworks and Public Perception

For synthetic gene networks, regulation frameworks and public perception become increasingly interrelated. Regulation frameworks inform the boundaries of operation within which synthetic biology is facilitated to evolve, impacting not only the science but also public acceptability. These frameworks must prioritize public interest, with a focus on values of ethics and risk management in developing new technologies. Specifically, as argued by Chaturvedi et al. (2015), the institutionalization of ethical discourses in regions like Europe, India, and China demonstrates how diverse cultural contexts can determine different approaches to governance. By incorporating ethics into policy-making, regulatory bodies can design more trustworthy and transparent policies and thus balance scientific innovation with public needs. Public opinion plays a significant role in the success of such regulatory policies. Since biotechnology is evolving at a rapid pace, public opinion towards synthetic biology is generally biased due to overall fears for security and ethical considerations. For instance, the growth in human genome sequencing and neurodevices has triggered public debate at the top level regarding privacy, autonomy, and vulnerability to exploitation, noted N/A (2011). These issues require regulatory bodies to engage with the public on an active level, so that the public is informed and engaged in the process of governance. This dialogue not only serves to de-mystify cutting-edge technology but also fosters a co-operative culture, wherein science and society cooperate in charting through the ethics web which comes with synthetic genetic advancement. The effectiveness of regimes in synthetic biology can be gauged by observing governance issues achieved in contemporary bioengineering. According to N/A (2011), the reverse trend of biology becoming technology and technology becoming biology calls for a changing regulatory platform that can keep pace with unexpected developments. Building law that both promotes innovation and addresses social issues is essential. This needs substantial public participation processes combining scientific education with participatory governance. By incorporating diverse perspectives into regulatory policy, stakeholders have the ability to help mitigate public concern and enhance the credibility of synthetic gene circuits, thereby paving the way for responsible scientific advancement and promoting a better-informed societal understanding of these groundbreaking technologies.

SUMMARY

In finalizing an examination of synthetic gene circuits, it is important to consider the broad implications of this rapidly evolving field of science. Synthetic biology, being a discipline that marries engineering concepts with biological information, takes advantage of intricate gene networks, enabling the design and assembly of novel circuits. The technique not only enhances the comprehension of biological functions but also facilitates the assembly of complex systems capable of executing pre-defined functions. The worth of a structured, engineering-type approach can be observed in numerous applications, including those in pharmaceutical manufacturing and metabolic engineering. The potential of synthetic biology to revolutionize these fields is underscored by its strict adherence to design principles, which are essential to advancing our understanding of biological function and design (Heinemann et al., 2006). Moreover, the dynamical function of artificial circuits, in the form of hybrid metabolic-genetic circuits known as metabolators, reflects the complexity that lies with synthetic gene circuit engineering. The research indicates that the accurate modeling of such circuits is fundamental to the understanding of their kinetic behavior, the direct influence of which on functional outcome is significant. The application of generalized modeling (GM) allows scientists to identify important kinetic parameters that govern circuit behavior, including retrieving primary bifurcation phenomena in circuit response. These results not only enhance our theoretical models but also allow for optimization of circuit performance in practical use. Thus, comprehension of these complex dynamics is essential in guiding the iterative design process at the core of effective synthetic circuitry (Alon et al., 2013). In the end, research in synthetic gene circuits is a frontier of scientific investigation of limitless promise. As researchers progress and probe further into the interactions and dependencies that characterize synthetic systems, the dividends extend far past the laboratory. These innovations have the capability to address major global challenges, from sustainable production and bio-manufacturing to new therapeutic strategies. The engineering-inspired scaffolds created can have an unprecedented impact on health, energy, and environmental conservation and illustrate the capability for change of synthetic biology. With the creation of a better understanding of synthetic gene circuits, we lay the groundwork for additional research and practical applications that will be turned into widespread practices someday, thus justifying the necessity for interdisciplinary collaboration in this new technology (Heinemann et al., 2006).

Synthetic gene circuit construction is a significant departure from conventional genetic engineering, particularly in plant biotechnology. Using programmable logic gates, these circuits allow the targeted modulation of gene expression, a far cry from traditional always-on transgenes. The ability to input multiple signals allows complex spatiotemporal modulation of target genes with minimal disruption of host cellular processes. Novel advances offer a variety of platforms for constructing such circuits, ranging from bacterial transcription factors to site-specific recombinases to CRISPR/Cas components. Thus, these technologies enable researchers to control plants with added characteristics, opening the door to farming technology that is both sustainable and beneficial to food security (Lloyd et al., 2025). Despite the potential of synthetic gene circuits, there are challenges associated with their application in plant systems. The slow generation times inherent in plant agriculture slow down the rapid design-build-test cycles critical to successful testing and iteration of circuits. Furthermore, the limited availability of characterized genetic elements in a range of species presents another significant barrier. All these challenges necessitate new approaches to circuit construction and reliable methods for stable plant transformation in plant genomes. With continued research, addressing these issues will be crucial in optimizing the performance of synthetic gene circuits and expanding their use in agricultural and ecological contexts (Lloyd et al., 2025). Synthetic gene circuits are also making breakthroughs in cancer treatment, particularly in targeting mutated RAS genes, which are prevalent in most human cancers. It has been demonstrated in the recent past that such circuits are efficient in sensing and integrating cancer-specific biomolecular signals to selectively express therapeutic proteins in cancer cells. The selectivity of such circuits, to avoid healthy cells from potential toxicity, is still a challenging issue. To overcome this, a new method was developed to integrate multiple RAS sensors into gene circuits designed for RAS-driven cancers, yielding unparalleled selectivity. This module-based design strategy allows the circuit to be tailored for certain cell lines, thereby enhancing its targeting capacity. Such advancements hold a favorable future for therapeutics in oncology, illustrating the revolutionary potential of synthetic gene circuits in medicine.

REFERENCES

1. Agarwal, S., Choi, S., Dohyun, J., Jeong, F., et al. (2019). Cell-free synthetic biology platform for engineering synthetic biological circuits and systems. https://core.ac.uk/download/210536168.pdf (accessed on 3rd October 2025).
2. Allenby, N., Hallinan, J. S., James, K., Misirli, G., et al. (2013). BacillOndex: An integrated data resource for systems and synthetic biology. https://core.ac.uk/download/287613592.pdf (accessed on 3rd October 2025).
3. Alon, U., Åström, K. J., Segrè, D., Reznik, E., Izhikevich, E. M., & Kaper, T. J. (2013). The dynamics of hybrid metabolic-genetic oscillators. http://arxiv.org/abs/1305.2954 (accessed on 3rd October 2025).
4. Anikeeva, P., Bao, Z., Benfenati, F., Bezanilla, F., et al. (2018). Roadmap on semiconductor-cell bio interfaces. https://core.ac.uk/download/287623639.pdf (accessed on 3rd October 2025).
5. Arnold, F. H., & Haseltine, E. L. (2007). Synthetic gene circuits: Design with directed evolution. https://core.ac.uk/download/4876350.pdf (accessed on 3rd October 2025).
6. Ayesa, B., Begum, M. S., Iqbal, F., Urmi, B. R., Malakar, R. K., Hossain, S., & Rashed, M. J. R. (2025). Advances in CRISPR-Cas9 gene editing: Biochemical applications and future directions. *International Journal of Science and Research Archive*. https://www.semanticscholar.org/paper/f828b1ac421b613eb2b95e57b2f19042b7feead5 (accessed on 3rd October 2025).
7. Baek, S., Kim, B., Pathiraja, D., & Choi, I. G. (2025). Development of a transposon-based genome engineering toolkit for efficient and adaptable genetic modifications in *Wolfiporia cocos*. *ACS Synthetic Biology*. https://www.semanticscholar.org/paper/8e5bf646fc20cc176a4028684cd87cd37725356e (accessed on 3rd October 2025).
8. Barnes, C., Sheng, X., Silk, D., Stumpf, M. P. H., et al. (2011). Bayesian design of synthetic biological systems. http://arxiv.org/abs/1103.1046 (accessed on 3rd October 2025).
9. Bervoets, I., Charlier, D., De Mey, M., Maertens, J., et al. (2018). A sigma factor toolbox for orthogonal gene expression in *Escherichia coli*. https://core.ac.uk/download/147045784.pdf (accessed on 3rd October 2025).
10. Bivona, T. G., Chang, S., Hernández, A. C., Keasling, J. D., et al. (2019). Overcoming the challenges of cancer drug resistance through bacterial-mediated therapy. https://core.ac.uk/download/287622369.pdf (accessed on 3rd October 2025).

11. Boehm, C. R., Chen, J. H., Lienert, F., Silver, P. A., et al. (2014). Rapid construction of insulated genetic circuits via synthetic sequence-guided isothermal assembly. https://core.ac.uk/download/28947102.pdf (accessed on 3rd October 2025).
12. Broedel, A. K., Isalan, M., & Jaramillo, A. (2016). Engineering orthogonal dual transcription factors for multi-input synthetic promoters. https://core.ac.uk/download/77020343.pdf (accessed on 3rd October 2025).
13. Brune, K. D., & Bayer, T. S. (2012). Engineering microbial consortia to enhance biomining and bioremediation. https://core.ac.uk/download/pdf/82866319.pdf (accessed on 3rd October 2025).
14. Cao, J., Gao, H., Liu, X., Wang, X., et al. (2008). On delayed genetic regulatory networks with polytopic uncertainties: Robust stability analysis. https://core.ac.uk/download/335782.pdf (accessed on 3rd October 2025).
15. Chaturvedi, S., Ladikas, M., Stemerding, D., Zhao, Y., et al. (2015). *Science and Technology Governance and Ethics: A Global Perspective from Europe, India and China*. https://core.ac.uk/download/42136809.pdf (accessed on 3rd October 2025).
16. Chen, Y. Y., Galloway, K. E., & Smolke, C. D. (2012). Synthetic biology: Advancing biological frontiers by building synthetic systems. https://core.ac.uk/download/9123958.pdf (accessed on 3rd October 2025).
17. Chong, L. S., Elowitz, M. B., Gao, X. J., Ince, T., et al. (2020). Engineering multiple levels of specificity in an RNA viral vector. https://core.ac.uk/download/323312290.pdf (accessed on 3rd October 2025).
18. Ding, Y., Tan, C., & Wu, F. (2014). Synthetic biology: A bridge between artificial and natural cells. https://core.ac.uk/download/323068426.pdf (accessed on 3rd October 2025).
19. Du, M., Gao, H., Li, Y., Pan, X., et al. (2010). Robust H∞ feedback control for uncertain stochastic delayed genetic regulatory networks with additive and multiplicative noise. https://core.ac.uk/download/337147.pdf (accessed on 3rd October 2025).
20. Elani, Y. (2016). Construction of membrane-bound artificial cells using microfluidics: A new frontier in bottom-up synthetic biology. https://core.ac.uk/download/77011265.pdf (accessed on 3rd October 2025).
21. Freemont, P., Kelwick, R., MacDonald, J. T., Webb, A., et al. (2014). Developments in the tools and methodologies of synthetic biology. https://core.ac.uk/download/76995845.pdf (accessed on 3rd October 2025).
22. Gilad, A. A., & Shapiro, M. G. (2017). Molecular imaging in synthetic biology, and synthetic biology in molecular imaging. https://core.ac.uk/download/84314121.pdf (accessed on 3rd October 2025).

23. Habib, R., Jacquemard, J. C., Nicolas, D., Omont, M., et al. (2007). *Commodities in a Global Vision for Agriculture Towards 2010.* http://agritrop.cirad.fr/546116/1/document_546116.pdf (accessed on 3rd October 2025).

24. Hall, G. M., & Howe, J. (2012). The impact of synthetic biology in chemical engineering: Educational issues. https://core.ac.uk/download/6112945.pdf (accessed on 3rd October 2025).

25. Heinemann, M., & Panke, S. (2006). Synthetic biology—Putting engineering into biology. https://core.ac.uk/download/pdf/12944386.pdf (accessed on 3rd October 2025).

26. Hell, D. (2018). Self-adjusting cytokine neutralizer cells as a closed-loop delivery system of anti-inflammatory biologicals. *ACS Synthetic Biology, 7*(11), 2518–2528. https://www.semanticscholar.org/paper/a0104c6bd7a3cd141de56a2a89046830630fb022 (accessed on 3rd October 2025).

27. Jefferson, C., Lentzos, F., & Marris, C. (2014). Synthetic biology and biosecurity: Challenging the "myths." https://core.ac.uk/download/42629805.pdf (accessed on 3rd October 2025).

28. Kallergi, A., Asin-Garcia, E., Martins dos Santos, V. A. P., & Landeweerd, L. (2020). Context matters. *EMBO Reports, 22*. https://doi.org/10.15252/embr.202051227 (accessed on 3rd October 2025).

29. Khan, A., & Lister, R. (2025). Synthetic gene circuits in plants: Recent advances and challenges. *Quantitative Plant Biology, 6*. https://www.semanticscholar.org/paper/0aaeeaf36c730ea93fb632ba7655c9a550e96e31 (accessed on 3rd October 2025).

30. Laubichler, M., Loettgers, A., & Maienschein, J. (2008). How can history of science matter to scientists? https://core.ac.uk/download/4879106.pdf (accessed on 3rd October 2025).

31. Leal-Alves, C., Deng, Z., Kermeci, N., & Shih, S. C. C. (2024). Integrating microfluidics and synthetic biology: Advancements and diverse applications across organisms. *Lab on a Chip*. https://www.semanticscholar.org/paper/7483c6d2faa3f9a09a2ec95093babbb37bbd19bd (accessed on 3rd October 2025).

32. Lloyd, J. P. B., Khan, A., & Lister, R. (2025). The switch-liker's guide to plant synthetic gene circuits. *The Plant Journal, 121*. https://www.semanticscholar.org/paper/005c3ea2d2dfe70f2883475517b7a5299e0d7c4d (accessed on 3rd October 2025).

33. Marchal, K., Vanderleyden, J., & Venken, L. (2013). Synthetic biology and microdevices: A powerful combination. https://core.ac.uk/

download/55705827.pdf (accessed on 3rd October 2025).

34. Marder, E., DeLong, M. R., Gazzaniga, M. S., & Wichmann, T. (2008). *2008 Progress Report on Brain Research*. https://core.ac.uk/download/71349141.pdf (accessed on 3rd October 2025).
35. Mukherjee, M., & Cao, B. (2020). Engineering controllable biofilms for biotechnological applications. *Microbial Biotechnology, 14*, 74–78. https://doi.org/10.1111/1751-7915.13715 (accessed on 3rd October 2025).
36. N/A. (2011). *European Governance Challenges in Bio-Engineering: Making Perfect Life: Bio-Engineering (in) the 21st Century: Final Report*. https://core.ac.uk/download/577016149.pdf (accessed on 3rd October 2025).
37. Senn, G., Nissen, L., & Benenson, Y. (2024). Synthetic gene circuits that selectively target RAS-driven cancers. https://www.semanticscholar.org/paper/51f32597c4aa4f2a9295599f20b89af450418867 (accessed on 3rd October 2025).
38. Vecchione, S. (2019). Development of novel orthogonal genetic circuits, based on extra cytoplasmic function (ECF) σ factors. https://core.ac.uk/download/294757775.pdf (accessed on 3rd October 2025).
39. Yu, L., Zhang, Y., & Marchisio, M. (2022). Gene digital circuits based on CRISPR-Cas systems and anti-CRISPR proteins. *Journal of Visualized Experiments, 188*. https://www.semanticscholar.org/paper/811bd570e9bd038550471927492f187a8b29f3cf (accessed on 3rd October 2025).
40. Zeng, J., & Zeng, X. (2023). Systems medicine for precise targeting of glioblastoma. *Molecular Biotechnology*. https://www.semanticscholar.org/paper/7676b93bf56763f477c18c5fdd8396bbb2e9c459 (accessed on 3rd October 2025).
41. Zhang, J., Chen, J., Chen, Y., Dai, L., & Dai, Z. (2023). [Engineering microbial consortia through synthetic biology approach]. *Sheng Wu Gong Cheng Xue Bao = Chinese Journal of Biotechnology, 39*(5), 2517–2545. https://www.semanticscholar.org/paper/646a895b5ee6304ff96e1a63978aaf482f726948 (accessed on 3rd October 2025).

CHAPTER 7

METABOLIC ENGINEERING

Contents

INTRODUCTION TO METABOLIC ENGINEERING

Metabolic engineering is a central field of biotechnology that entails the re-engineering of metabolic pathways of organisms. Using genetic and biochemical tools, researchers attempt to over-produce targeted products, from biofuels to medications. This cross-disciplinary field fuses genetics, molecular biology, and systems biology to create organisms with maximally optimized metabolic pathways specifically for industrial applications. The innovation of microbial fermentation processes, specifically with regard to bioethanol production, presents a first-rate example of applying metabolic engineering. Of the organisms used, yeasts have been a long-established part of biotechnological industries, successfully forming the bridge between renewable energy sources and traditional technologies through metabolic improvement and synthetic biology (Basso et al., 2019). Understanding the concept of metabolic engineering is therefore central to understanding how to confront the challenges presented by modern bioindustrial needs. With continued progress in metabolic engineering, the construction of genome-scale metabolic models has also been a major challenge in this field. In the past, their manual reconstruction used to be a lengthy and intricate process that required gigantic resources and time. Researchers have to analyze vast amounts of genomic and metabolic information, something that involves careful scrutiny of physiological information as well as established metabolic pathways.

Figure 7.1. Some of the steps in engineering a metabolic process.

Source: https://upload.wikimedia.org/wikipedia/commons/7/73/Computational_design_and_evaluation_of_DNA_circuits_to_achieve_optimal_performance.svg.

All these texts have contributed to the automation of this process of reconstruction, significantly enhancing efficiency and precision. Implementations such as the development of computational platforms like COPABI allow the development of metabolic models through probabilistic algorithms, thus changing how biologists build metabolic networks (de Córdoba et al., 2012). This shift not only reduces labor intensity in model construction but also facilitates more systematic exploration of metabolic potential across various organisms. The effects of metabolic engineering extend well beyond enhanced product titers; they also involve sustainable development and environmental stewardship. Engineered microbes, by means of metabolic pathway fine-tuning, can produce biofuels and biochemicals with a lower carbon footprint and low inputs compared to traditional methods. In addition, knowledge of metabolic pathways facilitates the reutilization of waste products as useful resources and, in the process, brings industrial procedures in line with nature. Such a shift in paradigm identifies metabolic engineering as a tool to combat global challenges like energy shortages and environmental degradation. The more scientists learn and develop in this area, the more possibilities the future of metabolic engineering has to integrate biotechnological advancement with sustainable processes for society and the world (Figure 7.1).

Definition and Scope of Metabolic Engineering

Metabolic engineering is a cross-disciplinary branch of science that focuses on the manipulation of the metabolic pathways of an organism for the optimal production of target products. The discipline employs the principles of biochemistry, molecular biology, and synthetic biology to engineer organisms for the production of desirable products such as biofuels, medicines, and biochemicals. The essence of metabolic engineering is the systematic alteration of metabolic routes in a manner that researchers use gene editing and pathway design to enhance yields and convert substrates better. As issues related to sustainable production grow, the role of metabolic engineering becomes increasingly important in providing novel solutions to meet society's demands while reducing environmental impact. The use of metabolic engineering is wide-ranging, particularly in biofuel production. For instance, yeasts, formerly critical to alcoholic fermentation, are presently leading the bioethanol industry due to breakthroughs spurred by metabolic engineering. Through addressing critical challenges in first and second-generation biofuels, researchers are designing yeasts that effectively convert lignocellulosic biomass into bioethanol. Furthermore, the introduction of synthetic biology opens the door to further fine-tuning of metabolic pathways, with the potential to produce bioethanol from a broad range of carbon substrates. This aligns with the broader vision of creating economically viable and sustainable renewable energy systems, with the enormous potential of metabolic engineering in fulfilling global energy needs

(Basso et al., 2019). At the intersection of metabolic engineering and human health is the herculean effort of reconstructing the human metabolic network. Reconstruction of the human genome-scale metabolic network provides a collective map of metabolic processes within the body and has huge potential to decode the intricacies of illness. Recent advancements, such as the consensus version Recon 2.2, illustrate how this discipline continues to evolve, adding comprehensive metabolic pathways into the model as well as ensuring correct mass and charge balances in reactions. These enhancements not only increase the model's predictive capacity toward ATP production from various substrates but also permit integration with genomics, transcriptomics, and proteomics data. This integrative approach significantly enhances the understanding of cellular metabolism in disease and health, emphasizing the central role of metabolic engineering in deciphering intricate biological pathways and designing novel therapeutic interventions (Ang et al., 2016) (Figure 7.2).

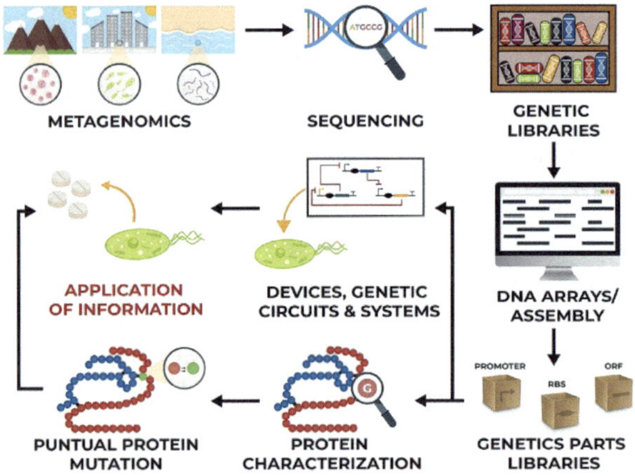

Figure 7.2. Different stages for the development of synthetic biology and metabolic engineering projects.

Source: https://www.frontiersin.org/files/Articles/439905/fbioe-07-00036-HTML/image_m/fbioe-07-00036-g001.jpg.

Historical Development and Milestones

The evolution of metabolic engineering traces its roots back to the early discoveries of molecular biology and biochemistry, a significant point in the history of the field. The discovery of recombinant DNA technology during the 1970s provided the basic tools that were employed to manipulate genetic information and enabled scientists to modify metabolic pathways in microorganisms. This era saw milestone achievements, such as the first synthesis of insulin using recombinant bacteria, which confirmed the feasibility of engineered metabolic systems to produce useful medicines. These achievements, besides advancing scientific insights, opened the door to downstream innovation in metabolic engineering, driving applications to reach agriculture, biofuels, and bioremediation. As researchers were learning about metabolic pathways, their integration gave rise to the period when metabolic engineering emerged as a critical area of study within biotechnology, and eventually became a sophisticated and multidisciplinary science. When the field transitioned into the 1990s and early 2000s, successes such as the development of synthetic biology radically changed metabolic engineering methodology. Scientists began to employ modular genetic building blocks in constructing and engineering metabolic pathways, thus enabling unprecedented control of cellular function. A case of particular interest was the assembly of bio-based pathways for producing complex molecules like artemisinin in genetically modified yeast, demonstrating the viability of bio-based alternatives for drug compounds that had typically been of plant origin. This was also the period when systems biology advanced, permitting increased understanding of cellular regulation and metabolism, and permitting scientists to predict and control metabolic fluxes more accurately. The marriage of metabolic engineering and synthetic biology created a series of innovations that revolutionized how biological systems could be utilized for industrial and environmental purposes. Not long ago, the integration of cutting-edge computing technologies with genome-scale engineering has encouraged further advances in metabolic engineering, such as the construction of highly sophisticated microbial factories. Technologies such as CRISPR-Cas9 genome editing have revolutionized the precision of gene editing, enabling more targeted intervention in metabolism. These have resulted in significant improvements in yield and efficiency for such bioproducts as biofuels, bioplastics, and fine chemicals. Additionally, the advent of artificial intelligence and machine learning in metabolic engineering is catalyzing fast iterations in strain development, expediting the optimization process. As the discipline advances, an increasing focus on sustainability and environmental factors pervades research, signifying that the classic history of metabolic engineering not only emphasizes notable technological achievements but also portends a future in which engineered systems make significant contributions to solving global problems.

Importance in Biotechnology and Industry

Biotechnology is now a fundamental element of contemporary industrial processes, primarily due to the concepts of Metabolic Engineering. This new science endeavors to engineer microbial metabolic pathways for scaling up the production of valuable biochemicals. Through the utilization of renewable raw materials, industries can not only reduce their dependence on fossil resources but also significantly reduce their environmental footprint. Such industrial biotechnology innovations are facilitating the transition towards more sustainable production methodologies. Through such biological systems, which are becoming more optimized with these genetic improvements, the window opens for improving better strains of organisms producing high-quality products required by various industries, including pharmaceuticals, foods, and biofuels. The integration of robust mathematical and computational models at the core is what enables predictions about how genetic changes affect microbial phenotypes, an aspect core to the industry's development (Rocha et al., 2013). One illustration of a key application of Metabolic Engineering is the production of bioethanol through yeast fermentation, which symbolizes the point of intersection between traditional practices and sophisticated biotechnology. Beginning in ancient history, the fermentation of yeast for producing alcohol has developed into a sophisticated industrial process today.

The problem points in the production of biofuels—first and second-generation bioethanol—hinge on increasing yield and efficiency and addressing problems involving the recalcitrance of substrate utilization. With the application of advanced Metabolic Engineering techniques, scientists are exploring new avenues that can be employed to enhance yeast strains for bioethanol production more efficiently, thus fortifying the role of engineered microbes in designing sustainable energy solutions. The possibility of the same underlines the importance of ongoing research and development of metabolic pathways that could possibly meet forthcoming demands for renewable energy resources (Basso et al., 2019). Besides, the greatest impact of Metabolic Engineering in biotechnology is evident since it facilitates interdisciplinary integration between systems biology, molecular biology, and synthetic biology. All these scientific disciplines provide their own approach that enhances understanding as well as manipulation of metabolic networks. All this is required to create a synergistic system for designing microorganisms for a given industrial application, e.g., microbial-engineered strains for chemical synthesis or waste minimization. Moreover, open-source software development tools such as OptFlux provide opportunities to researchers in academia and industry by making tools available to simulate microbial behavior and evaluate genetic alteration cost-effectively. Such technological advances not only make strain development easier but also expand innovation potential in the biotechnology sector, which provides Metabolic Engineering as a platform for industrial applications in the future (Rocha et al., 2013).

7.1. FUNDAMENTAL PRINCIPLES OF METABOLISM

Metabolism is the intricate biochemical activity maintaining life, converting substrates to energy and cell components necessary for growth and maintenance. Metabolism is ultimately rooted in the thermodynamic principles, namely Gibbs free energy and enthalpy principles, which dictate the spontaneity and feasibility of metabolic reactions. The relevance of such thermodynamic principles is evident in the various biochemical processes wherein enzymes catalyze reactions, thereby reducing the energy barrier necessary for conversion. The knowledge of the thermodynamic landscape enables scientists to anticipate possible metabolic fluxes and efficacy in terms of substrate utilization, a fundamental aspect that must be considered in metabolic engineering aimed at optimization of production processes. As researchers adopt a systems-level analysis of metabolic networks, sophisticated models yield information on pathway interactions that facilitate the design of efficient strategies for the manipulation of metabolism toward desired goals in biotechnology (Nielsen et al., 2025). In order to utilize metabolism efficiently towards biotechnological applications, an in-depth comprehension of metabolic networks and their regulators is crucial.

This complexity arises from thousands of interacting reactions and metabolites that constitute these networks and makes traditional study methods typically suited to single pathways inadequate. Mathematical modeling methodologies and systems biology new approaches enable scientists to map and analyze such complex interactions in greater depth. Further, the ability to integrate artificial intelligence tools provides unprecedented potential for predictive metabolic engineering outcome evaluation. By applying algorithms to plan the various dynamics of metabolic pathways, scientists can find potential areas of efficiency to improve and streamline processes for the production of drugs, biofuels, and other useful biochemicals. Computational strategies are being employed to fill in the gaps of what is known regarding metabolism so that innovative manipulation tactics will be in accordance with biological principles (Nielsen et al., 2025). Recent developments in metabolic engineering reinforce the importance of thermodynamic principles, particularly in optimizing genome-scale metabolic networks (GEMs). The development of tools such as the graph neural network-based model for predicting Gibbs free energy improves metabolic pathway analysis accuracy, which determines major reactions that regulate metabolic flux and efficiency. This view emphasizes the integration of thermodynamic principles with traditional metabolic modeling strategies to provide better constructs with a more realistic model of biological systems. The identification of thermodynamic driver reactions not only enhances our knowledge on metabolic control but also enlightens us about evolutionary pressures operating on metabolic pathways. As metabolic engineering advances, prioritizing thermodynamics in model curation will ensure the creation of efficient frameworks for the evaluation of new metabolic strategies towards industrial application and improving general biotechnological efficiencies.

7.1.1. Overview of Metabolic Pathways

Metabolic pathways are complex networks that facilitate the conversion of nutrients into energy and biomolecules required for cellular functions. These biochemical reactions are categorized into anabolic and catabolic pathways whose functions in cell metabolism are exactly opposite. The catabolic pathways break down complex molecules into simple molecules while liberating energy during the process, while the anabolic pathways construct complex molecules from simpler molecules using energy. An understanding of the pathways is crucial for metabolic engineering because one can redesign the metabolic networks to result in increased levels of production of the products of interest. The interaction of these pathways can lead to a finely balanced metabolic state, which is essential to organismal health and productivity. Information on these pathways also enables the identification of key enzymes and reaction steps that can be manipulated for metabolic alteration, thus providing acccss to sophisticated biosynthetic capabilities in various microorganisms. Genome-scale metabolic models are an integral part of an improved comprehension of metabolic pathways. These models mimic the entire repertoire of metabolic reactions in an organism and provide a full system for studying metabolic networks. For instance, Heinemann et al. (2005) documented a genome-scale model of *Staphylococcus aureus*, its use in the elucidation of the connections between bacterial metabolism and antibiotic resistance. These models are created through the integration of genomic information, experimental information, and physiological information and can be exhaustively tested against actual biological behavior. This computationally demanding modeling procedure not only facilitates deciphering complex interactions between metabolic pathways but also aids in the prediction of physiological response to gene alterations, thereby informing experimental design and conduct in metabolic engineering approaches. Comparative modeling of metabolic pathways across diverse organisms may also offer significant insights into evolutionary adaptation and functional improvement. Techniques such as the Multi-Metabolic Evaluator (MultiMetEval) enable multi-metabolic model evaluation and therefore allow scientists to verify how alterations in metabolic structure might affect flux distributions between biochemical pathways. Freitag et al. (2012) brought out the importance of multi-objective analysis for the study of trade-offs in natural product production and biomass production, which is particularly beneficial for biotechnological applications. Using such tools as MultiMetEval, scientists are not only able to identify optimal routes of overproduction of compounds, but also uncover hidden strategies implemented by other organisms in the biosynthesis of valuable metabolites. This holistic approach to unravelling metabolic pathways eventually leads to improved strategies in metabolic engineering, ideally enabling the possibility of industrial and pharmaceutical applications (Figure 7.3).

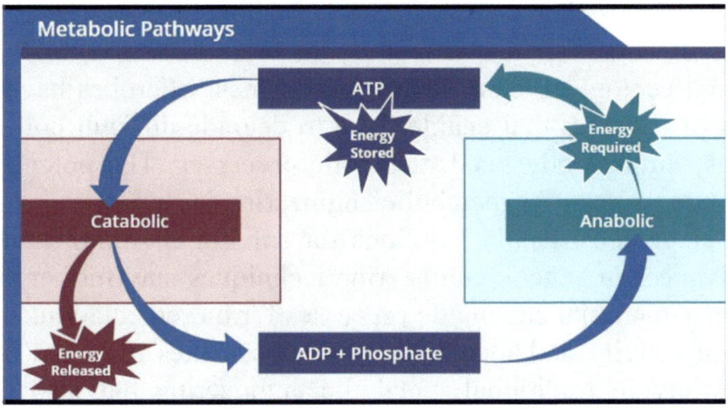

Figure 7.3. Metabolic pathways.

Source: https://i.ytimg.com/vi/m61bQYio9ys/maxresdefault.jpg.

7.1.2. Role of Enzymes in Metabolic Processes

Enzymes play a vital role in regulating metabolic processes as biological catalysts that facilitate biochemical reactions important to cell function. The manipulation of enzyme activity and specificity in metabolic engineering holds vast potential for optimizing the efficiency of metabolic pathways. This can lead to improved levels of precious products, such as biofuels, pharmaceuticals, and biodegradable plastics. By optimizing the levels of expression of enzymes and changing their substrate affinities, engineers are able to produce microbial strains that can efficiently convert raw materials into specific products. More knowledge regarding enzyme kinetics and regulatory mechanisms is key to these accomplishments because it makes possible the rational design of metabolic pathways that can change dynamically in response to environmental conditions and substrate availability. The function of enzymes transcends catalysis; they are part of the complete metabolic networks that define life. Central carbon metabolism (CCM), comprised of pathways such as glycolysis and the tricarboxylic acid cycle, is a fundamental template for energy production and biosynthetic precursors in living organisms. Current studies in metabolic engineering have employed the unique regulatory roles of enzymes in CCM towards facilitating cellular output maximization. The addition of heterologous pathways or overexpression of key regulatory enzymes has made it possible for scientists to alter metabolic fluxes such that production of specific products is increased. These strategies are the key intersection of enzymology and metabolic engineering and demonstrate how the subtle modulation of enzyme functions can have significant impacts in microbial biotechnology (Wu et al., 2023). Furthermore, the complementarity between enzymatic processes and microbial mechanisms of degradation, particularly in bioremediation, underscores the flexibility and ecological significance of enzymes. Microbes have developed diverse sets of enzymes that enable them to degrade difficult pollutants, such as pesticides, through advanced metabolic processes. The potential in using such enzymatic systems for metabolic engineering is enormous, particularly as industries pursue eco-friendly solutions for solving environmental problems. As seen, advances in genetic engineering techniques may further enable us to leverage these microbial enzymatic processes for bioremediation. This overlap of enzymatic activity and metabolic processes makes enzymes increasingly crucial not only in traditional metabolic engineering but also in ensuring ecological sustainability through innovative biotechnology interventions (José Ramírez et al., 2023) (Figure 7.4).

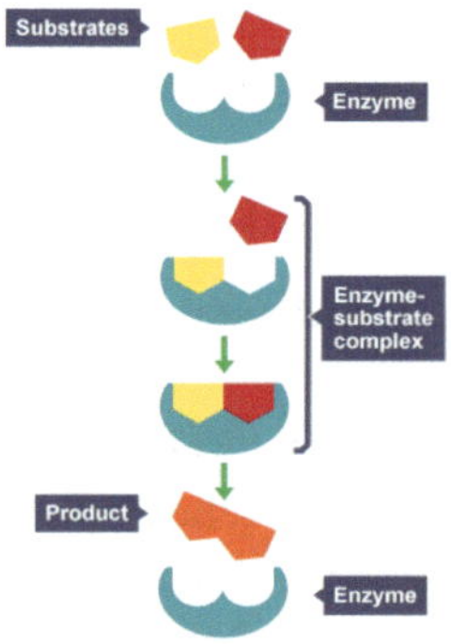

Figure 7.4. Control of metabolic pathways using enzymes.

Source: https://bam.files.bbci.co.uk/bam/live/content/z66rwmn/large.

7.1.3. Energy Production and Consumption in Cells

Energy production and utilization within cells is the fundamental component of metabolic engineering, more or less dictating the degree to which microorganisms can be designed for various applications. Cellular metabolism consists of separate pathways that convert substrates into energy, primarily through processes like glycolysis, the tricarboxylic acid (TCA) cycle, and oxidative phosphorylation. Effective energy utilization is crucial to maintaining cellular growth and function, especially in those engineered microorganisms intended for industrial applications. Metabolic engineering aims at manipulating these pathways to enhance energy production and optimize biosynthetic pathways to allow organisms to efficiently generate worthwhile bioproducts from cheap substrates. With the achievement of the intricate interaction of the metabolic pathways, researchers can reorganize cellular functions, thereby enhancing microbial processes of production, making microbial production processes more efficient, promoting greater sustainability in energy consumption, and promoting ecological balance. Microorganisms' application in the generation and utilization of energy has been a topic of significant interest in the area of metabolic engineering, especially for the production of biofuels. Yeasts, for instance, have played a key role in fueling the bioethanol industry by efficiently fermenting sugars into alcohol.

The shoal engineering of metabolic pathways of such organisms has unveiled new potential for increased bioethanol yield via metabolic engineering and synthetic biology, overcoming challenges faced in producing biofuels using conventional biofuel production methods, such as substrate specificity and metabolic bottlenecks. Furthermore, new technologies in engineering techniques have stimulated research in developing yeast strains that can resist higher alcohol and toxicity levels, ultimately coming up with more sustainable and cheap biofuel production systems with the potential to compete with fossil fuels in efficiency and yield (Basso et al., 2019). Further, more recent developments in microbial electrosynthesis are a paradigm shift in the production of energy, shifting the direction towards the new use of electricity to drive metabolic reactions in bacteria. Certain acetogenic bacteria, such as *Clostridium autoethanogenum*, can utilize electrical current as a supplemental source of energy, making a major diversion in metabolic pathways possible. By depositing electrons into such bacteria under *in vitro* conditions, researchers have observed drastic changes in the distribution of carbon, resulting in increased production of industrially relevant biochemicals like lactate and 2,3-butanediol at the expense of the conventional acetate. This new approach not only demonstrates the scope of microbial systems for energy transformation but also highlights the potential of the integration of electrocatalytic reactions into the topology of metabolic engineering. These findings set the stage for optimizing fermentative processes and enhancing production efficiency through electrochemical methods, ultimately contributing to the development of sustainable chemical production technologies (Bernhardt et al., 2016).

7.2. TECHNIQUES IN METABOLIC ENGINEERING

Metabolic engineering entails a series of sophisticated methods to control the metabolic processes in living things to maximize the production of desirable chemicals. Among the most apparent methods in this regard is the usage of genome editing technologies, such as CRISPR-Cas9. This innovative instrument allows for the specific editing of an organism's genetic material, which subsequently permits the knockout of specific genes that are responsible for competing with pathway-specific biosynthesis. For instance, in research carried out on *Chlamydomonas reinhardtii*, scientists selectively edited LCYE and ZEP genes to enhance zeaxanthin yield by significantly large margins. Through the use of gene editing combined with overexpression methods of rate-limiting enzymes, as well as cultivation optimization, metabolic engineers have significantly enhanced the yield of zeaxanthin by significant margins, demonstrating the flexibility as well as strength of genome editing as a tool in metabolic engineering (Jang et al., 2025).

Another key approach to metabolic engineering is synthetic pathway design in microbial hosts. By incorporating heterologous genes that participate in the biosynthesis of target compounds, researchers have successfully engineered microorganisms for the production of valuable metabolites in an eco-friendly fashion. Saccharomyces cerevisiae is one such example of engineered microorganisms that produce salidroside and hydroxytyrosol, demonstrating the potential of yeast modification to yield bioactive compounds more efficiently compared to traditional extraction strategies. Researchers achieved significant yield improvements by optimizing precursor supply and amplifying cofactor availability. Through multiple rounds of metabolic pathway optimization and genetic manipulation, significant strides were achieved toward microbial production levels, thereby addressing industrial requirements and demonstrating the relevance of synthetic biology in metabolic engineering (Sun et al., 2025). Aside from genome editing and pathway building, fermentation optimization is required to maximize the production yields of metabolites.

The technique seeks to alter various bioprocess parameters such as pH, temperature, and nutrient concentration to achieve a good growing environment for microorganisms and metabolite formation. Strategic management of fermentation conditions can lead to significant product yield enhancement, as evidenced by the advancements of zeaxanthin and salidroside production from engineered strains in controlled bioreactors. Through maximizing biomass accumulation and metabolite content, these approaches not only increase productivity but also improve the sustainability of production processes. Lastly, the complementarity of these varied methods proves the dynamic nature of metabolic engineering and the potential of the discipline to counteract the increasing global demand for naturally occurring molecules (Jang et al., 2025; Sun et al., 2025).

7.2.1. Genetic Modification and Synthetic Biology

Synthetic biology and genetic modification play central roles in the domain of metabolic engineering, offering new solutions for the production of biologically relevant compounds. Thanks to advanced technologies such as CRISPR-Cas technology, it is now possible for scientists to precisely modify microbial genomes for the fine-tuning of metabolic pathways. This precision enables improvements in metabolite biosynthesis, with vast implications for the biotech and pharmaceutical industries. By integrating synthetic biology methods, researchers are not only revolutionizing how we engineer microorganisms towards a desired purpose but also making processes of bioproduction optimal. Innovative technologies and strategies apply to broaden the scope of metabolic engineering, accelerating the pipeline for new medicines and biologically active metabolites. With these advances come their set of challenges, which include regulatory compliance and maintaining the sustainable production norms in the industrial setting (Sadanov et al., 2025).

Among the signature uses of metabolic engineering within synthetic biology is the deployment of *Corynebacterium glutamicum*, a biotechnology-wide ordinary organism. *C. glutamicum* has been engineered to produce extra amounts of amino acids, biofuels, and other target chemicals and thus represents a highly significant biomanufacturing device. Development of *C. glutamicum* strains involves an array of methodologies, including genomic alteration and synthetic biological device-aided metabolic flux optimization. Also, directed evolution strategies have shown potential in creating strains with improved performance during bioprocesses. The engineered strains form effective cell factories used to develop sustainable methods for producing essential bioproducts, such as N-acetylglucosamine and hyaluronic acid. The ongoing identification of *C. glutamicum* and other similar organisms is a testament to the revolutionary potential and versatility offered by synthetic biology in metabolic engineering (Chai et al., 2021). Despite the phenomenal advances in synthetic biology and genetic engineering, the field has some significant challenges that must be overcome. The first significant challenge is the engineering of the metabolism pathways for large-scale production on a financially viable scale. Additionally, regulation of genetically modified organisms (GMOs) is an issue that hinders innovation and utilization of new technologies. Addressing these challenges requires multidisciplinary efforts, for example, the integration of artificial intelligence and machine learning to predict metabolic interactions and improve bioprocesses. Through the design of proper biocontainment measures and compliance with international standards, the activity can sidestep potential risks associated with genetic modification. With consistent research efforts and advancements in technology, the prospect of metabolic engineering is vast with enormous opportunities to address world demands in health, energy, and sustainable materials (Sadanov et al., 2025) (Figure 7.5).

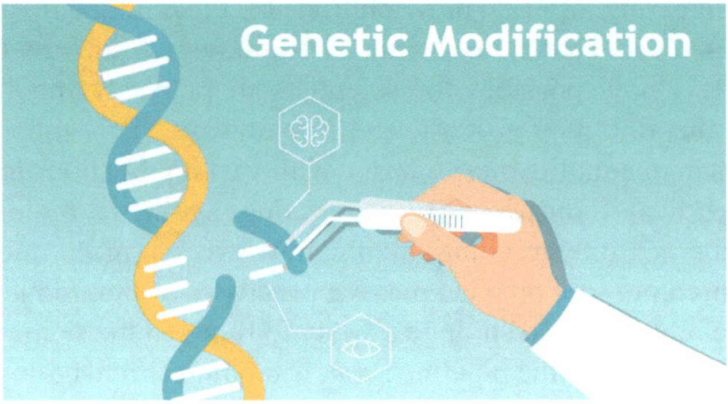

Figure 7.5. Genetic modification.

Source: https://cdn.educba.com/academy/wp-content/uploads/2024/08/Genetic-Modification.jpg.

7.2.2. Metabolomics and Systems Biology Approaches

The integration of metabolomics and systems biology is a cutting-edge development in metabolic engineering that creates immense insight into the biochemical pathways that govern cellular processes. Metabolomics is the comprehensive measurement of metabolites in biological systems and provides a snapshot of the physiological state at any given moment. This is significant since metabolites represent the end products of cellular functions, linking genetic information and environmental influences to cellular phenotypes. By employing advanced analytical techniques, such as mass spectrometry and nuclear magnetic resonance spectroscopy, scientists are able to profile levels and fluxes of metabolites between various conditions. Understanding these metabolic fluxes is valuable not only for unraveling complex biological networks but also for directing them with high efficiency in engineered organisms, particularly to maximize the optimization of production processes for medicines or biofuels. The cooperative paradigm provided by systems biology contributes to the utility of metabolomics by situating metabolic information in integrated models that constrain cellular behavior. This integrative architecture allows for the study of how metabolic networks interact and respond to varied stimuli, including genetic changes and environmental variations. Kinetic modeling, stressed in recent publications, remains a core weapon for unveiling plant metabolism and other complex systems. By unraveling the regulatory mechanisms that control metabolic fluxes, researchers are able to better predict the implications of specific interventions and thus design more accurate metabolic engineering plans. This modeling not only facilitates the understanding of plant cellular compartments but also transcends inherent shortcomings in the quantification of fluxes, thereby driving the field forward toward the achievement of target phenotypic traits in engineered species (Fernie et al., 2013). Moreover, the role of deciphering metabolic networks extends into essential areas such as cancer research, where altered metabolism is a preeminent force in disease progression. Interfacing of systems biology and metabolomics enables an improved comprehension of malignancy-specific metabolic changes, which has been discovered as one of the hallmarks of cancer. New advances in omics tools and computational modeling are yielding valuable information regarding dysregulated states of cancer metabolism. Such information offers opportunities to discover new drug targets and biomarkers that can translate into therapeutic interventions. By shedding light on the complex metabolism of cancer, the integration of these methods not only advances scientific insight into tumor biology but also contributes significantly to metabolic engineering with the capability to provide a method to develop targeted therapies through engineered cell systems (Chaneton et al., 2015) (Figure 7.6).

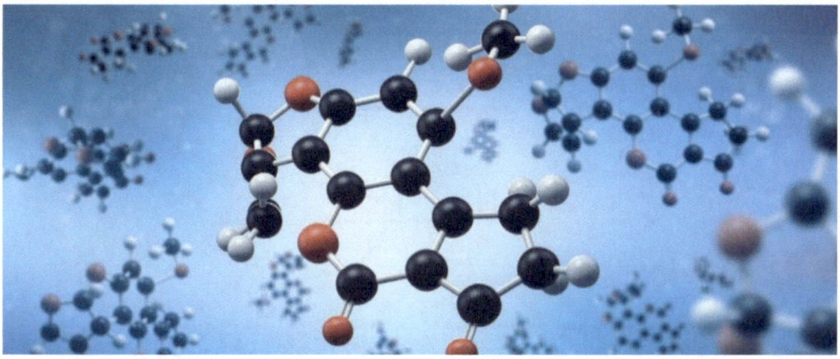

Figure 7.6. Metabolomics.

Source: https://www.agilent.com/cs/publishingimages/L2_HeroBanner_1280x545 px.jpg.

7.2.3. CRISPR and Gene Editing Technologies

The emergence of CRISPR and gene editing tools has brought about a revolutionary age in metabolic engineering. Using techniques such as CRISPR/Cas9, scientists can directly alter the genomes of organisms to target specific genes that control metabolic processes. This specificity not only increases the knowledge about metabolic processes but also facilitates the creation of optimized metabolic characteristics in organisms. For instance, CRISPR has been used to enhance algae biofuel production by removing some energy-intensive genes, thus channeling metabolic processes toward increased biomass accumulation. With continued advancement in research, the integration of such gene editing is likely to yield crops with better nutritional content, improved disease resistance, or greater ability to thrive in alternative environmental conditions, hence addressing global food security and sustainability concerns. In addition, the applications of CRISPR technologies extend beyond the traditional to areas that require a close understanding of population-specific genetic variations. The existing literature discussed here highlights the potential of advanced genome-editing technologies to explore such variations in depth. By investigating population-specific point mutations through functional genome editing, scientists can move beyond associative studies in isolation, instead to create empirical evidence connecting genetic variation and particular phenotypic outcomes.

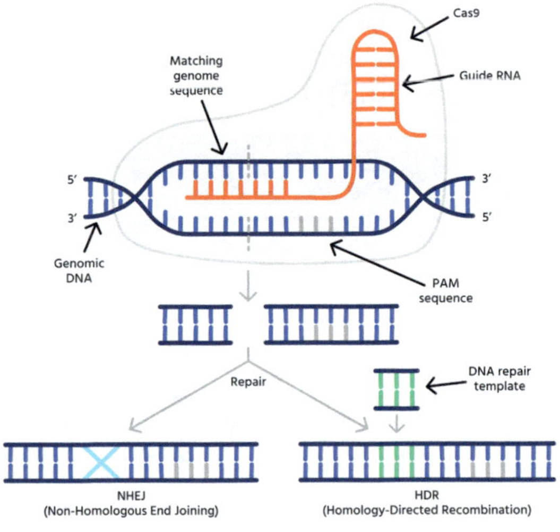

Figure 7.7. CRISPR-Cas9 gene editing.

*Source:*https://iotasciences.com/wp-content/uploads/2024/06/CRISPR-Cas9_mechanism.jpg.

The case of the CREBRF gene, associated with thrifty mutations, shows how the application of CRISPR can open up further understanding of evolutionary pressure and metabolic adaptation within a population, illuminating the relationship between genetics and environmental selection. These results not only contribute to the metabolic engineering body of knowledge but also place a spotlight on the necessity for interdisciplinary convergence between population genetics and genomics (Fox et al., 2020). Enhanced genomic technologies also translate to ethical issues, with their application increasingly coming to the forefront. The issue of human germline editing and its long-term consequences for future generations poses significant questions about natural genetic diversity and uniqueness. Similarly, when gene-edited plants are developed, regulatory frameworks must weigh whether such advancements will be labeled as GMOs. These labels affect the public's willingness to accept and market forces, which are key to the successful integration of these technologies into agriculture. Emerging developments in mitochondrial replacement therapy, for instance, demonstrate the potential and pitfalls of gene editing in medicine. These debates not only reflect the variety of CRISPR applications but also demonstrate the ongoing need to be in the ethics space as the world copes with the future consequences of engineering life itself, underlining the role of regulated innovation in metabolic engineering (Beccari et al., 2017) (Figure 7.7).

7.3. APPLICATIONS IN MEDICINE

Combining metabolic engineering and medicine has unleashed revolutionary applications with the potential to fundamentally change healthcare. The most astonishing of these is the production of flavonoids, molecules whose antioxidant activities have been well documented and whose anti-chronic-disease-inducing properties are suspected. While their therapeutic potential is considerable, nature's limitation of flavonoids in plants presents a barrier to global usage in pharmaceutical and nutraceutical industries. The most recent research aims at new strategies to enhance flavonoid biosynthesis through the use of metabolic engineering, including the manipulation of transcription factors regulating biosynthetic pathways. Through the integration of synthetic biology and biotechnological tools, scientists are building plant cell factories capable of manufacturing these premium compounds at higher yields, eventually developing sustainable, high-production systems that meet the demands of research and industry (Wang et al., 2025). Another significant area of research in metabolic engineering is the application of adipose-derived mesenchymal stem cells (ADSCs) in regenerative medicine. These multipotent cells, abounding in adipose tissue, possess the unique ability to differentiate into many varied cell types and possess immense therapeutic potential. Their plasticity allows ADSCs to influence systemic metabolism and immune response, thus rendering them extremely valuable for the treatment of metabolic syndromes and inflammatory disorders. Besides, their interaction with the tumor microenvironment presents not only obstacles but also possibilities for treating cancer, revealing their character as modulators during tumor development. Genetic modification of the metabolic pathways of ADSCs not only enhances their therapeutic potential but also enables targeted delivery of therapeutic molecules, which can complement traditional treatment strategies and pave the way for next-generation regenerative treatments (Wang et al., 2025). Secondly, the reach of metabolic engineering goes beyond direct production of therapeutic compounds to cure complex health conditions such as antibiotic resistance. By reprogramming microbial metabolism, researchers are developing engineered strains that can produce novel antimicrobials. This biotechnology approach addresses the need for alternative medicine amid rising rates of antibiotic-resistant infections. Improved microbial platforms, optimized by metabolic engineering, have the potential to produce higher levels of antibiotics and more varied types of compounds, thereby widening the range of resistant pathogens that can be combated. With advancing technology in metabolic engineering, its impact on medicine not only has the potential to address current healthcare needs but also to provide new modes of solution that enhance patient outcomes and public health overall (Wang et al., 2025) (Figure 7.8).

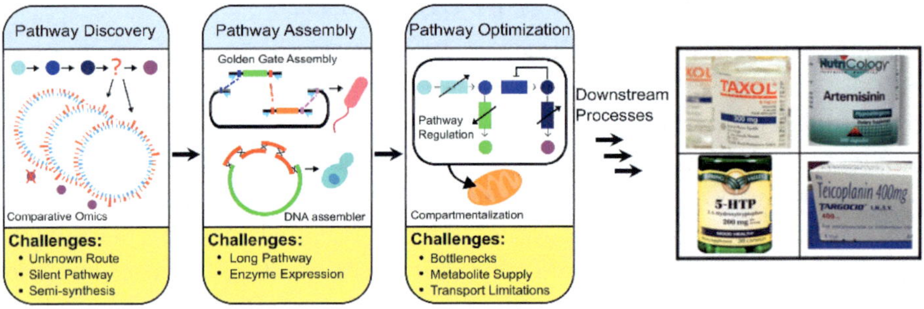

Figure 7.8. The three major steps in metabolic engineering for production of small molecule drugs, including pathway discovery, pathway assembly and pathway optimization.

Source: https://www.mdpi.com/fermentation/fermentation-02-00004/article_deploy/html/images/fermentation-02-00004-g001.png.

7.3.1. Production of Pharmaceuticals and Vaccines

Pharmaceutical and vaccine manufacturing has increasingly aligned with advances in metabolic engineering to exploit the capacity of microorganisms and plant systems to manufacture high-value chemicals on a sustainable scale. Metabolic engineering techniques have revolutionized the production of critical drug intermediates so that active pharmaceutical ingredients can be produced in an economical and efficient way. For instance, squalene, a critical precursor for steroid biosynthesis and certain vaccines, can be produced via genetically modified microbial systems such as yeast and bacteria. By regulating metabolic pathways, researchers have been able to redirect flux into squalene production, offering an environmentally friendly alternative to traditional sources like shark liver oil, with ecological connotations dependent on over-exploitation of marine resources (Patel et al., 2022). The change not only enhances the sustainability of drug production but also meets increasing demand in an evolving health climate. In the process of vaccine production, metabolic engineering has made significant contributions, particularly in terms of creating safer and more effective adjuvants. Adjuvants, chemicals that enhance the body's immune reaction to vaccines, have been obtained for a long time from natural sources. That has changed with the use of synthetic biology methods coupled with metabolic engineering, where these chemicals can now be produced through microbial fermentation.

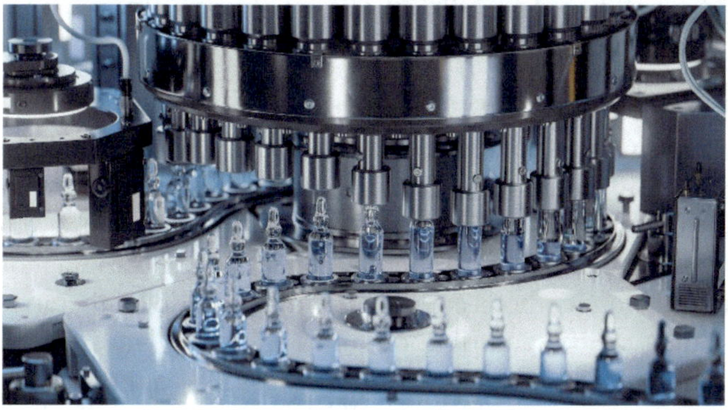

Figure 7.9. Vaccine production.

Source: https://www.news-medical.net/images/Article_Images/ImageForArticle_1247_174368561873655.jpg.

Not only does this make production easier, but it also decreases dependency on animal-derived materials, thereby eliminating ethical and regulatory problems associated with traditional adjuvant sources (Shi et al., 2020). The ability to engineer microbes for the manufacturing of adjuvants is a tangible evidence of the use of metabolic engineering in biotechnology that is groundbreaking and pivotal towards the attainment of public health objectives. Besides, plant systems engineering, in the form of hairy root cultures, presents a promising route to the production of sophisticated pharmaceutically active metabolites. Hairy roots, which are generated by transformation by *Agrobacterium rhizogenes*, maintain the major characteristics of the host plant with high growth rates and high yields of secondary metabolites. These modified roots may be optimized to enhance specific biosynthetic pathways, leading to higher yields of therapeutically valuable compounds such as alkaloids or terpenoids that play a crucial role in drug discovery (Shi et al., 2020). By integrating traditional plant biotechnology and metabolic engineering, researchers can enhance drug production efficiency while leveraging the vast chemical diversity of medicinal plants. This synergistic approach holds a rosy future for green pharma manufacturing, where prompt and reactive remedies for emerging health conditions are achieved rapidly (Figure 7.9).

7.3.2. Gene Therapy and Metabolic Disorders

Gene therapy is a breakthrough in the treatment of metabolic disorders, particularly genetically mutation-induced ones. Metabolic disorders are caused by an enzymatic deficiency that disrupts metabolic pathways, which has serious health implications. Gene therapy tries to repair these deficits by introducing, deleting, or modifying genes responsible for metabolic disease. Among the methods being explored, CRISPR-Cas9 stands apart because it is capable of delivering precision in editing genomic sequences and allowing the possibility to correct pathogenic mutations at specific locations within the genome. Besides, other cutting-edge techniques such as base editing and prime editing widen the repertoire of gene therapy, offering treatments for most metabolic diseases. These groundbreaking methods not only promise additional therapeutic benefits but also provide a central contribution to the evolving landscape of metabolic engineering, paving the way for new paths to disease treatment and management (Suthar, 2025). While the exciting possibility of gene therapy is exhilarating, the pitfalls inherent in these genomic editing methods cannot be overlooked. Off-target effects, efficiency of editing, and immunogenicity are still worrisome issues. In addition, the need for stable and secure delivery systems renders the translational process of these therapies complicated. The current state of gene therapy also lacks robust regulatory systems, raising concerns about its safety and ethics. Successful longevity of gene treatments is obtained through research in identifying the cellular environment and mechanisms that regulate the temporal and amplitude of gene expression. One promising direction is the application of synthetic gene circuits, which hold the promise for precise control of therapeutic intervention and making personalized treatments that have the potential to significantly enhance patient outcomes while minimizing the risk of side effects (Carneiro et al., 2024). In the field of metabolic disorders, synthetic gene circuits are an emerging frontier in which the advanced technology and precision required for therapeutic applications meet. These gene circuits could perform logical computations on genetic inputs to yield tailored responses based on individual patient needs. With the use of multiple genetic components such as promoters and repressors, synthetic gene circuits can regulate protein expression with accuracy, which improves the efficacy of therapy. The capacity of these circuits as very sensitive biosensors allows for real-time monitoring of metabolic markers, where it becomes possible to implement proactive intervention in treatment protocols. As research advances, the possibility of developing implantable biodevices to produce bioactive molecules on-demand grows increasingly possible. Whilst development in this area is challenging through concerns about commercial feasibility and optimization, the prospect of revolutionizing metabolic disorder therapy is an area worth investigating in metabolic engineering (Carneiro et al., 2024) (Figure 7.10).

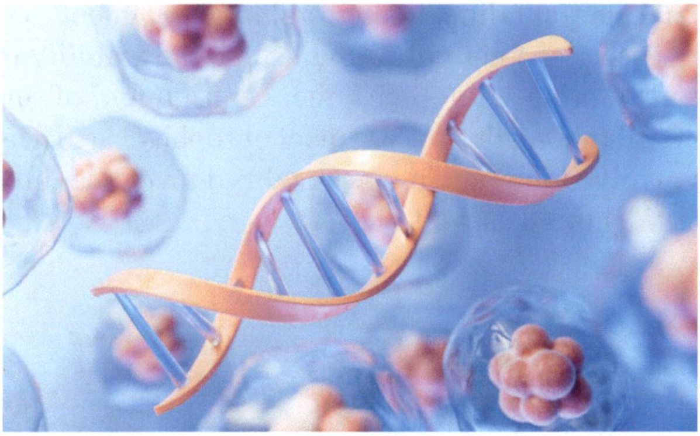

Figure 7.10. Scientists and clinicians are collaborating on the development of a clinical trial to test a gene therapy for a rare metabolic disease.

Source: https://ncats.nih.gov/sites/default/files/2024-11/shutterstock_2352632967.jpg.

7.3.3. Personalized Medicine and Metabolic Profiling

The development of personalized medicine, more specifically through metabolic profiling, is a monumental step in metabolic engineering. With healthcare moving increasingly from a one-size-fits-all strategy to treatment plans that cater to the individual, it is becoming increasingly important to know a patient's metabolic profile. Metabolic profiling utilizes the specific biochemical markers in an individual's metabolism and can tell them much about their health status, susceptibility to disease, and response to therapies. By studying metabolites, which are small molecules involved in metabolic processes, practitioners can tailor therapies based on the very specific metabolic pathways and status observed in each patient. This paradigm not only enhances treatment efficacy but also minimizes side effects associated with generalized treatments, thereby emphasizing the revolutionary nature of individualized approaches to modern medicine. Integrating multi-omics data, such as genomics, proteomics, and metabolomics, empowers personalized medicine even more by giving a systems-level view of a person's biological terrain. This makes the complex diseases better understood and facilitates the development of personalized treatments accordingly. For instance, alterations in the gut microbiota composition have also been linked to metabolic diseases like gallstone disease, where imbalances in cholesterol metabolism and inflammatory signaling pathways are primary drivers of health hazards (Tan et al., 2025). Personalized therapy through metabolic profiling can thus screen out at-risk patients and apply targeted therapies such as dietary modification or probiotic treatment to restore balance in metabolism. These plans not only demonstrate the viability of personalized healthcare but also bring into focus the contribution of interdisciplinary collaboration towards aiding the management of sophisticated health conditions. The challenges involved in realizing personalized medicine through metabolic profiling must be identified as well.

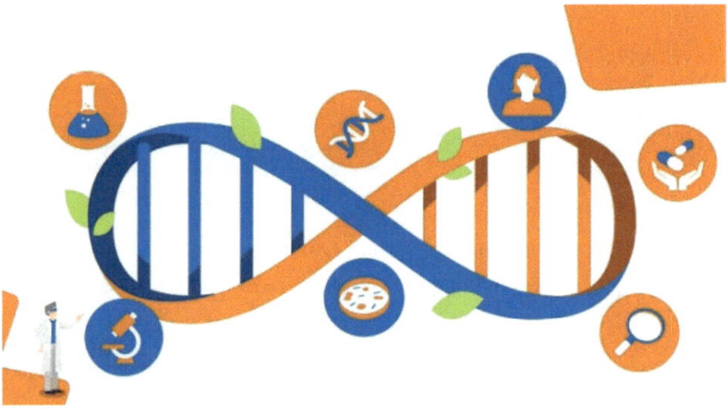

Figure 7.11. Personalized medicine.

Source: https://ispe.org/sites/default/files/styles/site_750_x_420/public/2021-10/1121_PE_SO_Valley_01.jpg?itok=YK-_K8_8.

Although more treatment regimens are promised, a variety of hindrances still exist, including the utilization of advanced technologies for precise scrutiny of metabolic profiles and integrating these findings in clinical decision-making. Furthermore, ethical concerns about data privacy and the production of biological material must be honored. Microbial sources for active pharmaceutical ingredient production is one-of-a-kind example of how metabolic engineering can drive innovation in drug development (Md. Fakruddin et al., 2024). By means of genetic manipulation, microbial metabolism can be tailored so that scientists can design personalized therapeutic treatment by aligning it with specific patient needs, thus furthering the agenda of personalized medicines. In doing so, metabolic profiling is at once key to disease management and to shaping the therapeutic development of the future (Figure 7.11).

7.4. APPLICATIONS IN AGRICULTURE

Applications in agriculture are widely seen to be the next imperative for the improvement of food security amid growing population pressure worldwide. Through metabolic pathway optimization, researchers can genetically tailor plants to enhance yield, nutritional value, and defense against biotic and abiotic stresses. These alterations are especially critical as conventional practices fail to provide enough to match the demand of a growing population. Modern technology employs gene modifications to generate crops that not only yield more but also conserve resources more efficiently, which contributes to making agricultural production sustainable. It is an all-encompassing solution to food production issues that combines scientific discovery with economic and social factors to drive a new form of agriculture that is innovative and also aligned with ethical values (Montagu et al., 2020). Besides crop yield increases, the role of biotechnology further involves plant improvement with improved nutritional content and resistance to environmental stresses such as drought and salt. For instance, metabolic engineering has the ability to drive endeavors in biofortification, enhancing the micronutrient quality of staple foods to address the challenge of malnutrition globally. Moreover, utilization of the latest technologies, such as genome editing, has led to developing crops with increased climate change resilience and reduced chemical inputs and, thereby, lesser environmental damage. Globalization of these biotechnologies is essential to meet future food security needs since they have been found to provide potential remedies to the constraints of resources and uncertainty in the climate (Gartland et al., 2018). Ultimately, the union of metabolic engineering with agriculture is a leap of revolution towards environmentally friendly management of agricultural systems. This paradigm consists of a multidisciplinary platform in which scientific progress, economic viability, and social acceptability are unified with a view to restating food systems. As the demand for sustainable agriculture increases, the ethical implications of biotechnological strategies gain prominence. By developing higher transparency and stakeholder interaction, the agriculture sector can build a solid foundation for the implementation of metabolic engineering concepts in harmony with environmental stewardship and social responsibility. The intersection of these elements, well integrated, is necessary to build a robust and strong agricultural future (Montagu et al., 2020).

7.4.1. Enhancing Crop Yield and Resistance

The enhancement of crop yield and resistance by metabolic engineering is increasingly important to address the food security challenges of the world. With an estimated population of nearly 9.6 billion in 2050, traditional agriculture will fail to satisfy the food requirements. The emphasis is now on not only improving crop yields but also developing crop varieties resistant to environmental stresses, including salinity, drought, and temperature stresses. The importance of such advancements is that the farming sector is under double pressure of increasing demand and decreasing arable land. New metabolic engineering strategies are able to assist in realigning germplasm and strategic genetic manipulations for aiding the resistance traits in moving towards a higher yield under adverse situations. These strategies are central to enabling staple foods, particularly crops like rice, to nourish a growing world population in the coming decades. In the development of stress-tolerant crops, salinity has emerged as an important issue, ranking second only to drought as a determinant of agricultural productivity. Salinity-tolerant rice varieties discovery and development are essential to regions with soil salinity, which results in crop damage and reduces the yield potential.

The integration of conventional and newer biotechnological methods, such as Quantitative Trait Loci (QTL) mapping and Marker Assisted Selection (MAS), provides breeding options for resistant crops. These techniques facilitate the precise selection of traits associated with salinity tolerance. Moreover, transgenic and haploid breeding technology holds promising ways of accelerating the generation of cultivars that are not only amenable to saline environments but also yield competitively. With these novel methods, scientists are well-equipped to address salinity stress issues, resulting in increased crop yield as well as sustainable agriculture in the long run. Despite the promise of metabolic engineering to alter crop abiotic stress resistance, there exist significant obstacles in our understanding of the genetic and molecular characteristics of abiotic stress tolerance.

Limited knowledge is hindering the effective development of stress-resistant cultivars that can survive environmental stress. A harmonistic approach integrating conventional breeding with advanced biotechnology strategies could balance these obstacles to ultimately develop more stress-tolerant crop varieties. By testing rice germplasm for desirable traits, such as stress tolerance, researchers can select and reproduce strains with a higher likelihood of surviving stressful conditions. In addition, ongoing research in somaclonal variation and other new technologies continues to supplement the set of tools at the disposal of crop improvement. Closing the gap between basic sciences and real-world applications is essential because increased crop yield and resistance depend on our capacity to innovate and evolve agricultural methods in an increasingly dynamic world (Figure 7.12).

Figure 7.12. Crop yield

Source: https://brandonbioscience.com/wp-content/smush-webp/2024/04/getty-images-r_rXoOYAvy4-unsplash-1024x681.jpg.webp.

7.4.2. Biofortification and Nutritional Improvements

The demand for nutritional enhancement in food sources has never been so urgent in the context of global malnutrition, which impacts around one-third of the world's population. Biofortification, the increase in the nutrient quality of staple food crops through metabolic engineering and deliberate breeding, is a working option for alleviating micronutrient deficiencies. Incorporating gene stacking techniques, scientists have been able to substantially increase key vitamins and minerals in widely consumed crops, thus offering an environmentally friendly way to combat hunger and improve overall public health. These advances are well aligned with the United Nations' Sustainable Development Goal 2 of eradicating all forms of hunger and malnutrition by 2030. The utilization of biofortified crops not only can improve economic growth among vulnerable populations but also can contribute significantly to global social development by reducing health inequities and improving food availability (Straeten et al., 2020). Agronomic biofortification is an important method of enriching crops with essential micronutrients like selenium. As selenium deficiency poses a significant threat to health on a global scale, particularly in nations with selenium-poor soils, elevating the concentration of this micronutrient in the edible parts of crops is a necessity. Through metabolic engineering, researchers aim to elucidate the mechanisms of selenium uptake and assimilation in crops and thus optimally accumulate it in the edible parts of crops. This process not only changes the nutritional value of staple food but also increases the pathways for controlling micronutrient deficiency through diet. Furthermore, studies indicate that selenium-enriched food has beneficial effects on human health, for instance, as a potential intervention against most diseases and strengthening the immune system. Thus, agronomic biofortification is a visionary step towards food security and nutritional well-being worldwide (Schiavon et al., 2020). The future of biofortification has immense possibilities with the integration of different micronutrient improvement methods and new methods, such as nanosized biofortification. These technologies aim at simplifying the complexities of nutrient deficiencies in various types of environmental conditions, in addition to rendering agriculture sustainable and viable. By exploring the nexus between microbial interaction and plant health, researchers can further enhance nutrient delivery in plants to create crops that work actively towards dietary diversity and resistance. The potential of biofortified crops to address current health challenges, such as recent global pandemics, underscores their relevance as nutrition interventions. With the help of sophisticated metabolic engineering techniques, production of biofortified food with improved quality can not only be utilized to enhance nutritional levels but also to build the human body strong against new health hazards, integral growth towards a healthy and well-fed population (Schiavon et al., 2020; Straeten et al., 2020) (Figure 7.13).

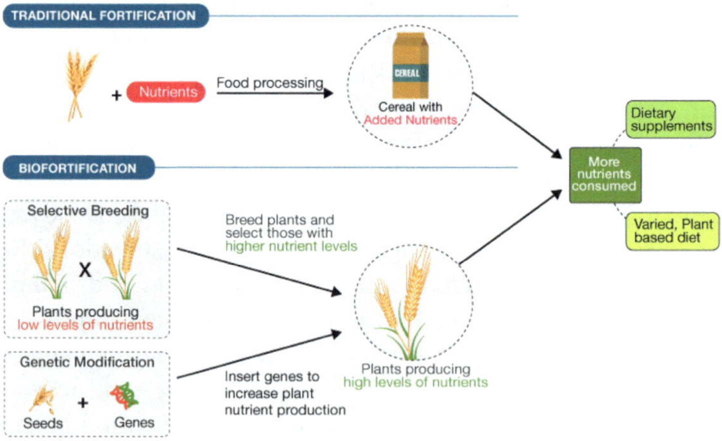

Figure 7.13. Methods of biofortification.

Source: https://cdn1.byjus.com/wp-content/uploads/2018/11/biology/2018/05/11072538/Methods-of-Biofortification.png.

7.4.3. Sustainable Agricultural Practices

With growing global food demands, alongside environmental degradation, conventional cultivation practices need to be adopted on a large scale for future food security and ecological equilibrium. Application of biosurfactants—surface-active compounds naturally produced during biosynthesis—can be one such promising avenue to this end. Biosurfactants, upon metabolic engineering, can enhance the nutrient bioavailability, stimulate plant growth, and enhance soil structure. Through their decrease in reliance on synthetic pesticides and fertilizers, biosurfactants offer environmentally friendly alternatives in keeping with sustainable agricultural practices. That they are biodegradable and of low toxicity helps protect the environment even further by promoting sustainable production, and their incorporation into standard agricultural systems cannot happen soon enough (Abdoli et al., 2025).

Figure 7.14. Sustainable agricultural practices.

Source: https://www.envpk.com/wp-content/uploads/2021/04/Sustainable-Agriculture-Practices-and-their-Advantages-1.jpg.

Other than biosurfactants, studies on natural fibers from waste crops offer a further component of sustainability. Pseudostem fibers from Alpinia galanga offer a new approach to addressing waste management challenges and the demand for biodegradable resources. Traditionally discarded, they are rich in cellulose and lignin content, making them good substitutes as reinforcement material for composites. With the use of metabolic engineering techniques to enhance their characteristics through treatments such as alkaline and permanganate processing, it is possible to actualize their potential as green resources. This approach not only promotes the circular economy with recycling of agricultural residues but also discourages non-renewable input dependence and thereby promotes ecological sustainability for agriculture (Babu et al.,

2025). Incorporation of these new approaches creates a new paradigm of sustainable agriculture production, whereby metabolic engineering emerges as a key feature. By enhancing the production and functionality of biosurfactants and natural fibers, agricultural systems can be enhanced to become resilient and efficient. Such resilience not only optimizes crop and soil quality but also minimizes environmental impact, hence serving the general agenda of sustainability. With ongoing research on these biotechnological advances by scientists and agricultural experts, the integration of metabolic engineering with agricultural procedures is promising to revolutionize traditional agriculture. This advancement accentuates the need for interdisciplinary studies in designing solutions that reconcile agro-productivity and environmental conservation, leading to a more sustainable and resilient food system (Figure 7.14).

7.5. INDUSTRIAL APPLICATIONS

Metabolic engineering has facilitated significant enhancement of industrial applications, particularly in the production of organic acids. Microbial platforms have been engineered to efficiently produce pyruvate, lactic acid, and succinic acid, thereby serving pivotal roles in the biochemical industry. Pathway reconstruction and gene editing enhance the metabolic capabilities of microbes, leading to higher titers of these organic acids. Moreover, implementation of new strategies, including by-product pathway knockouts and major enzyme overexpression, addresses production efficiency issues. With increased focus on sustainability, the utilization of non-food biomass resources like lignocellulose and algal feedstocks not only reduces competition with food but also serves to connect industrial processes with environmentally friendly goals (Wang et al., 2025). Metabolic engineering is also leading to the introduction of specialty plant oils for industrial applications to the marketplace, in addition to organic acids. For instance, genetic transformation techniques have been employed to enhance the erucic acid and wax esters in oilseed plants, such as *B. carinata*. By introducing genes like BnFAE1 and LdPLAAT, researchers have been able to drastically enhance the content of erucic acid, with some lines having up to 52.7% content. The innovation not only optimizes the composition of fatty acids for specific industrial uses but also encourages the development of bio-derived materials and products. In addition, production of wax esters through metabolic engineering has the potential to be utilized in surfactants and lubricants, proving how crops engineered previously can be designed to yield specific commercial demands while maintaining environmental footprints at a minimum (Tesfaye et al., 2024). Applications of metabolic engineering in the industrial sector extend beyond single compounds, with the design of a paradigm shift for biomanufacturing operations. By employing omics-driven approaches and system-wide optimization techniques, industries are favorably placed to respond to imbalances in metabolic flux and substrate inhibition problems at a faster rate. Production systems benefit from the use of stress-resistant strains being a significant leap, enabling microorganisms to survive under very harsh industrial conditions. The use of biorefinery concepts also enhances overall economic viability by promoting multi-product output and resource efficiency. With the continued development of metabolic engineering research, the potential to fundamentally transform a variety of industries, including chemical manufacturing, fuels, and biodegradable products, becomes more apparent, further establishing it as a cornerstone of sustainable industrial processes (Wang et al., 2025).

7.5.1. Biofuels and Renewable Energy Sources

The role of biofuels in the transition to renewable energy has become increasingly more significant as global awareness regarding climate change continues to grow. The effective production of biofuels, particularly bioethanol, has seen tremendous breakthroughs due to the expanding science of metabolic engineering. The growing discipline is driving innovation in the conversion processes, permitting improved utilization of both first-generation and second-generation feedstocks. The former one depends almost entirely on food crops such as corn and sugarcane, while the latter one aims to surpass ethical challenges by using lignocellulosic biomass and crop residues. With the integration of metabolic engineering strategies, researchers are overcoming the limitations associated with the depolymerization of lignocellulosic feedstock and fermentation to make bioethanol a promising alternative to fossil fuels. As the promise of biofuels is increasingly realized, being aware of the scientific advancements in this area is more important for fueling a low-carbon economy (Walker et al., 2011). Bioethanol production is founded on high-tech biochemical reactions in which yeasts are pivotal. Yeasts have played an essential part in alcoholic fermentation throughout history, and have established themselves as key players in the world of biofuels. Metabolic engineering is a viable path to enhance yeast strains to increase their efficiency and viability in bioethanol production. This involves bridging barriers to fermentation by maximizing metabolic routes and thus ethanol production and tolerance. Moreover, the integration of synthetic biology into yeast culture offers possibilities for the genetic modification of organisms to exploit different substrates more efficiently, leading to a diversified biofuel production system. Advancements here not only add to the sustainability of bioethanol as a sustainable source of green energy but also form the basis for a sustainable energy policy that can reduce greenhouse gas emissions substantially (Basso et al., 2019). Along with the expansion of the biofuels sector, it is worth noting the prospects of metabolic engineering and current research projects. New technology for second-generation biofuels, such as utilizing lignocellulosic feedstocks, incorporates both economic and environmental factors. These technologies can effectively transform waste streams, closing food security and energy supply gaps. Furthermore, by cracking open non-food biomass resources, the industry can follow a cleaner paradigm that reduces reliance on conventional food crops. Thus, these efforts not only reinforce the significance of biofuels in meeting world energy requirements but also highlight the imperative necessity for ongoing innovation and regulatory support that promotes metabolic engineering operations. The future of biofuels is really based on this intersection of technology and sustainability and presents a viable path to a cleaner energy future (Walker et al., 2011) (Figure 7.15).

Figure 7.15. Types of biofuels exploring the future of renewable energy.

Source: https://www.coherentmarketinsights.com/blogimg/1742907999.png.

7.5.2. Bioplastics and Biodegradable Materials

Escalating environmental concerns regarding traditional petroleum-based plastics have triggered great interest in bioplastics and biodegradable materials. These new-generation substitutes, made from renewable biological resources such as plants and microbes, offer a promising solution to the ever-growing problem of plastic waste and fossil fuel dependence. The bioplastics can be manufactured from a range of different organic feedstocks, and as such are highly attractive not only for application in the packaging sector but also as feedstocks into durable goods manufacturing. The materials' flexibility allows potential alteration of their properties through advanced processing techniques, which would enhance their mechanical performance and make them suitable for specific market applications. Such versatility is the key to the potential of bioplastics to harmonize sustainability and use, advancing the argument that metabolic engineering has a central position in creating such sustainable alternatives (Cecchini et al., 2017). Turning food waste into biodegradable products is likely the strongest argument in favor of bioplastics. Utilization of food processing industry byproducts, i.e., peels and shells, has the added benefit of reducing wastage and creating valuable material. With an estimated third of the world's food going to waste, these materials made into biodegradable plastics can contribute substantially to a reduced environmental impact of food production.

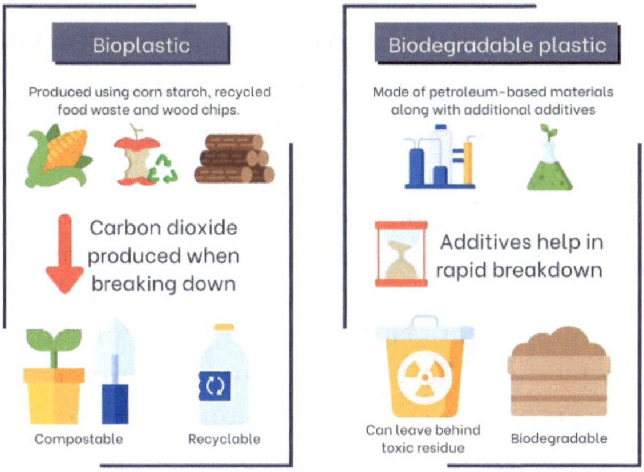

Figure 7.16. Bioplastics and biodegradable plastic.

Source: https://packmojo.com/cms_assets/optimized/large_bioplastic_vs_biodegradable_plastic_graphic_f97cf3a8a5.webp.

Such a process not only saves economic expenses on food waste disposal but also encourages innovative research and design in metabolic engineering. Researchers are increasingly engaged in the optimization of extraction pathways as well as the discovery of lucrative sources of high-yielding biopolymers, hence demonstrating the untapped prospects of these otherwise underestimated sources. The convergence of biorefinement techniques with metabolic pathways enlightens the promising possibilities of sustainability within the domain of material science (Cecchini et al., 2017). Research on bioplastics has been of particular interest, particularly concerning the microbial process of production. For instance, some bacteria have shown remarkable capabilities in producing polyhydroxybutyrate (PHB), a biodegradable polymer, under varying conditions. Bacillus cereus SE-1's ability to efficiently utilize various carbon sources for PHB production is just one example demonstrating the application of metabolic engineering to stimulate bioplastic production. Through the optimization of growth media and substrate utilization, researchers aim for higher yield and lower costs, paving the way for the commercialization of bioplastics. In addition, blending biopolymers like PHB with thermoplastic starch can facilitate improved material properties, widening their potential for use. The development of these technologies not only serves the immediate purpose of addressing the need for substitutes to traditional plastics, but it also highlights the importance of sustainable practices in the cycle of production and consumption (Figure 7.16).

7.5.3. Bioremediation and Environmental Applications

Combining bioremediation and metabolic engineering offers promising opportunities for the remediation of environmental pollutants and disasters. Metabolic engineering, or manipulating microbial metabolic pathways, enhances organisms' capacity to break down toxic substances such as heavy metals, petroleum, and persistent organic pollutants. By genetically engineering microbes to synthesize specific enzymes or pathways, researchers can significantly increase the degradation rates of such pollutants, thus providing a measure for solving oil spills and other similar environmental catastrophes. Not only does this procedure restore contaminated environments, but it also conforms to sustainability and environmental balance and allows us to address environmental issues in a more holistic manner. Hence, the development of metabolic engineering techniques has the potential to propel the move towards more efficient bioremediation processes that support the revitalization of degraded environments and foster sustainable environmental well-being. Despite the vast potential of bioremediation, several challenges hinder its widespread application. Conventional methods are unable to cope with low levels of contaminants such as asphaltenes and polycyclic aromatic hydrocarbons (PAHs), which slow down rapid clean-up operations during spillage such as oil spillage. Other difficulties in low nutrient bioavailability and bioaugmentation viability interfere with the effectiveness of microbial treatments. Evidence shows that the utilization of biosurfactants and thermophilic emulsifiers has the potential to enhance the biodegradation of complex hydrocarbons, but these processes must be explored further to gain maximum field applicability (Macaulay et al., 2013). Thus, overcoming these issues necessitates an interdisciplinary strategy that, in addition to the development in metabolic engineering, encourages collaboration between governmental and nongovernmental organizations in environmental research. Institutional participation in designing bioremediation techniques is critical in guaranteeing innovation and the right implementation of metabolic engineering. The Marine Board is one important agency in promoting collaborative marine science and developing common priorities among European marine research organizations (Boyen et al., 2025). By the application of scientific principles in policy and practice, such organizations can guide the development of a sustainable and effective bioremediation strategy. Encouraging public-private partnerships is central to financing field studies to validate new technologies and methods on-site. In addition, the engagement of various stakeholders can contribute to the enhancement of the dissemination of successful bioremediation practices, making the public more educated and promoting the involvement of the public in environmental remediation processes. With concerted efforts and mutual expertise, the successful implementation of bioremediation strategies can significantly minimize the magnitude of environmental degradation (Figure 7.17).

Figure 7.17. Bioremediation.

Source: https://microbenotes.com/wp-content/uploads/2022/01/Bioremediation-Definition-Factors-Types-Advantages-Disadvantages.jpeg.

7.6. ETHICAL AND REGULATORY CONSIDERATIONS

The landscape of metabolic engineering is significantly influenced by ethical as well as regulatory issues, especially with the advancements in genetic engineering tools like CRISPR-Cas systems. The potential for altering microbial genomes has led to debates on the ethics of such actions in natural systems. Some of the areas of concern from an ethical perspective are the potential risks of unintended consequences of genetic modification that may translate into ecosystems and human health. For instance, the application of genetically modified microbes in agriculture could lead to the annihilation of local biodiversity or the emergence of pathogen resistance. As is provided in recent summaries, solving these ethical issues is of utmost importance in facilitating public trust in biotechnological developments while ensuring the safe advancement of metabolic engineering applications (Yang et al., 2024). Regulation is also an essential aspect that accompanies advancements in metabolic engineering. The governance framework must adapt to accommodate the speed of biotechnological development, requiring a balance between innovation and safety. In virtually all countries, existing laws do not cover the specific challenges of genetic modification of organisms introduced into the environment. The deficiency can compromise the commercial viability of microbe strains generated by metabolic engineering. The GMO case illustrates the merits of comprehensive regulatory systems assessing potential hazards while facilitating useful strategies to enter the market. The integration of ethical considerations into such regulatory frameworks will facilitate the creation of policies that protect both the environment and public interest and encourage safe application of metabolic engineering techniques (Dhokane et al., 2023). Public sentiment will finally shape the future of metabolic engineering and regulation. The more informed society becomes of the implications of genetic engineering, the more stakeholders must engage with the public to have an enlightened discussion on the benefits and risks of these technologies. Such engagement can demystify the metabolic engineering technology and alleviate misconceived fears. There is a need for interdisciplinary, multidisciplinary collaboration among scientists, ethicists, policymakers, and society to develop stable guidelines to guide the application of these advanced approaches. By airing ethical and regulatory concerns freely and inclusively, the field of metabolic engineering can enhance its ability to solve problems and develop solutions to worldwide challenges in health, agriculture, and sustainability (Dhokane et al., 2023; Yang et al., 2024).

7.6.1. Ethical Implications of Genetic Modifications

The ethical consequences of genetic modifications, and particularly in the case of metabolic engineering, are multifaceted and must be carefully evaluated. As genetic technologies advance to allow for more precise changes to organismal genomes, one must weigh the potential benefits with the ethical dilemmas entailed. For example, although the potential of crop engineering to increase nutritional value or pest resistance for food security is promising, it at the same time also poses questions regarding biodiversity and environmental sustainability. This dual nature of innovation only mirrors a larger conversation into which stakeholders such as scientists, policymakers, and the general public need to enter in a critically engaged manner. The controversy surrounding the use of GMOs and their regulation underscores the need for an open system that encompasses not only the scientific and economic aspects but also society's values and the ethical considerations that inform such modifications. In addition, the interaction between the genetic modifications and equity makes the ethical scenario even more complex. As metabolic engineering technologies become increasingly accessible, unequal access to such technologies can increase health and agricultural inequities. It is only chosen populations that will reap the rewards of engineered crops or drugs, while marginalized communities are likely to be subjected to economic and health inequalities. There should therefore be a guarded approach wherein further progress in genetic engineering works not to consolidate cleavages in society but to bring about fair access and distribution. A careful ethical consideration of how these technologies are shared and deployed is necessary to avoid a situation in which the advantages of metabolic engineering are not equally shared, in effect undermining the very purpose of bettering health and wellbeing in a wide range of populations. Last, the ethical considerations with respect to the use of genetic alterations carry over into the arena of safety and long-term effects. With the employment of GMOs and the utilization of novel biotechnologies, there are concerns regarding the possible unforeseen ecological repercussions and long-term durability of designed traits. For instance, gene flow between transformed and untransformed organisms can lead to uncontrollable and unpredictable ecological transformation. Moreover, informed consent is imperative, as consumers are entitled to know about engineered products they may be consuming. These problems are to be tackled by both high-level scientific analysis and ethical commitment to transparency and accountability in metabolic engineering applications. Fostering an open platform where scientists, ethicists, and the public converge to share will be instrumental in creating policies that balance both the promise of genetic modifications and the ethical issues that surround them, so that future growth is for the greater good (Chehelgerdi et al., 2023; Malik et al., 2023).

7.6.2. Regulatory Frameworks Governing Metabolic Engineering

Regulation of metabolic engineering is of increasing significance as the science is progressing towards real-life use in healthcare, agriculture, and environmental conservation. Varying regulatory systems guide the development and utilization of engineered organisms for safety and efficacy. For instance, within the United States, organizations such as the Environmental Protection Agency (EPA) and the Food and Drug Administration (FDA) oversee the assessment of biotechnological products to prevent risks associated with GMOs. The agencies take a case-by-case approach, taking into account the specific nature and functions of engineered systems. This structure allows for a deregulated environment in a flexible manner, but raises concerns about the comprehensiveness and consistency of evaluation, affecting the progress being made in metabolic engineering at a time when rapid advancement demands unmistakable guidance and proper oversight. In addition to national regulations, international guidelines as well as agreements also have a significant role in regulating metabolic engineering. The Convention on Biological Diversity and the Cartagena Protocol on Biosafety emphasize prudence and international collaboration among states in the management of living modified organisms (LMOs) to foster cooperation and consensus on standards. These frameworks aim at mitigating transboundary impacts of metabolic engineering in consideration of potential environmental consequences and health risks generated by engineered organisms. The involvement of diverse stakeholders, from the government to industry players and non-governmental organizations, reflects the complexity in establishing a harmonized regulatory system. Therefore, despite these systems being designed to provide guidelines, divergence is possible, complicating global trade and the progress of scientific research in metabolic engineering. The ethical considerations in metabolic engineering contribute to the complexity of regulatory systems and necessitate an interdisciplinary system of governance. Ethical implications have a tendency to overlap with the beliefs of society, and therefore, public discussion and transparency are required components of the regulation process. As metabolic engineering becomes advanced, particularly in areas such as gene editing, societal impacts need to be constantly evaluated. The emergence of such biotechnologies as CRISPR/Cas9 has raised controversy regarding the risks and benefits of gene editing, necessitating thorough debates around the ethical and moral boundaries of scientific practice. Public dialogue and reorienting regulatory policies toward ethical values are stepping stones to establishing a stable framework that ensures public trust and encourages innovation. This interconnection of ethics with regulation is central to the ethical advancement of metabolic engineering, where societal values are maintained while adversaries progress.

7.6.3. Public Perception and Societal Impact

Public perception significantly influences the course and adoption of metabolic engineering technologies. Societal fear for these approaches is typically a reflection of previous experiences with GMOs. The majority of consumers express resistance towards genetically engineered products, which will lead to potential rejection of applications of metabolic engineering. In order to navigate this situation, a comprehensive risk and benefit analysis ought to include health and environmental, as well as societal and economic considerations. One needs to recognize that consumer engagement is not merely a product of scientific breakthrough but also the product of ethical considerations and public values. Thus, innovation must speak to social needs, and implication of public opinion at the early development phase is inherent. Research into the discovery of consumer priorities is required to ensure that the progress in metabolic engineering has societal approval and expectations, thereby enhancing the prospect of integration into everyday use (Coles et al., 2016). Also, the controversy surrounding metabolic engineering is often fueled by misinformation about its potential and limitations. An experiment with different scholars proved how words associated with synthetic biology are able to evoke different emotions in the public. For instance, metabolic engineering vocabulary, such as genetic engineering versus GMOs, is greatly capable of determining public opinion and instigating or lowering concerns. Furthermore, incorrect assumptions for what the scientists have done to create life further complicate the ability of the public to comprehend. Engaging in constructive discussion is essential to discrediting myths and outlining the actual scientific protocols under metabolic engineering. The discussion should be aimed at bridging the difference between scientists and society and emphasizing that scientists' ability is always overrated. Interdisciplinary meetings that encourage discussion can help heal the disability of communication, resulting in an informed public debate on metabolic engineering (Acevedo-Rocha et al., 2016). The social impact of metabolic engineering extends to policy-making and regulation. As developments are made, policymakers must contend with the ethical and pragmatic considerations of these technologies. Complete insight into metabolic engineering is what regulators must have in order to successfully create policies that promote innovation as well as safeguard public interests. The current dialogue between policymakers, scientists, and the general public can lead to more intelligent and equitable legislation that captures the potential and also the ethical complexities presented by metabolic engineering. In an environment of open communication, stakeholders can collaborate to create policies that reflect society's values, ensuring that metabolic engineering serves the public while advancing scientific understanding. Such interdisciplinarity will ultimately shape the future of metabolic engineering to enable its valuable contributions to society to flourish (Acevedo-Rocha et al., 2016).

SUMMARY

Overall, achievements in metabolic engineering have produced groundbreaking findings that are revolutionizing biotechnological and pharmaceutical industries. Through the synergy of computational modeling and experimental validation, metabolic engineers can now design cellular pathways for enhanced product yield and specificity today. For instance, one of the models presented in more current research demonstrates a keen understanding of continuous cell culture systems, i.e., how the ratio of cell density to dilution rate is a useful control parameter for metabolic properties across a broad class of perfusion systems (Fernandez-de-Cossio-Diaz et al., 2017). This insight not only paves the way for better bioproduction methods but also reveals the underlying complexities of cell metabolism that can be exploited to attain maximum efficiency in bioreactors. In addition, the study of tissue-specific metabolic profiles provides validation of the potential to translate metabolic engineering beyond traditional microbial systems. The importance of spatial heterogeneity of metabolic processes within tissues, particularly in the liver, highlights the necessity of modifying metabolic engineering strategies to meet the unique biological settings. The recent introduction of platforms such as the Metabolic Patterning on a Chip (MPOC) enables researchers to engineer intricate metabolic gradients, simulating *in vivo* scenarios and facilitating the investigation of tissue-specific metabolic functions (Eo et al., 2018). This level of sophistication in engineering techniques is an indication not only of the promise of metabolic engineering but also of its essential value in addressing issues met in drug development, personalized medicine, and regenerative therapies. Finally, with continuing advances in metabolic engineering, interdisciplinary strategies will be key to future developments. Collaborations among biologists, engineers, and data scientists will be able to use state-of-the-art analytics and machine learning to predict metabolic pathways and outcomes with levels of precision not previously thought possible. Emphasizing the interconnectedness of metabolic networks will also enhance our understanding of biological systems so that more efficient and resilient microorganisms or bioengineered tissue can be engineered. As the field advances, the impact of enhanced metabolic processes should permeate across an array of applications within sustainability, medicine, and biotechnology, making metabolic engineering an essential area of study in addressing some of humanity's most significant challenges.

Metabolic engineering is a new approach in biotechnology where metabolic pathways are manipulated and leveraged for a specific application. At the core of this field is controlling microbial hosts to optimize production of the valuable compounds and thereby address global issues, including the demand for renewable energy as well as sustainable production methods. Engineering strategies typically involve genetic control that alters pathways of carbon and nitrogen metabolism, enabling microorganisms to produce valuable products more efficiently. Notably, understanding the regulatory networks that manage these metabolic pathways is important as it directs the development of such interventions towards the development of improved-performing and productive microbes. Studies have indicated that extensive analysis of *E. coli* can result in significant innovations in rational metabolic engineering strategies, providing a foundation for sustainable chemical and biofuel production, hence the use of metabolic engineering in contemporary science (Zhang et al., 2025). Moreover, finding alternative sources of energy calls for new methods, most significantly for the production of hydrogen from lignocellulosic biomass using microbial systems. The complexity involved in this production process, including the requirement for effective pretreatment and enzymatic hydrolysis, is the challenge facing scientists. However, work has been done with organisms such as *Caldicellulosiruptor bescii*, which can produce hydrogen without the use of the conventional pretreatment methods. This consolidated bioprocessing reduces the process complexity, which may reduce costs and mitigate the impact of inhibitory toxic byproducts that suppress microbial growth during hydrogen production. The metabolic engineering of *C. bescii* to further augment its hydrogen production further attests to the potential of metabolic engineering in solving energy problems in a manner that is ecologically sustainable (Cha et al., 2024). Such results emphasize the importance of microbial metabolic knowledge to create effective biotechnological remedies. In general, the science of metabolic engineering not only facilitates the customization of microbial metabolism but also resolves greater ecological and economic problems. By close scrutiny of metabolic pathways and the application of genetic engineering techniques, scientists can unlock the potential of microorganisms to serve as biofactories for producing fuels and chemicals. Synergistic integration of metabolic engineering and microbial physiology knowledge facilitates a better understanding of optimizing production and overcoming new solutions for energy and resource sustainability. With ongoing research accelerating in this rapidly evolving field, interdisciplinary collaboration is more and more evident. This collaboration will be the key to surmounting the current technical challenges, paving the way for the implementation of renewable energy applications and for biotechnology's contributions to resolving the energy crisis (Cha et al., 2024; Zhang et al., 2025).

REFERENCES

1. Abdoli, S., Asgari Lajayer, B., Bagheri Novair, S., & Price, G. W. (2025). Unlocking the potential of biosurfactants in agriculture: Novel applications and future directions. *Sustainability.* https://www.semanticscholar.org/paper/2dfd1107fbb1ac11720e462dfd4a804e019e8d87 (accessed on 3rd October 2025).
2. Acevedo-Rocha, C. G., Chizzolini, F., Rothschild, L. J., Verseux, C., et al. (2016). Misconceptions of synthetic biology: Lessons from an interdisciplinary summer school. https://core.ac.uk/download/pdf/76424512.pdf (accessed on 3rd October 2025).
3. Ang, K. S., Borth, N., Brewer, J., Dobson, P., et al. (2016). Recon 2.2: From reconstruction to model of human metabolism. https://core.ac.uk/download/323064087.pdf (accessed on 3rd October 2025).
4. Babu, S. S., Jiyas, N., Sasidharan, I., Kumar, K. B., Thomas, V. P., Thomas, B., & Dan, M. (2025). Sustainable valorization of *Alpinia galanga* pseudostem fibers for characterization and materials engineering applications. *Scientific Reports, 15.* https://www.semanticscholar.org/paper/470b0f970d96f7bddd5067837cd1124b28daa7f7 (accessed on 3rd October 2025).
5. Basso, L. C., Basso, T. O., Walker, G. M., Walker, G., et al. (2019). The future of bioethanol. https://core.ac.uk/download/196581185.pdf (accessed on 3rd October 2025).
6. Beccari, T., Dundar, M., Gartland, J. S., Gartland, K., et al. (2017). Advances in biotechnology: Genomics and genome editing. https://core.ac.uk/download/293859551.pdf (accessed on 3rd October 2025).
7. Bernhardt, P. V., Kracke, F., Kromer, J. O., Rabaey, K., et al. (2016). Redox dependent metabolic shift in *Clostridium autoethanogenum* by extracellular electron supply. https://core.ac.uk/download/84043721.pdf (accessed on 3rd October 2025).
8. Boyen, C., Børresen, T., Dobson, A., Höfle, M., et al. (2025). Marine biotechnology: A new vision and strategy for Europe. https://core.ac.uk/download/pdf/29241634.pdf (accessed on 3rd October 2025).
9. Carneiro, D. C., Rocha, V. P. C., Damasceno, P. K. F., Barbosa, J. D. V., & Soares, M. B. P. (2024). Therapeutic applications of synthetic gene/genetic circuits: A patent review. *Frontiers in Bioengineering and Biotechnology, 12.* https://www.semanticscholar.org/paper/0a28758d2e1abda7760e2ebff6edf14572cb9b5f (accessed on 3rd October 2025).
10. Cecchini, C. (2017). Bioplastics made from upcycled food waste: Prospects

for their use in the field of design. https://core.ac.uk/download/98337583.pdf (accessed on 3rd October 2025).

11. Cha, M., Kim, J. K., Lee, W. H., Song, H., Lee, T. G., Kim, S. K., & Kim, S. J. (2024). Metabolic engineering of *Caldicellulosiruptor bescii* for hydrogen production. *Applied Microbiology and Biotechnology, 108.* https://www.semanticscholar.org/paper/ae9bfefb02eeb19eaa86220d75b803826b382e96 (accessed on 3rd October 2025).
12. Chai, M., Deng, C. X., Chen, Q., Lu, W., Liu, Y., Li, J., Du, G., et al. (2021). Synthetic biology toolkits and metabolic engineering applied in *Corynebacterium glutamicum* for biomanufacturing. *ACS Synthetic Biology.* https://www.semanticscholar.org/paper/99fefbdb325579911e1ed38b4cbe38f84567c7e9 (accessed on 3rd October 2025).
13. Chaneton, B., Gottlieb, E., Ruppin, E., Yizhak, K., Kiuru, M., Li, L., Lobo, C., et al. (2015). Modeling cancer metabolism on a genome scale. https://core.ac.uk/download/42363494.pdf (accessed on 3rd October 2025).
14. Chehelgerdi, M., Chehelgerdi, M., Allela, O. Q. B., Cosme Pecho, R. D., Narayanan, J., Rao, D. P., Thamaraikani, T., et al. (2023). Progressing nanotechnology to improve targeted cancer treatment: Overcoming hurdles in its clinical implementation. *Molecular Cancer, 22.* https://doi.org/10.1186/s12943-023-01865-0 (accessed on 3rd October 2025).
15. Coles, D., Dijkstra, A. M., Frewer, L., et al. (2016). Synthetic biology applied in the agrifood sector: Societal priorities and pitfalls. https://core.ac.uk/download/74206381.pdf (accessed on 3rd October 2025).
16. de Córdoba, P. F., Fuente, D., Gamermann, D., Montagud, A., et al. (2012). Automation on the generation of genome scale metabolic models. http://arxiv.org/abs/1206.0616 (accessed on 3rd October 2025).
17. Dhokane, D., Shaikh, A., Yadav, A., Giri, N., Bandyopadhyay, A., Dasgupta, S., & Bhadra, B. (2023). CRISPR-based bioengineering in microalgae for production of industrially important biomolecules. *Frontiers in Bioengineering and Biotechnology, 11.* https://www.semanticscholar.org/paper/d162fe509a0fea200c5228315fcda7edf1d19ff4 (accessed on 3rd October 2025).
18. Divekar, P. A., Srinivasa, N., Divekar, B. A., Kumar, R., Gowda, G. B., Ray, A., Singh, A. K., et al. (2022). Plant secondary metabolites as defense tools against herbivores for sustainable crop protection. *International Journal of Molecular Sciences, 23,* 2690–2690. https://doi.org/10.3390/ijms23052690 (accessed on 3rd October 2025).

19. Eo, J., & Kang, Y. B. (Abraham), Mert, S., Usta, M., et al. (2018). Metabolic patterning on a chip: Towards *in vitro* liver zonation of primary rat and human hepatocytes. https://digitalcommons.georgefox.edu/cgi/viewcontent.cgi?article=1056&context=mece_fac (accessed on 3rd October 2025).

20. Fakruddin, M., Prima, M. J., Chowdhury, T., Ferdous, U. T., Afroz, J., Shishir, M. A., & Fakruddin, M. (Correspondence). (2024). Nature's tiny chemists: Microorganisms as sources of next-gen active pharmaceutical ingredients (APIs). *Journal of Primeasia.* https://www.semanticscholar.org/paper/dde856e1f7b38f60a874b55950bbafde62aee036 (accessed on 3rd October 2025).

21. Fan, W., Ding, C., Huang, D., Zheng, W., & Dai, Z. (2024). Unraveling principles of thermodynamics for genome-scale metabolic networks using graph neural networks. https://www.semanticscholar.org/paper/44a37915ce910138caaf3edfadcbcbacfcce37a3 (accessed on 3rd October 2025).

22. Fernandez-de-Cossio-Diaz, J., León, K., & Mulet, R. (2017). Characterizing steady states of genome-scale metabolic networks in continuous cell cultures. http://arxiv.org/abs/1705.09708 (accessed on 3rd October 2025).

23. Fernie, A., & Morgan, J. (2013). Analysis of metabolic flux using dynamic labeling and metabolic modeling. https://core.ac.uk/download/210620021.pdf (accessed on 3rd October 2025).

24. Fox, K., Komor, A. C., & Rallapalli, K. L. (2020). Rewriting human history and empowering Indigenous communities with genome editing tools. https://core.ac.uk/download/323064893.pdf (accessed on 3rd October 2025).

25. Freitag, A., Gevorgyan, A., Mithani, A., Gevorgyan, A., Feist, A. M., Feist, A. M., Feist, A. M., et al. (2012). MultiMetEval: Comparative and multi-objective analysis of genome-scale metabolic models. https://core.ac.uk/download/9652011.pdf (accessed on 3rd October 2025).

26. Gartland, J. S., & Gartland, K. (2018). Contributions of biotechnology to meeting future food and environmental security needs. https://core.ac.uk/download/293884179.pdf (accessed on 3rd October 2025).

27. Guerrero Ramírez, J. R., Ibarra Muñoz, L. A., Balagurusamy, N., Frías Ramírez, J. E., Alfaro Hernández, L., & Carrillo Campos, J. (2023). Microbiology and biochemistry of pesticides biodegradation. *International Journal of Molecular Sciences, 24,* 15969–15969. https://doi.org/10.3390/ijms242115969 (accessed on 3rd October 2025).

28. Heinemann, M., Kümmel, A., Panke, S., et al. (2005). *In silico* genome-scale reconstruction and validation of the *Staphylococcus aureus* metabolic

network. https://core.ac.uk/download/pdf/12944401.pdf (accessed on 3rd October 2025).

29. Jahan, N. (N.), Javed, M. A. (M.), Khan, S. (S.), Manan, A., et al. (2016). A short review on the development of salt tolerant cultivars in rice. https://core.ac.uk/download/480549642.pdf (accessed on 3rd October 2025).

30. Jang, J., Baier, T., Kneip, J. S., Kruse, O., & Jin, E. (2025). High-yield zeaxanthin production in *Chlamydomonas reinhardtii* via advanced metabolic pathway engineering. *Biotechnology for Biofuels and Bioproducts, 18.* https://www.semanticscholar.org/paper/0362b12583c36430640894bc6cb0ea6f0aade84d (accessed on 3rd October 2025).

31. Karimov, N., Sarybaev, M. K., Kaipnazarov, A., Djumageldiev, N. R., Reymbaev, R., & Kholdarova, F. (2024). Historical development of construction techniques: From ancient architecture to modern engineering. *Archives for Technical Sciences.* https://www.semanticscholar.org/paper/09d41a9b4b7af69bc578d4e96deec3ce6e115c3a (accessed on 3rd October 2025).

32. Li, H., Yang, Y., Hong, W., Huang, M., Wu, M., & Zhao, X. (2020). Applications of genome editing technology in the targeted therapy of human diseases: Mechanisms, advances and prospects. *Signal Transduction and Targeted Therapy, 5.* https://doi.org/10.1038/s41392-019-0089-y (accessed on 3rd October 2025).

33. Macaulay, B. M., & Rees, D. (2013). Bioremediation of oil spills: A review of challenges for research advancement. https://core.ac.uk/download/42392380.pdf (accessed on 3rd October 2025).

34. Malik, S., Muhammad, K., & Waheed, Y. (2023). Emerging applications of nanotechnology in healthcare and medicine. *Molecules, 28,* 6624–6624. https://doi.org/10.3390/molecules28186624 (accessed on 3rd October 2025).

35. Nielsen, J., & Petranovic, D. (2025). Modeling for understanding and engineering metabolism. *QRB Discovery, 6.* https://www.semanticscholar.org/paper/0c507b0c532853db517a38c0798e56257058572c (accessed on 3rd October 2025).

36. Patel, A., Bettiga, M., Rova, U., Christakopoulos, P., & Μάτσακας, Λ. (2022). Microbial genetic engineering approach to replace shark levering for squalene. *Trends in Biotechnology, 40,* 1261–1273. https://doi.org/10.1016/j.tibtech.2022.03.008 (accessed on 3rd October 2025).

37. Rocha, I. (2013). Systems biology for the development of microbial cell factories. https://core.ac.uk/download/55627681.pdf (accessed on 3rd October 2025).

38. Sadanov, A., Baimakhanova, B., Orasymbet, S., Ratnikova, I., Turlybaeva, Z., Baimakhanova, G., Amitova, A. A., et al. (2025). Engineering useful microbial species for pharmaceutical applications. *Microorganisms, 13.* https://www.semanticscholar.org/paper/6b83d4412d5aa7c05bbf760883d7a24fc7cb8520 (accessed on 3rd October 2025).

39. Schiavon, M., Nardi, S., Dalla Vecchia, F., & Ertani, A. (2020). Selenium biofortification in the 21st century: Status and challenges for healthy human nutrition. *Plant and Soil, 453,* 245–270. https://doi.org/10.1007/s11104-020-04635-9 (accessed on 3rd October 2025).

40. Shi, M., Liao, P., Nile, S. H., Georgiev, M. I., & Kai, G. (2020). Biotechnological exploration of transformed root culture for value-added products. *Trends in Biotechnology, 39,* 137–149. https://doi.org/10.1016/j.tibtech.2020.06.012 (accessed on 3rd October 2025).

41. Sun, J., Zhu, L., Duan, L., Li, F., Guo, S., Bian, J., & Yang, L. (2025). Metabolic engineering of *Saccharomyces cerevisiae* for *de novo* biosynthesis of hydroxytyrosol and salidroside. *Applied and Environmental Microbiology.* https://www.semanticscholar.org/paper/8e18c0dc7944f4e7a8b6f26771d06c23e198b6fb (accessed on 3rd October 2025).

42. Suthar, P. (2025). Advances in gene editing techniques. *International Journal for Research in Applied Science and Engineering Technology.* https://www.semanticscholar.org/paper/9a208d4102c7d759cb1e4b72a93991ed050b76c4 (accessed on 3rd October 2025).

43. Tan, L., Jia, F., & Liu, Y. (2025). Advances in research on the role of gut microbiota in the pathogenesis and precision management of gallstone disease. *Frontiers in Medicine, 12.* https://www.semanticscholar.org/paper/9f9f256da328366432684c59c7898da920a5b80e (accessed on 3rd October 2025).

44. Tesfaye, M., Wang, E., Feyissa, T., Herrfurth, C., Haileselassie, T., Kanagarajan, S., Feussner, I., et al. (2024). Enhancing erucic acid and wax ester production in *Brassica carinata* through metabolic engineering for industrial applications. *International Journal of Molecular Sciences, 25.* https://www.semanticscholar.org/paper/7e332447eb67cc91454896f7c06be7cdb9a5ce68 (accessed on 3rd October 2025).

45. Van Der Straeten, D., Bhullar, N. K., De Steur, H., Gruissem, W., MacKenzie, D. J., Pfeiffer, W., Qaim, M., et al. (2020). Multiplying the efficiency and impact of biofortification through metabolic engineering. *Nature Communications, 11.* https://doi.org/10.1038/s41467-020-19020-4 (accessed on 3rd October 2025).

46. Van Montagu, M. (2020). The future of plant biotechnology in a

globalized and environmentally endangered world. https://core.ac.uk/download/287941311.pdf (accessed on 3rd October 2025).

47. Walker, G. M. (2011). 125th anniversary review: Fuel alcohol: Current production and future challenges. https://core.ac.uk/download/228176769.pdf (accessed on 3rd October 2025).

48. Wang, L., Jiang, X., Zhao, F., Duan, P., Li, Z., & Luo, Y. (2025). A review of adipose-derived mesenchymal stem cells' impacts and challenges: Metabolic regulation, tumor modulation, immunomodulation, regenerative medicine and genetic engineering therapies. *Frontiers in Endocrinology, 16.* https://www.semanticscholar.org/paper/e2985dfd36d5aa3903c9a926491c9abdbf4367b1 (accessed on 3rd October 2025).

49. Wang, T., Xue, H., Liu, H., Yuan, H., Huang, D., & Jiang, Y. (2025). Advancements in metabolic engineering: Unlocking the potential of key organic acids for sustainable industrial applications. *Frontiers in Bioengineering and Biotechnology, 13.* https://www.semanticscholar.org/paper/2a9e923a5b62408fc5dc624d48bea317c787b5f8 (accessed on 3rd October 2025).

50. Wang, Y., Chen, J., He, G., Yin, L., & Liao, Y. (2025). Unlocking the potential of flavonoid biosynthesis through integrated metabolic engineering. *Frontiers in Plant Science, 16.* https://www.semanticscholar.org/paper/5824e611bf70dc18a926d256a941fd52b7048301 (accessed on 3rd October 2025).

51. Wu, Z., Liang, X., Li, M., Ma, M., Zheng, Q., Li, D., An, T., et al. (2023). Advances in the optimization of central carbon metabolism in metabolic engineering. *Microbial Cell Factories, 22.* https://doi.org/10.1186/s12934-023-02090-6 (accessed on 3rd October 2025).

52. Yang, P., Condrich, A., Lu, L., Scranton, S., Hebner, C., Sheykhhasan, M., & Ali, M. A. (2024). Genetic engineering in bacteria, fungi, and oomycetes, taking advantage of CRISPR. *DNA.* https://www.semanticscholar.org/paper/40220f7501e061f1752d8a029d45b6b018eb2847 (accessed on 3rd October 2025).

53. Zhang, J., Fang, H., Du, G., & Zhang, D. (2025). Metabolic regulation and engineering strategies of carbon and nitrogen metabolism in *Escherichia coli. ACS Synthetic Biology.* https://www.semanticscholar.org/paper/004e5c4af9ac852ecc5ba72c1dd22c49d896984d (accessed on 3rd October 2025).

CHAPTER 8

SYNTHETIC CELLS AND MINIMAL GENOMES

Contents

INTRODUCTION

Synthetic cell and minimal genome research constitutes a cutting-edge new frontier within modern biotechnology, in which the fundamental principles of biological engineering intersect with advanced scientific research. Synthetic biology aims to construct organisms with specific functions employing techniques such as artificial gene network design and computational protein design. By taking an engineering-oriented approach, scientists can examine the complexities of biological systems, removing the underlying scientific principles that enable them. This science not just enables artificial forms of life to be manufactured but also provides us with greater capability to answer real-world challenges, such as the manufacturing of medicines. With the ability to synthesize de novo DNA becoming greater, the capacity to construct intricate biological entities is racing ahead at a rapid speed, which marks a milestone for synthetic biology's utility in other sectors (Heinemann et al., 2006). One of the most promising advancements made so far in this field is the capacity to engineer minimal genomes, which involves dramatically simplifying the genetic complexity of organisms but retaining major functions. This endeavor is a test to filter out and keep only those genes necessary for bare cellular survival, laying the groundwork for the assessment of life's basic necessities. It has been demonstrated by recent studies that it is feasible to reorganize the genomic makeup of organisms like *Escherichia coli*, holding the promise of having fewer operons but also exhibiting steadfast functionality across variable environments. Such research points to a central principle of synthetic biology: how to rewire transcriptional networks without losing cellular flexibility. Further, the ability to increase regulatory response rates by more than 20% in such redesign strains is the proof of concept for minimal genome designs (Carrera et al., 2013). The implications of these results extend far beyond academic interest. By advancing our understanding of minimalist genomes and synthetic cell technology, researchers can innovate groundbreaking applications in medicine, biotechnology, and the environmental sciences. For instance, genetically engineered microorganisms could be employed for the bioremediation of polluted environments or for the creation of renewable biofuels, drastically reducing some of the world's key challenges. In addition, synthetic cell research holds the potential for revolutionary approaches to disease therapy, including the engineering of targeted therapies for complex diseases. As synthetic biology continues to mature, the integration of minimal genomes into practical applications may very well redefine our interaction with living organisms and biotechnological processes, heralding a new era of innovation and discovery in the life sciences (Carrera et al., 2013; Heinemann et al., 2006).

Definition of Synthetic Cells

The concept of synthetic cells is a part of synthetic biology, the discipline that endeavors to engineer and construct new life forms from biological components. Synthetic cells are essentially man-made cellular systems that have been re-engineered or assembled to perform specific biological functions. Synthetic cells are designed to perform functions that range from the production of drugs to biorecognition and sensing the environment. Synthetic biologists use genetic engineering to create organisms with selected functionalities that might not be present naturally, and therefore expand the field of the native characteristics of life and biological systems. As scientists explore the components of living cells, they seek to disassemble the major components of life itself, raising concerns about the potential for new functions and the ethics of their construction and use. Central to cell synthesis is the concept of minimal genomes, or the identification and application of the smallest set of genes needed to sustain life processes. This concept addresses the possibility of designing cells with less genetic content, making it therefore possible for researchers to learn about the fundamental building blocks of cellular life.

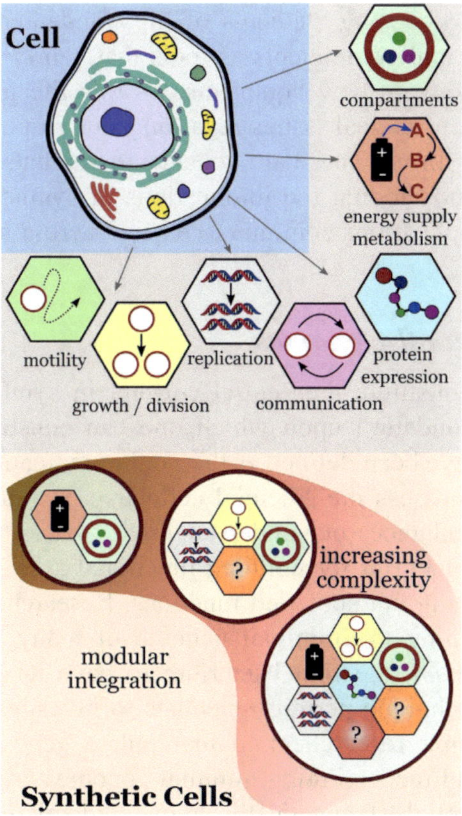

Figure 8.1. Modular approach for building synthetic cells with cell-like properties. The integration of functional modules creates synthetic cells with increasing complexity.

*Source:*https://onlinelibrary.wiley.com/cms/asset/915bd9d0-f653-4045-841b-5255ce547c42/anie202110855-fig-0001-m.jpg.

Recent work has indicated that one can design surrogate transcriptomic models with retained crucial behaviors of wild-type organisms under environmental stressors, affirming the adaptability and plasticity of synthetic cells (Carrera et al., 2013).

Genetic construct reengineering not only avoids identifying genes essential for survival but also opens the possibility of creating organisms with optimized traits tailored to specific industrial or clinical applications. Furthermore, synthetic cells can be viewed as a toolbox in the framework of life, as described in the synthetic biology literature (Deplazes-Zemp et al., 2018). This analogy suggests that living things can be understood as a collection of various functionalities, each equivalent to tools for specific functions. Synthetic biologists' attention is redirected toward choosing elements of life, fundamentally working with biological processes and components as tools that can be tailored, amplified, or fitted together to create new organisms. As this life philosophy emerges, it introduces important ethical issues regarding our responsibility in altering and engineering life forms. These are considerations that emphasize the need for careful debate surrounding the implications of synthetic cell technology, particularly as its applications continue to expand across society and research (Figure 8.1).

Overview of Minimal Genomes

Minimal genomes constitute the central concept in synthetic biology, to be employed as the foundation upon which one can construct synthetic cells. Minimal genomes have been defined as the minimum quantity of genes needed for an organism to possess the essential cellular processes and reproduction. This idea not only enhances our understanding of basic biological processes but also presents a blueprint for building functional artificial cells capable of replicating and carrying out specified functions. Researchers have progressed significantly in defining the minimal genome of many organisms, such as *Mycoplasma genitalium*, the first bacterium to have its genome engineered synthetically. The technique permits scientists to investigate gene interaction and function, allowing researchers to manipulate gene pathways in novel ways. Identifying and reconstituting minimal genomes are allowing scientists to tap into the potential of genetically engineered microorganisms that can be built for biotechnological applications and medicine. Understanding what limits minimal genomes is critical to the successful construction of synthetic cells. The functioning of minimal genomes illuminates the fine line needed to sustain life. For instance, while a compressed genetic apparatus upholds the simplicity of the organism, it could also threaten their capacity to adapt and recover from environmental stress. This inherent tension between minimalism and resilience reinforces the complexity entailed in genetic engineering for synthetic purposes. Furthermore, the characterization of minimal genomes allows identifying essential genes that can be employed as drug targets and in metabolic engineering. As researchers delve further into the metabolic pathways of minimal genomes, they discover intricate interconnects that reinforce the evolutionary significance of various gene functions, enabling better design of artificial life forms endowed with economically valuable features, like the synthesis of biofuels or detoxification. The use of minimal genomes within synthetic biology presents a confluence of potential for innovation across various scientific fields. As genome synthesis continues to advance, researchers can pair synthetic biology with other fields like computer science to further develop the production of intricate biological circuits and devices. Using this cross-disciplinary method, functionalized organisms may be created to address global challenges, including disease regulation and environmental conservation. Apart from this, as science is progressing, more relevant ethical concerns about synthetic genomes also crop up. Dual-use issues, environmental impacts, and equitable access to advancements all require a robust ethical foundation so that synthetic biology benefits all human beings. As the science of minimal genomes progresses, such controversies will chart the future course of bioengineering, influencing subsequent advances capable of altering society's benefits while respecting the subtle interaction of natural processes (Figure 8.2).

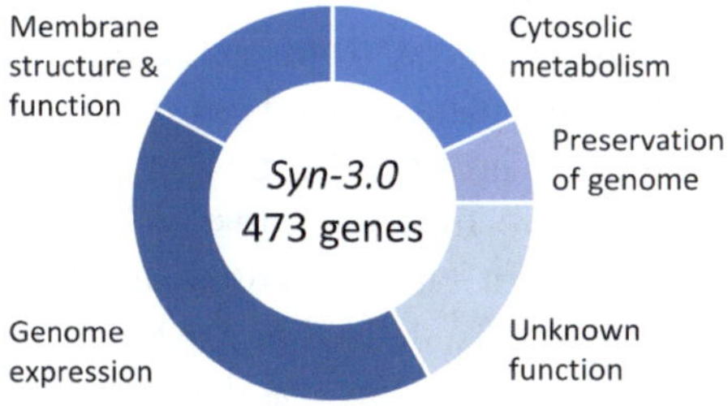

Figure 8.2. Gene functions in Syn 3.0, a minimal-genome variant of *Mycoplasma genitalium*.

*Source:*https://upload.wikimedia.org/wikipedia/commons/5/5b/Syn3_genome.svg.

Importance of Studying Synthetic Biology

The question of synthetic biology is at the core of the understanding of the essential characteristics of life that define life itself, particularly in the case of synthetic cells and minimal genomes. These researchers aim to boil down the complexities of living systems to their core factors with a focus on survival-critical processes. Through the engineering of minimal cells, researchers can understand the intricacies of cell mechanisms like DNA segregation. For instance, the assembly of a minimal segrosome, which is necessary for faithful separation of DNA during cell division, is a fine example of how synthetic biology can produce information regarding fundamental processes of life. Not only do such synthetic systems reveal knowledge regarding rules of life, but they also have practical applications in biomimetic systems that could utilize these fundamental processes towards technologies (Hürtgen et al., 2018). In addition, synthetic biology research plays an important role in the current debate regarding the origin of life, synthesizing erstwhile contradictory theories into one harmonious understanding. The union of replicator-first and metabolism-first theories is revolutionizing our understanding of life's evolution, bringing a more sophisticated perspective that allows for diverse temporal dynamics found in biological processes. This perspective lays specific emphasis on intermediate timescales—e.g., those at which metabolism and evolution occur—describing how life is able to emerge from less complex chemical interactions evolving towards more sophisticated self-organizing systems (Froese et al., 2013). Through exploring these processes in artificial environments, scientists can simulate and dissect the emergence of life-like processes, expanding our knowledge of biological complexity and the likely paths by which life gave rise to the first living organism. Besides, synthetic biology is also crucial in addressing global challenges, from medicine to environmental sustainability, through innovative solutions derived from fundamental biological research. Synthetic cell design and engineering expand our toolkit for drug development, biofuel, and bioremediation. Through the design of minimal genomes that can perform essential metabolic functions, researchers can potentially create organisms optimized for specific tasks, such as the degradation of toxins or the production of pharmaceuticals. It is this aspect of synthetic biology that bridges the gap between abstract science and practical application, showing how an understanding of life's building blocks can become revolutionary breakthroughs. The scope of such research is immense, potentially offering solutions to the most daunting problems associated with health, energy, and conservation of the environment, this topic has to be learned in contemporary society.

8.1. HISTORICAL BACKGROUND

The history of synthetic biology dates back to basic genetic studies that began to arise in the mid-20th century. Early on, the finding of the DNA structure and the deciphering of the genetic code formed the foundation for comprehending how data was stored in living things. Such knowledge helped spur breakthroughs in molecular biology, especially with the development of tools such as recombinant DNA and gene editing. Such strategies paved the way for synthetic biology since they made it possible to manipulate genetic material with precision. While researchers were exploring the minimal genome strategy, they began searching for critical genes necessary for life, laying the foundation for making synthetic cells. This development not only deepened the understanding of biological phenomena but also illuminated the prospect of constructing life structures with a certain functionality and thus towards providing an affluent historical context to contemporary advancements in synthetic cells and minimal genomes. The paradigm of thinking shift towards achieving living systems as a dynamic rather than fixed structure has been a major development in the field. Initial studies focused not only on the significance of coding regions but also of various cellular functions required to sustain life.

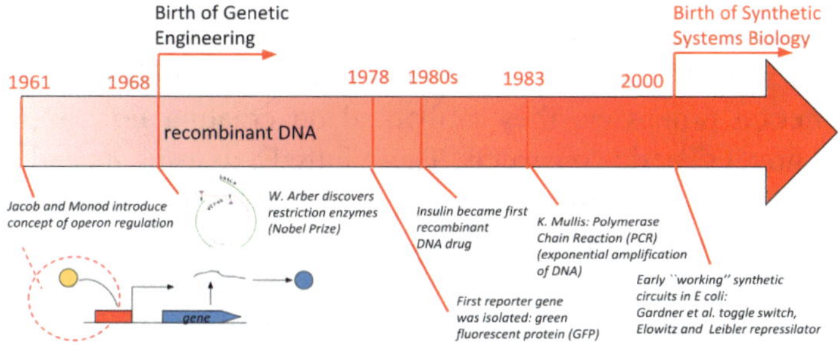

Figure 8.3. History of synthetic biology.

*Source:*http://www.sontaglab.org/images/history_synthetic_biology_adapted_from_many_sources_rev1.png.

Research into minimal genomes established that such life-sustaining functions, similar to theoretical Maxwell's demon, play an essential role in the management of information and energy within cells. Besides fundamental cell processes, minimal genomes incorporate intricate processes like nutrient transport and ribosomal assembly that are essential for cellular independence and efficiency. As minimal genome research has been conducted, it emphasizes the intricate dependence of genomic information on cellular behavior. These findings have utterly transformed the way scientists now approach the design and assembly of synthetic cells, illustrating an evolution in the science that is itself historical and pedagogic. Moreover, the visions of systems biology are a breakthrough in the history of synthetic biology. The integration of voluminous amounts of molecular information to model biological processes is a paradigm for predictive and controlled approaches to understanding life. Systems biology, since its inception, has aimed at integrating traditional genetic understanding with computational paradigms towards a comprehensive understanding of life's intricacies. This integration, though, has created ethical concerns with special fear surrounding possible consequences of life form tampering. With the evolving field, it is increasingly necessary to consider the socioethical dimensions of creating synthetic life, which can transform life and biotechnology forever in people's minds. Examining these transdisciplinary connections allows researchers to better anticipate and control the societal impacts of advances in minimal genomes and synthetic cells, gaining responsible and informed guidance for the future (Calvert et al., 2013; Danchin et al., 2019) (Figure 8.3).

8.1.1. Early Experiments in Synthetic Biology

The beginning of synthetic biology as a science originated from early experiments that aimed to control biological systems at the genetic level. The pioneering studies focused on learning about the fundamental building blocks of life, which were DNA, RNA, and proteins. As researchers delved deeper into cell function processes, they embarked on creating new approaches to synthesizing novel genetic constructs. These initial experiments had a tendency to imply the building of miniature genetic circuits that were designed to produce new phenotypes in model organisms such as *Escherichia coli*. This was no experimental benevolence; it laid important foundations for genetic engineering that would enable subsequent progress in the build-up of synthetic cells and minimal genomes. The implications of such experiments were substantial, pointing to the beginning of a more directed and engineered period of life sciences. One of the experiments that bears witness to the promise of synthetic biology is the design of high-throughput genetic interaction studies, and this was beautifully illustrated by the approach which is known by the name *GIANT-coli*. This new technique allowed the controlled generation of double mutants in *E. coli*, thereby enabling researchers to untangle complex genetic interactions significant in biological contexts. By tracking variation in the size of colonies, researchers could quantify cellular fitness and interaction dynamics. Such high-throughput screening not only accelerated genetic interaction data generation but also facilitated the easy detection of new synthetic interactions. Thus, early experiments like *GIANT-coli* established a firm ground that has now taken the center stage in shedding light on our understanding of gene function as well as on synthetic manipulation of living organisms (Roguev et al., 2008). Also, the search for minimal cells necessitated the exploration of minimal cellular components, prompting studies of DNA segregation systems. The minimal segrosome theory emerged out of the need to account for how genetic information could reliably be distributed throughout these engineered organisms. The first experiments took into consideration a number of different prokaryotic segregation systems, those that were capable of conferring genetic stability in minimized cellular models. An example is the type II segregation system, R1-ParMRC, which proved to be a prime candidate. Researchers managed to reconstitute and characterize this system *in vitro* successfully, demonstrating its potential applications in the design of biomimetic systems, e.g., synthetic cells (Hürtgen et al., 2018). This innovation represented a bigger objective in synthetic biology: developing life-like objects that have minimal genetic baggage yet retaining critical functions, thereby creating opportunities for future synthesis in synthetic and minimalist genomes.

8.1.2. Key Milestones in the Development of Synthetic Cells

The way to the construction of artificial cells has been one of milestone discoveries that signal advances in biological engineering and the molecular manipulation of life itself. Synthesis of minimal genomes is among the earliest breakthroughs in this area, a milestone step of prime significance in determining what are the minimum requirements for cellular life. Researchers focused on deconstructing the genome of the bacterium to identify the minimum genes needed for life, leading to the creation of minimal life forms that have the capability to continually grow. Breakthroughs were made in 2010 when the J. Craig Venter Institute created a complete bacterial genome and inserted it into a host cell to initiate the first synthetic life form. This early achievement set a precedent for the field, demonstrating the potential for making life from scratch and the promise that synthetic cells could hold to advance medicine and biotechnology. Advances in gene-editing technologies, including the development of CRISPR-Cas9, have further driven progress in the field of synthetic biology more recently. The ability to edit genomes with unprecedented precision has enabled researchers to alter the genetic makeup of synthetic cells to accomplish something particular or produce a material of interest. The technology has ushered in the possibility for synthetic cells to new frontiers in applications such as drug development, biosensing, and bioremediation. As synthetic cells mature, the equipment and methods employed to build them advance further, opening up the possibility of increasingly intricate systems that simulate natural cellular processes. The synergy between synthetic biology and cutting-edge technologies promises to improve our capacity to design living systems that can react to environmental stimuli or perform particular therapeutic functions, extending beyond traditional uses. Another critical advancement in the synthesis of synthetic cells was the establishment of standard protocols for biocontainment and assessment of engineered biosystems. Secure development and use of synthetic cells depend upon the use of safety mechanisms against unwanted ecological impacts or biosecurity threats. Global workshops and expert panels, such as the MicrobiomeSupport project ones, are places where institutional frameworks for the regulation of synthetic biology research are being debated, thereby facilitating responsible innovation. The argument is proof of growing consensus on the need for public engagement and regulation of synthetic biology, and this underlines the social implications of creating synthetic life. As scientists continue to push the boundaries of what synthetic cells can accomplish, guidelines are paramount for addressing ethical challenges while ensuring that experiments are effectively translated into practical applications in health, industry, and environmental stewardship (Figure 8.4).

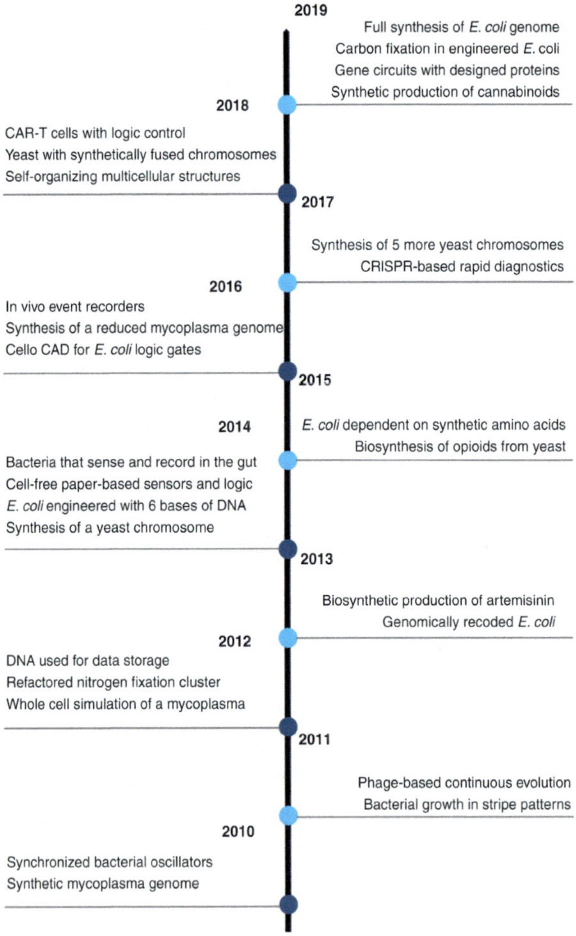

Figure 8.4. Synthetic biology and minimal cells: an historical perspective.

Source: https://media.springernature.com/lw1200/springer-static/image/art%3A10.1038%2Fs41467-020-19092-2/MediaObjects/41467_2020_19092_Fig1_HTML.png

8.1.3. Evolution of Minimal Genome Research

The minimal genome idea has come a long way since its inception, leveraging progress in genetics, molecular biology, and synthetic biology. Scientists have been driven by the need to understand the essence of life and thus explored the absolute minimum to sustain cellular functions. Through minimal genome construction, scientists attempt to determine the minimum genes required for reproduction and survival, which would map out the most primitive life blocks. This reductionist approach not only enhances our perception of cellular machinery but also enables us to design and engineer synthetic cells with specific functions. One of the main areas of this endeavor is the synthesis of cellular structures that contain biological processes and therefore merge the limitations of biological and synthetic systems. Findings from minimal genome research will resonate across various fields of science, from cell biology and biotechnology to astrobiology, inspiring greater interest in life's underlying mechanisms (Rothschild et al., 2024). With ongoing development in minimal genome research, researchers encounter a range of challenges that inform synthetic cell advancement.

At the core of the success of this field is the identification and reconstitution of gene sets that can replicate core processes of life, such as metabolism, growth, and stress response. Although great strides in synthesizing the minimal cellular components have been made, the task of reconstituting these components into working systems remains. The development of such methodologies to create minimal genomes in an efficient way remains a serious bottleneck towards the final aim of synthesizing a living cell de novo. Future studies will need to concentrate on scaling up these synthetic processes as well as examining the interactions and behaviors of these minimal organisms in various environments, enabling them to adapt and respond as natural cells (Rothschild et al., 2024). The promise of the successful assembly of synthetic cells with minimal genomes extends far beyond basic biological research. The possibility of engineering living organisms from the very components of molecules has the prospect to radically transform disease therapy, environmental management, and bioengineering applications. Minimized genome microbes can be employed as factories for biopharmaceuticals, environmental biosensors, or even new materials, and thus their use is multifaceted. Moreover, the experience accumulated in the production of reduced cellular structures reinforces the idea that it is essential to understand evolutionary processes to build life-like systems. With researchers increasingly making headway in this field, biology-bio-synthetic system interactions will likely usher in revolutionary advances in medicine and industry capable of redesigning societal expectations and interactions in nature. Thus, the advancement of minimal genome science must be framed not only as a scholarly pursuit but also as an entry into groundbreaking applications leveraging the principles of synthetic biology (Rothschild et al., 2024).

8.2. THE SCIENCE OF SYNTHETIC CELLS

The science of synthetic cells and minimal genomes is a revolutionary interface of biology and engineering that pushes the boundaries of our understanding of life itself. By synthesizing the components of cells and reengineering genomes, scientists are not only exploring the fundamental nature of cellular function but also unlocking the gates to novel applications. Recent research demonstrated that by employing advanced computational design, it is possible to construct substitute transcriptomic models maintaining the functional characteristics of cells under various environmental contexts. These findings demonstrate that altered organisms can maintain critical processes of life even as their genetic blueprints are radically rearranged. These are capabilities that have very wide-ranging implications in industries from medicine to biotechnology, where microorganisms might be tailored on a bespoke basis for particular functions, such as pollution remediation or targeted therapies (Carrera et al., 2013). Introducing even more complexity into synthetic biology is that scientists are currently applying engineering principles to design and construct biological systems with precise engineering, including artificial cells. This discipline calls on an engineering-style philosophy that marries purposefulness with deep scientific understanding. The modular structure of biological components enables researchers to apply strategies like electronic engineering, i.e., toggles or oscillators, in designing dynamic biological reactions. Additionally, the advances made in de novo DNA synthesis also enable the creation of more advanced synthetic systems that can switch theoretical ideas into practical reality. The synthesis of molecular modularity and computational design significantly enhances the potential of synthetic biology to design new living organisms, driving the development of engineered organisms that can sense and respond to selected environmental stimuli or operational conditions (Heinemann et al., 2006). The more advanced synthetic cells are, the more potential there is for designing minimal genomes—lean, mean versions of conventional genomes that support life processes essential for existence. These small-scale structures provide new insights into the fundamental requirements for life, enabling scientists to outline the components necessary for viability. This recent work in redesigning genetic networks in organisms like *Escherichia coli* demonstrates this effort, demonstrating that it is possible to cut down on genetic information by reducing the number of operons while preserving function. This study defines a new approach to genomic reorganization, in which reengineered genomes react faster to environmental stimuli, thereby creating an improved responsive cellular framework. Such works confirm the revolutionizing capability of synthetic cells in reshaping our understanding of life's complexity and versatility in a continuously evolving world (Carrera et al., 2013) (Figure 8.5).

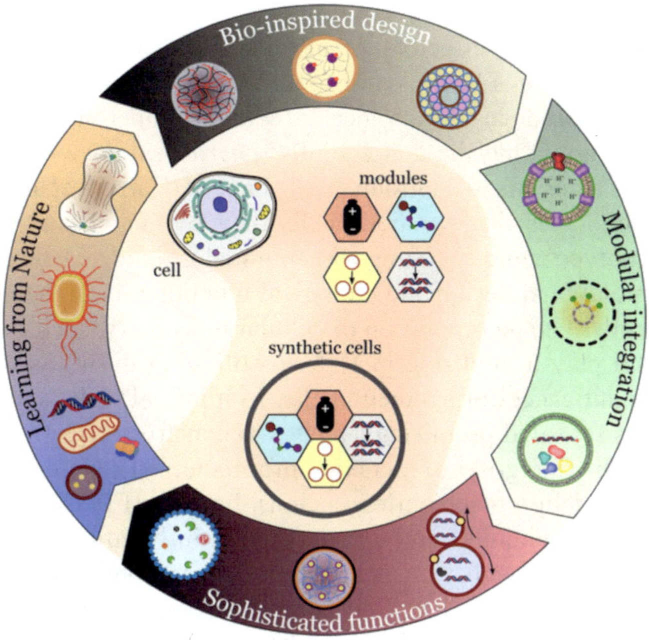

Figure 8.5. Synthetic cells: From simple bio-inspired modules to sophisticated integrated systems.

Source: https://onlinelibrary.wiley.com/cms/asset/833a376c-b5ba-4212-97b0-9d567a6b7f57/anie202110855-toc-0001-m.jpg.

8.2.1. Components of Synthetic Cells

Synthetic cells in synthetic biology are constructed with a focus on minimal genomes upon which life at the most fundamental level is understood. Synthetic cells are constructed through staged assembly of essential cell components, including nucleic acids, proteins, and cell membranes, to create viable forms that are similar to natural organisms. By adopting a minimalist approach to design, scientists hope to maintain the delicate balance between simplicity and functionality, ultimately leading to an even deeper understanding of cellular functions and the underlying dynamics of life. As has been observed through recent studies on evolution, engineered genomes can maintain their basic metabolic function while adapting a more efficient organizational framework, providing new avenues for the development of synthetic life forms that can be designed to address various environmental stresses (Carrera et al., 2013). Key features of synthetic cells are their gene expression and genetic design, in addition to regulatory functions. Synthetic genome design has a preference for utilizing coding sequences ordered in a way so as to maintain significant transcriptional behaviors, effectively rearranging genes such that functionality is not impacted. This has been demonstrated by experiments where scientists have been able to reduce the number of regulated operons by a significant percentage by rearranging the genome of *Escherichia coli* without disrupting suitable cellular reactions to environmental changes (Carrera et al., 2013). Such development indicates that synthetic cells may be engineered to produce a particular phenotype while operating with minimal genetic makeup, marking the union of computational design and biological functionality in the generation of synthetic biocells. The incorporation of cellular machinery, such as segregation systems, which play a vital role in DNA partitioning during cell division, is another important aspect of synthetic cells. In minimal cells, studies of DNA segregation has revealed promising avenues to construct effective segrosomes that maintain the integrity of genetic information. The study of multiple prokaryotic segregation machineries, such as the type II ParMRC system, has revealed that these subunits can be reconstituted in biomimetic environments to yield stable DNA transmission. Apart from that, by combining these systems with energy-scavenging and regenerating abilities, scientists can enhance their efficiency, thereby establishing the boundary for what can be achieved in replication and survival by synthetic cells. This encapsulation of complex molecular machinery is one step towards fully achieved synthetic biological systems.

8.2.2. Methods for Constructing Synthetic Cells

Construction of artificial cells, which is the theme running through synthetic biology, employs several cutting-edge methods based on the mechanisms of engineering sciences and biological sciences. DNA design and synthesis, or the blueprints of cellular function, underlie the very basis of this approach. De novo DNA synthesis advances allow scientists to create custom-designed sequences that specify specific proteins to allow them to build fundamental cellular components. The engineering-oriented strategy allows for the evolution of specific biological functions, such as gene regulation and signaling systems. In addition, artificial gene network integration enables cell design to execute complicated behaviors in terms of oscillations and toggle switches, hence modeling natural cell processes and behavior in an ordered and controlled manner. This capacity for design at the DNA level is the ultimate expression of the synergetic relationship between engineering and biology to enable enhanced synthesis of cells (Heinemann et al., 2006). Additionally, the process of construction is extended to protein engineering, a technique that supplements DNA synthesis with a focus on the functional assembly of proteins. This part of synthetic biology is defined by the rational design of protein interactions, with scientists being able to develop modular proteins that can be recombined like the interlocking blocks of Lego. By adopting this module concept, scientists can redesign protein functions to incorporate the intended cellular responses, making synthetic cells more flexible and functional. The connectivity of functional domains between proteins allows for the creation of entities with intricate signaling and metabolic functions, addressing the inherent uncertainty of biological systems, finally. This engineering design enables synthetic biologists not only to construct but also to improve and optimize protein design in synthetic cells, thereby pushing biological engineering and application in various fields (Freemont et al., 2014). Also, the combinatorial use of computational software with experimental methodology further improves the design framework for the assembly of synthetic cells. The implementation of a biological design cycle, characterized by iterative cycles of specification, modeling, analysis, assembly, and testing, reflects this integrated strategy. Computational models are employed at each step by scientists to predict and optimize cellular behavior, which is key in order to bridge the complexity that is typically present in biological systems. The combination of computational design and experimental verification renders the synthetic cells stronger, allowing the systematic exploration of their function. This systematic incorporation not only streamlines the construction process but also enables continuous optimization of synthetic cell designs for a variety of applications, like drugs or biofuels. In this way, the vast set of tools secured in synthetic biology constitutes an essential milestone in the pursuit of minimal genomes and synthetic cellular constructs (Heinemann et al., 2006).

8.2.3. Applications of Synthetic Cells in Research

Synthetic biology has also gone a long way, particularly in utilizing synthetic cells for scientific experiments. The artificially constructed biological machines, which have been created by deliberate construction of minimal genomes, can perform various kinds of functions that biological entities cannot perform. Artificial gene networks for performing certain tasks are among the most important areas of study that help understand genetic and biochemical processes in a more profound way. For instance, artificial cells can be employed as model systems to investigate the complex dynamics of cellular signaling networks so that scientists may dissect complex cellular behavior under controlled situations. With the implementation of modular components as envisioned in the concepts of synthetic biology, scientists can assemble functional components in a heterologous manner and regulate synthetic cells, promoting basic science and industrial applications in the fields of gene therapy and biomanufacturing (Heinemann et al., 2006). Another significant application is the generation of synthetic cells that can be tailored for drug delivery and therapeutics. By packaging drugs in artificial cell bodies, researchers are able to design the specificity with which these therapies journey to their locations, limiting the side effects of conventional delivery methods. The targeted delivery system leverages the capacity of synthetic cells to respond to specific biological signals or environmental cues, releasing therapies only where they are required. Additionally, the engineering of minimal segrosomes, as seen in recent investigations, showcases potential strategies for stabilizing genetic materials within synthetic cells. Such systems could revolutionize how drugs are formulated and administered, leading to more effective treatments for diseases while simultaneously deepening our understanding of cellular mechanisms in health and disease (Hürtgen et al., 2018). Moreover, synthetic cells offer fundamental platforms to investigate fundamental biological processes, permitting scientists to mimic life-like systems while reducing complexity. Their ability to incorporate minimal genetic elements allows researchers to examine essential functions necessary for life, such as replication, metabolism, and motility. This minimalist approach separates individual cellular activity so that the relationship between genetics and function in living systems may be more clearly understood. For instance, reconstitution of segregation systems within artificial cells sheds light on DNA partitioning mechanisms. By encasing core biological principles, such artificial structures not only enhance our insights into life sciences but also offer the gateway to new applications ranging from environmental biosensing to biofabrication of personalized biomolecules. Lastly, the current research on synthetic cells holds promise to reveal new frontiers in biological knowledge and application (Hürtgen et al., 2018).

8.3. MINIMAL GENOMES: DEFINITION AND CHARACTERISTICS

Minimal genomes refer to a term that captures a significant advance in synthetic biology, the possibility of understanding simple life processes by reducing organisms to their core genetic contents. Defined as the fewest number of genes necessary for a cell to sustain life and reproduce, minimal genomes give us a roadmap for building synthetic cells. Minimal genomes may be derived from a variety of organisms but are often obtained through extensive genomic reduction with retention of essential metabolic and replicative functions. By focusing on the minimal genes that support cellular viability, researchers aim to create minimal forms of life that can inform the boundary conditions of life and simplify the complexity of biological interactions. These experiments not only propel the evolution of minimal cells but also advance our knowledge of cellular structure and evolution, which has the potential to greatly impact biotechnology and metabolic engineering. One of the characteristics of minimal genomes is their ability to maintain fundamental biological processes at the cost of dispensable ones. By this reductionist approach, scientists can dissect the complexity of cellular metabolism and gene regulation into more manageable pieces. Constraint-based models of metabolic networks can be utilized by researchers to demonstrate how metabolic pathways are hierarchically arranged, defining the exchange of metabolites between organisms.

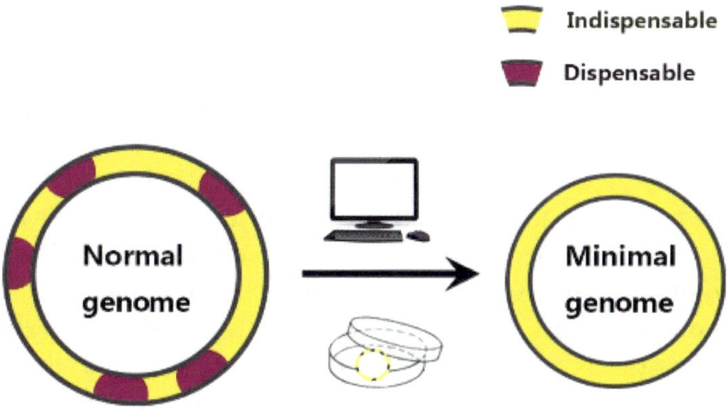

Figure 8.6. Synthetic biology used to create a minimal genome.

Source: https://bioethicsobservatory.org/wp-content/uploads/2016/06/minimal-genome.png.

The ability to selectively target metabolic subnetworks, as defined in research, allows core pathways supporting growth to be detected, hence giving rise to improved comprehension of the metabolic function embedded within minimal genomes (Wedmark et al., 2024). This pathway analysis further facilitates examining potential applications in metabolic engineering and synthetic biology, illustrating how cellular function may be controlled for specific biotechnological applications. Furthermore, identification of certain genetic elements that control critical cellular processes provides new prospects for therapeutic and engineering design. For instance, studies of immune response and cancer pathways have clarified the potential of decreased genomic blueprints in designing targeted therapies for diseases such as cancer, whose development is characterized by pervasive immune evasion (Muzzi et al., 2021). By leveraging information gathered from minimal genomes, researchers can design approaches for increasing immune activation and targeting unique biomarkers. The fundamental information gained from minimal genomes has the capability to revolutionize the way new organisms are designed, which would eventually find its way into new applications in the fields of healthcare, environmental stewardship, and biotechnology. The ongoing study of these fundamental properties not only underscores the significance of minimal genomes but also underscores their relevance to future synthetic biological advances (Figure 8.6).

8.3.1. What Constitutes a Minimal Genome

The minimal genome is the core concept of synthetic biology, which is the minimum amount of genetic code needed to sustain a living cell. In the stepwise process of simplifying cellular complexity, scientists have also made significant strides to identify which genes are essential for simple cell functions such as replication, metabolism, and response to the environment. These compact genomes not only yield blueprints for cellular syntheses but are also employed as laboratories to probe fundamental biological ideas. By creating organisms with fewer genes, scientists can analyze how much they can reduce a genome without annihilating the cellular organism. The minimal genome project highlights the fine balance between cellular function and genetic information, raising basic questions regarding the very nature of life itself and what constitutes life processes in cellular organisms (Ding et al., 2014). Research on minimal genomes also significantly enhances our understanding of cellular behavior and synthetic cell design. Because synthetic biology aims to make life forms that are tailor-made for specific functions, learning from minimal genomes is applied to guide the construction of synthetic equivalents. Artificial cells are a model system here, made up of cell parts that are required of them but still simpler to manipulate and understand. Through exploiting insights from minimal genomes, researchers can construct synthetic cells that utilize biological membranes, transcriptional apparatus, and functional genetic modules—all working to define the essential characteristics necessary for life (Heinemann et al., 2006). Pointing out this transition from abstract theories to concrete applications is a demonstration of how determining minimal genomes as the foundation for more advanced synthetic biology research makes it possible to advance in biomedical applications, biomanufacturing, and environmental biotechnology. Furthermore, the designation of a minimal genome raises significant questions regarding synthetic evolution and cellular adaptability. As researchers test the boundaries of minimal genomes, questions arise about how these synthetic organisms will evolve and adapt to environmental variation. The minimalist approach is interested in what genetic elements are absolutely required for minimal survival, but it also questions the investigation of how even the reduced forms of life attain dynamic qualities and evolution over the passage of time. Cellular communication, duplication, and evolutionary potential in a minimalist genome can provide important insights into the mechanisms of life at its most basic level. Ultimately, this study not only confirms the significance of minimal genomes in constructing synthetic cells but also challenges expectations of what life is, demonstrating that the essence of living things may be derived from a relatively short list of genetic code (Ding et al., 2014).

8.3.2. Examples of Organisms with Minimal Genomes

The smallest genome organism is extremely relevant to research, since it provides a glimpse into the very fundamentals of what life is based upon, specifically for synthetic biology. Such organisms usually have more compact genetic codes, which enable them to live in highly specialized, nutrient-rich environments. *Mycoplasma genitalium*, for instance, has one of the smallest described genomes of any free-living organism with approximately 580 kilobase pairs. Its minimal genome emphasizes the essential genes necessary for life, paving the way for the research of biological minimalism. By studying these minimal genomes, researchers can identify essential functionalities and metabolic processes, which will contribute to progress in synthetic cell engineering. Research on these organisms not only enhances knowledge about cellular life but also provides the platform for designing synthetic cells and organisms with genome features for specific applications in biotechnology (Bernstein et al., 2018). *Nanoarchaeum equitans* is another noteworthy organism with a compact genome and presents a unique perspective on genome reduction. This microorganism, with the record-breaking very small genome of approximately 491 kilobases, is an extreme obligate symbiont with host dependencies. Such mutual dependence underscores metabolic interaction, which has been a subject of keen interest for researchers. Studies on Nanoarchaeums metabolic adaptations, in the context of microbial society, demonstrate the opportunities for doing research with minimal genomes to analyze wider biological networks. Intersecting networks and interdependencies of such minimal organisms are able to yield critical information on evolutionary processes and ecological interactions that further inform designs for synthetic biological systems (Heinemann et al., 2006). Implications of minimal genomes extend to real-world applications in the field of synthetic biology. Synthetic biology's engineering approach seeks to construct complex biological systems from the knowledge gained through the analysis of organisms with compact genomes. For instance, the simplicity of these organisms may lead to efficient design principles for constructing artificial gene networks and metabolic pathways. By leveraging naturally occurring efficiencies in minimal genomes, researchers can leverage engineering principles that enable the assembly of new biological systems with particular functions. This approach is pioneering in biotechnology, including potential advancements in therapeutic compounds and bio-manufacturing. On a whole, minimal-genome microbes are an underlying material for building the field of synthetic biology, hence having beneficial uses and additional biological knowledge (Bernstein et al., 2018).

8.3.3. The Role of Minimal Genomes in Synthetic Biology

Minimal genome discovery is a foundation of synthetic biology since it provides researchers the ability to abbreviate cellular function to its essential constituents. By designing organisms that have smaller genomes, researchers are able to explore systematically basic biological processes necessary for life and discover complex biological mechanisms. These simple systems facilitate the breakdown of genetic interactions, which gives insight into the functions of individual genes and pathways within intricate cellular networks. This strategy is very much in line with the engineering principles of synthetic biology, in which goal-oriented designs are central. For example, the construction of such reduced genomes makes it possible to generate custom-made organisms that may be designed with desirable traits and behavior suitable for particular functions, like the production of biofuels or the synthesis of medicines, thus taking biological innovation to new limits beyond what nature can offer (Heinemann et al., 2006). Minimal genome application also involves metabolic pathway optimization through structural simplification of cell components, ultimately contributing to the progress in understanding required genetic interactions. With the growing genomic data, massive-scale genetic interaction studies have emerged that allow us to gain deep insights into gene functionality and the dynamics of relationships in cellular networks. The technology of genetic interaction analysis for *E. coli* (*GIANT-coli*) enable efficient production and screening of double mutants, revealing synergistic and antagonist interactions essential to map metabolic pathways. The design strategy maximizes our understanding of genetic dependencies and enables cellular behavior reprogramming. Using minimal genomes, researchers can distinguish specific genetic activities that are vital in synthetic organisms for practical purposes, building an ethos of rapid innovation based on empirical observation (Roguev et al., 2008). Moreover, minimal genomes serve as a model system for the development of the engineering of synthetic cells, which have been suggested to be platforms for complex biotechnological applications. The simplicity of these genomes allows for the integration of synthetic circuits and addition of new functions without being burdened by unnecessary biological components. This capability is consistent with the new trends in synthetic biology, emphasizing modular construction and scalability. Using minimal-genomic designs, researchers can deploy advanced biological behaviors, such as toggle switches and oscillations, thereby increasing the flexibility of synthetic organisms. By enhancing our understanding of minimal genomes, not only do we advance fundamental biological knowledge but also make significant progress in the design of synthetic cells to address key challenges in health, energy, and ecological sustainability and thus redefine the promise of biological engineering (Heinemann et al., 2006).

8.4. TECHNIQUES IN GENOME ENGINEERING

Technological advancements in genome engineering have revolutionized the field of synthetic biology to enable the manipulation of genetic information under controlled circumstances to create synthetic cells and minimal genomes. One of the principal approaches involves the use of computational design in restructuring genomic layouts without sacrificing cellular function. It has been demonstrated in recent years that different transcriptomic models can be constructed that produce similar environmental responses under varied conditions as their wild-type counterparts. It is a two-step process that starts with the development of a theoretical model and then transferring this model to a synthesized nucleotide sequence. Notably, research has shown that the genome of *E. coli* can be reorganized into a structure with up to 86% fewer regulated operons, an unparalleled improvement in the ability to engineer genomes that rely on optimized regulatory networks (Carrera et al., 2013). Apart from computational techniques, synthetic biology capitalizes on a range of engineering principles applied to design complex biological systems. The field uses engineering design standards in which hierarchies of abstraction and functionality direct the assembly and construction of the synthetic entities. Methods such as de novo DNA synthesis and manipulation of artificial gene networks are the engineering-inspired strategies that are prevalent in contemporary genome engineering. Such methods also depict the emergence of protein structure modularity that allows for customizable recombination similar to Lego block construction. Such module-based architectures may make it possible to specifically reprogram particular pathways or functions and thus extend the functionalities of synthetic organisms for use in a number of applications, such as biosynthesis and environmental monitoring (Heinemann et al., 2006). Last but not least, advances in genome engineering methods pave the way toward the realization of fully synthetic cells and minimal genomes that support life processes but are more streamlined than their natural equivalents. The unparalleled enablement of de novo DNA synthesis allows scientists to produce sequences unattainable through conventional means. This ability to design specific genetic blueprints forms the foundation of a new era of synthetic biology, wherein scientists are able to experiment with biological functions to levels of complexity and control previously unimaginable. As these techniques develop, they not only hold the promise to further illuminate genetic organization but also to give way to real-world applications in fields from medicine to agriculture and bioenergy. Ultimately, the union of computational design and engineered gene networks represents a quantum leap toward what synthetic biology strives to achieve: building life from scratch (Heinemann et al., 2006).

8.4.1. CRISPR-Cas9 Technology

The advent of CRISPR-Cas9 technology has significantly influenced the use of genetic engineering, particularly in synthetic biology and minimal genomes. The technology provides for precise manipulation of DNA, providing the potential for exploring genetic function and interaction with high precision. Compared to previous genome editing approaches that relied on complex methods, CRISPR-Cas9 is simpler to use and more efficient. Scientists can design special RNA sequences to guide the Cas9 endonuclease to target genomic locations, causing double-stranded DNA breaks. This is a principal component of synthetic cell and minimal genome creation since it will allow researchers to gain further insights into the essential genes for cellular metabolism and life, ultimately yielding advancements in bioengineering and treatment therapies. One of the several notable applications of CRISPR-Cas9 technology is a case in point with regard to its use in creating enhanced metabolic pathways, particularly in microorganisms such as *Saccharomyces cerevisiae*. Recent developments have included the utilization of a multiplex CRISPR (CRISPRm) approach, which enables multiple loci in the genome to be edited simultaneously. This approach has helped to improve some traits, for instance, cellobiose fermentation capabilities, by over tenfold, by selectively choosing optimized genetic variants through a single round of mutagenesis. Through the application of CRISPR-Cas9, researchers are not only improving our understanding of metabolic engineering but are also speeding up the selection process of beneficial mutations. These are significant to constructing minimal genomes that contain only the necessary components for life, thereby progressing to synthetic cells that are the height of simplicity and optimized efficiency in biological function (Arkin et al., 2014). With the further development of CRISPR-Cas9 technology, the alignment with next-generation sequencing (NGS) has provided genetic analysis with unprecedented opportunities. Through the integration of CRISPR genome editing with advanced sequencing techniques, researchers can assess their edit outcomes with high sensitivity and speed. Bioinformatics software development, such as

BATCH-GE, is an example that allows for comprehensive batch processing of NGS data derived from genome editing experiments. Beyond making it easier to detect indel mutations and quantify mutagenesis efficiencies, it also enhances the versatility of experimental designs. These innovations emphasize the value of CRISPR-Cas9 in synthetic biology, as they allow systematic dissection of genetic functions and construction of minimal genomes. These activities, in turn, allow for increased understanding of elementary biological processes and the construction of novel organisms for a variety of applications (Boel et al., 2016) (Figure 8.7).

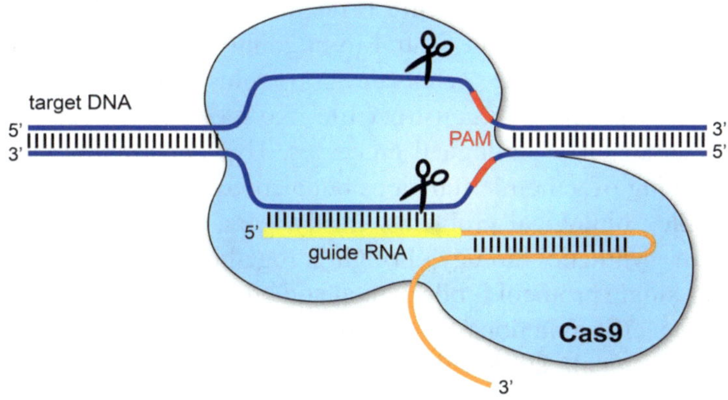

Figure 8.7. The CRISPR/Cas9 system.

Source: https://ep.bmj.com/content/edpract/101/4/213/F1.large.jpg?width=800&height=600&carousel=1.

8.4.2. Gene Synthesis and Assembly

Advancements in gene assembly and synthesis are core to the building of synthetic cells and minimal genomes, enabling researchers to create new biological entities with accuracy. By applying techniques such as de novo DNA synthesis, scientists can engineer and design specific sequences of genes that may not exist in nature. This level of control is central to synthetic biology because it allows genetic circuits to be designed so that they can function in a manner similar to electronic circuits, thus making it possible for new routes to be built within organisms. The ability to create sophisticated functions such as toggle switches and oscillators gives the full range of potential applications from drugs to biofuels and shows how gene synthesis is not a process of copying but an innovative design process based on engineering principles (Heinemann et al., 2006). Also, the assembly of synthesized genes into functioning genomes demands a disciplined strategy based on both biological and engineering practice. The combination of diverse genetic components, often described as a modular structure analogous to building blocks, enables the creation of flexible synthetic systems. Having standard components and procedures, scientists can simplify the assembly process to make it more efficient while reducing the randomness usually observed in biological systems. This engineering principle makes synthetic applications more functional and reliable. By building up a robust design cycle, synthetic biologists can specify, model, and test new biological designs, bringing minimal genome projects nearer to exploring the basic processes of life (Freemont et al., 2014). Overall, gene synthesis and assembly remain central to ongoing research on synthetic cells and minimal genomes. As researchers develop sophisticated tools and methods, it is increasingly possible to construct life-like models that summarize only the most fundamental of biological processes. This undertaking not only lifts our understanding of genomic functions but also holds out potential routes for solving complex biological problems. Lastly, gene synthesis is an important foundation of synthetic biology, bringing together numerous branches of science into a single protocol for designing and building biological systems that can perform specific functions across different applications (Freemont et al., 2014; Heinemann et al., 2006) (Figure 8.8).

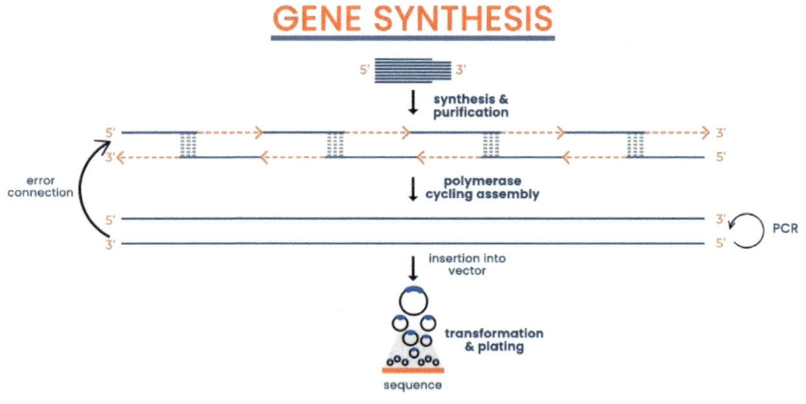

Figure 8.8. Gene synthesis and assembly.

Source: https://lh7-us.googleusercontent.com/-idBrZk6SX0L5-2ncgPsOha-OoLgerZrDcPEQ5oOKicqdyPDudZ-70co6EE1tiwq573r8-xhfMrWc4foLgc-7CLF3d95ENgEV9hmiiEWW2BOS3C9UbjJQJKUrCCmGfe79lmEp8jB8K-2GJhjdPiGjgAhXE.

8.4.3. Genome Editing Tools and Their Applications

Synthetic biology has brought forth a new generation of genome editing tools which have revolutionized our capacity for precise and efficient manipulation of genetic material. It is at the vanguard of these technologies that CRISPR-Cas9 and other nucleases stand, providing the possibility of targeted modification of an organism's DNA. These technologies allow not just the detection of a specific gene but also the addition of favorable characteristics using gene editing.

Such capability has tremendous potential to be applied to agriculture, as crops can be engineered for more yield and resistance to diseases from pests, hence alleviating food security concerns. In addition, the convergence of genome editing technologies offers an additional insight into the function of genes as well as mechanisms governing varied biological processes, a crucial awareness critical to synthetic biology (Freemont et al., 2014). The uses of genome editing extend beyond agriculture; they have profound effects on medical research and therapeutics as well. The ability to edit genes responsible for genetic diseases with precision opens revolutionary possibilities in gene therapy, with the hope of treatments that may correct genetic mutations for diseases such as cystic fibrosis and sickle cell anemia.

Moreover, genome editing technologies are used to generate models of human disease *in vitro*, allowing us to better understand pathogenesis and enabling the screening of new drugs. The efficiency in the analysis of gene function is improved through high-throughput approaches that integrate genome editing with data mining tools, making it easier to identify gene interactions and their functions in disease (Franco et al., 2015). These developments reaffirm the revolutionary possibility of genome editing in biomedical research. Alongside these technologies, however, arise ethical issues in relation to their implications and, therefore, an open debate regarding the place of genome editing. Genome editing and synthetic biology only continue to develop, and hence there exists an imperative to develop policies that offset the ethical concerns against the manipulation of the very constituents of life. Misuse concerns of such highly potent tools by possible bioterrorism or accidental ecological consequences underscore the necessity for cultivating responsible research practice. Moreover, the ability to create artificial organisms also challenges biodiversity and long-term impacts on natural ecosystems. Navigation of these seas of ethics is necessary as society embraces the possibilities presented by genome editing technologies, ensuring that their applications are aligned with ethical principles and advance human health and the environment (Franco et al., 2015; Freemont et al., 2014) (Figure 8.9).

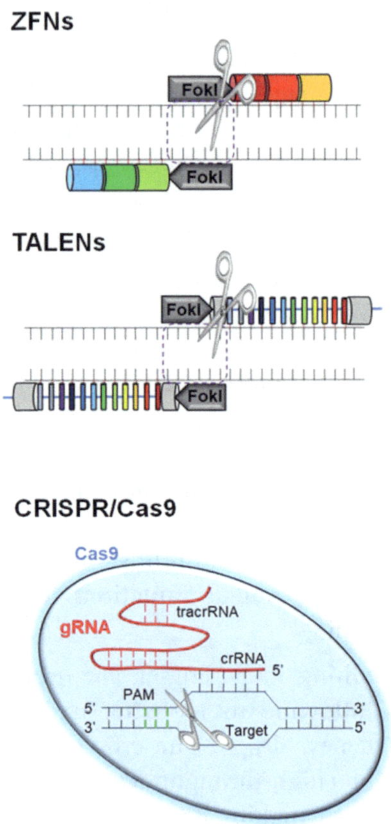

Figure 8.9. Genome editing tools. Three types of genome editing tools including ZFNs, TALENs, and CRISPR/Cas9 are shown.

Source: https://www.researchgate.net/publication/344222176/figure/fig1/AS:983366363000838@1611464219946/Genome-editing-tools-Three-types-of-genome-editing-tools-including-ZFNs-TALENs-and.ppm.

8.5. APPLICATIONS OF SYNTHETIC CELLS AND MINIMAL GENOMES

Synthetic cells and minimal genomes represent a paradigmatic change in synthetic biology, with immense consequences for biotechnological applications. By reducing cellular processes to the most elementary essentials, researchers are fabricating systems able to perform specialized biological activities more efficiently than ever before. This reductionist approach enables the programming of cells tailored for applications such as drug delivery, biosensing, and environmental remediation. Secondly, the ability to engineer and construct minimal cellular systems confirms the principle of modularity, under which components can be swapped or modified in a bid to create new functions. This versatility increases the potential for innovation across various fields, from medicine to agriculture, by enabling the development of highly specialized and powerful solutions to complex problems (Freemont et al., 2014). Synthetic cells can be used in biomimetic applications as tools for the study of cell processes in ways that cannot be achieved with naturally evolved organisms. As an example, scientists have created minimal cells that emulate the central processes of life, such as DNA replication and protein synthesis.

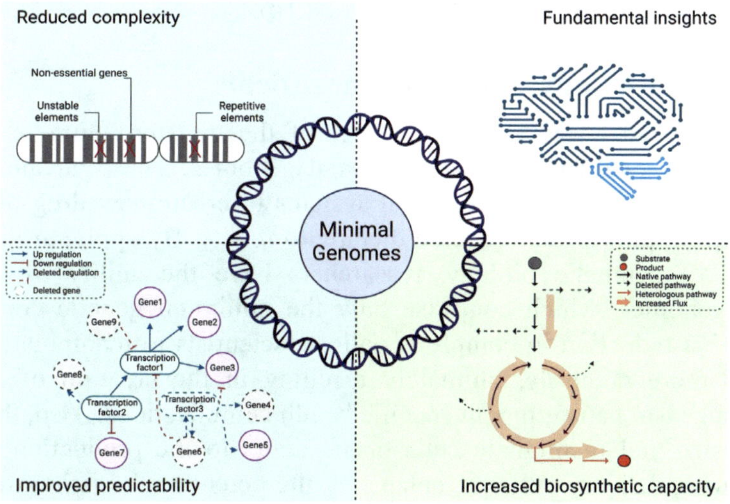

Figure 8.10. Applications of synthetic and synthetic-minimal genomes.

Source: https://media.springernature.com/full/springer-static/image/art%3A10.1038%2Fs41467-023-37748-7/MediaObjects/41467_2023_37748_Fig1_HTML.png?as=webp.

These synthetic systems not only offer additional insight into primitive biological processes, but they also create new avenues for engineering applications, such as the development of new therapeutic strategies. Minimal genomes have also been explored as vehicles to circumvent evolutionary constraints built into larger genomic contexts, allowing researchers to engineer production pathways for therapeutics, such as biofuels or pharmaceuticals. By exploring the functions and interactions of individually designed components, researchers are making new knowledge accessible to continue further advances in synthetic biology (Hürtgen et al., 2018). Synthetic cell use also finds its reach in biosafety and biohazard protection. Synthetic cells' engineered simplicity enables them to be rigorously tested and examined for their behaviors in a wide range of environments, such that they do not pose a containment issue before they are utilized. Minimal cells that have embedded safety switches, for instance, can ensure that they will not operate outside of their designed conditions, avoiding ecological hazards. In addition, synthetic biology can enable the solving of global issues like climate change and scarce resources by having microorganisms capture carbon or recycle nutrients. Ultimately, as synthetic biology continues to advance, the combined knowledge of controlling, utilizing, and deploying synthetic cells and minimal genomes will give answers that not only enhance human capacity but also establish harmonious interactions with nature (Freemont et al., 2014) (Figure 8.10).

8.5.1. Biotechnology and Pharmaceuticals

Biotechnology advances are at the heart of the transformation in pharma, particularly in synthetic cells and minimal genomes. These advances bring the potential of engineered biological systems to create new drug discovery, manufacturing processes, and even therapy to reality. By applying the design principles of synthetic biology, researchers have the ability to construct minimal genomes, which comprise only the minimum genetic components for life. This reduction in complexity allows scientists to manipulate cellular processes more precisely, ultimately resulting in the creation of specially crafted drugs that can surmount specific health complications. Also, the ability to synthesize and manipulate cells opens doors to the production of more efficient drug-delivery systems, enhancing the potency of existing treatments while hopefully reducing side effects. This intersection of biotechnology and pharmacology also has significant implications for public health and safety. The emergence of antibiotic-resistant bacteria has created a global health pandemic, necessitating innovative countermeasures to combat these virulent pathogens. Bacteriophages, or "phages," have been highly sought after as possible alternatives to traditional antibiotics. By leveraging the use of these viruses that selectively infect and lyse bacteria, scientists are exploring phage therapy as a potent antidote to antibiotic resistance. This push toward phage therapies underscores the importance of nimble regulatory frameworks with the ability to keep up with new biotechnological developments (Todd et al., 2019). Thus, the ability to align synthetic biology with drug innovation is not only critical to the development of new therapies but also crucial in fostering a viable and adaptive health ecosystem. To achieve the maximum potential benefits of these biotech advances, it is critical to address the intellectual property and commercialization issues. The pharmaceutical industry typically relies on strong patent rights to promote research and development; however, recent judicial uncertainty regarding the patent eligibility of biopharmaceuticals, including phage therapies, poses a serious threat to innovation. Absent robust protections, investment in the development of such new therapeutics will be reduced, as the risk of money deters entry by new players into the market. Thus, exploring new models of protection for biotechnology innovation—such as trade secrets or regulatory exclusivities—becomes increasingly essential. These structures attempt to achieve a balance that encourages research while making subsequent therapies available to those in greatest need, thereby bridging the space between scientific advancement and public utility (Boyen et al., 2025) (Figure 8.11).

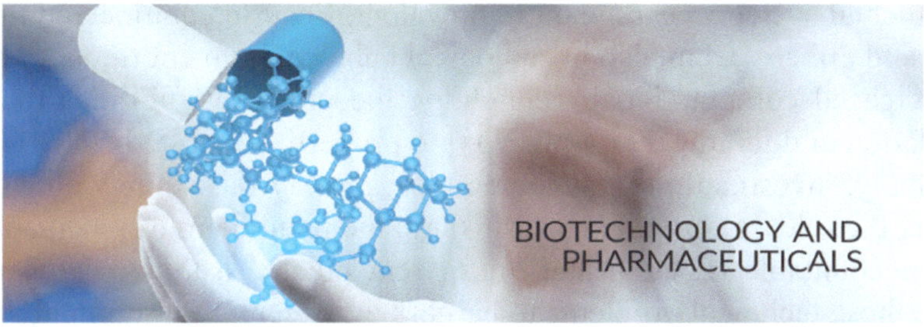

Figure 8.11. Biotechnology and pharmaceuticals.

Source: https://pbindustries.gov.in/static/assets/images/manufacturing/header-biotech.png.

8.5.2. Environmental Applications

The potential of minimal genomes and synthetic cells to fulfill environmental requirements is immense, and it will enhance our capacity for engineering biological systems to control resources and pollution. Through computational design approaches, researchers have been able to demonstrate the construction of synthetic organisms with engineered transcriptional networks that will engage harmoniously with environmental perturbations. A landmark research effectively rearranged the genome of *E. coli* to deliver a model that was more than 20% quicker in responding to external stressors than its wild-type equivalent. This revolutionary method reflects the potential of synthetic biology in engineering organisms to survive and adapt to changing conditions, thus enabling applications in bioremediation and strategies for waste management that are crucial for restoring ecosystems as well as ensuring sustainability (Carrera et al., 2013). Moreover, the innovation in synthetic biology has many uses in the production of biofuels and sustainable commodities. Through the engineering of microorganisms with compact genomes, scientists can adjust metabolic pathways to make renewable resources effectively converted into energy-dense compounds. Synthetic organisms can be engineered to convert agricultural residues or other organic substrates into biofuels, enabling a circular economy and reducing reliance on fossil fuels. This approach not only provides a green alternative energy source but also supports global efforts to manage climate change by reducing greenhouse gas emissions. The integration of computational models and synthetic biology laws underscores the revolutionary potential for the rational design of production mechanisms in environmentally friendly schemes (Heinemann et al., 2006). The regulation frameworks of these synthetic organisms are also crucial as their applications in environmental systems become more prevalent. Ensuring engineered microbes are safe and effective, but also in ecological balance, is an enormous challenge. With increased computational technology, the predictive abilities of genetic manipulation and its impact on phenotypic properties will be better enhanced. Additionally, investigation in synthetic biology may raise some ethical issues regarding the release of such organisms into the natural environment. With the help of accountable research and transparent practices, the researchers must navigate these technical problems to harness the potential of minimal genomes and synthetic cells for environmental applications in minimizing potential harms (Carrera et al., 2013; Heinemann et al., 2006).

8.5.3. Synthetic Cells in Bioengineering

Synthetic cells are a revolutionary advance towards the construction of bioengineering, and fresh avenues are opened up to address some of humanity's most important medical and environmental problems. Synthetic cells allow scientists to manipulate biological systems in new, unprecedented ways, with an arena being provided for the study of cellular processes and interactions. Synthetic cell engineering and design are motivated not only by the need for new uses but also by the need to unravel the complexity of life at its most fundamental. Synthesizing minimal genomes enables researchers to deconstruct cellular processes, examining how essential building blocks create the operations of life. This approach is the intersection of biology and engineering, aiming to apply design principles to the assembly of biologically integrated systems that can perform given tasks reliably and efficiently, ultimately advancing biotechnology and health (Freemont et al., 2014). One of the greatest challenges that synthetic biology has to overcome is to deal with the unpredictability inherent in biological systems. This uncertainty demands a good design cycle through which synthetic biologists delineate, model, analyze, build, and validate biological networks. The reason that the design cycle is iterative is that it allows the researchers to refine their synthetic cells recurrently, thus creating more stable and functional designs. Engineering principles are employed to create tools that are employed to battle the complexity of biology so that synthetic cells can be engineered that can mimic the natural counterparts. For instance, using systematic testing and optimization, it is possible for scientists to engineer synthetic cells with environmental responsiveness that can be useful in drug delivery and environmental remediation. This systematic process offers a window of opportunity to the engineer-driven design of biological systems that were earlier only under consideration in the field of theoretical studies and provides new opportunities for practical uses of synthetic life (Heinemann et al., 2006). Furthermore, the integration of computational modeling in the design of synthetic cells greatly enhances the ability to predict and regulate cellular behaviors. Sophisticated computational methods enable scientists to model various biological interactions, as well as insights for the potential synthesis of synthetic genomes. This ability to model complex interactions enables the establishment of synthetic cells that are able to be imbued with valuable traits, such as targeted drug production or biosensing capabilities. By combining computational design with experimental testing, researchers are pushing the boundaries of synthetic biology and developing functionalities that were previously unavailable. The more sophisticated the level of design of synthetic cells becomes, the more their applications will be in solving world problems, like sustainable production processes and personalized medicine. The crossing of engineering and biology thus holds much promise for the future of synthetic cells with respect to the importance of the contribution made by bioengineering in charting the way for biological research and application (Freemont et al., 2014).

8.6. ETHICAL CONSIDERATIONS

The invention of synthetic cells and minimal genomes has sparked so much interest in biotechnology but has simultaneously posed significant ethical issues. As geneticists design genes with tools such as CRISPR/Cas9, there is the risk of unforeseen consequences. As synthetic biology advances bring promises for breakthroughs in the cure of genetic diseases and pathogen attacks, so also do threats of unforeseen ecological effects and bioterrorism, and thus a severe examination of their applications. Concerns on the human interference level with natural organisms and the morality of creating life forms that may not exist naturally warrant much consideration. As cited by numerous researches, including those focused on CRISPR/Cas, addressing such ethical concerns is at the core of ensuring responsible stewardship of synthetic biology innovation (Zhou et al., 2024). Further, the ethical considerations regarding consent and accountability in the application of synthetic cells and minimal genomes need to be seriously examined. As technology advances, particularly in gene editing and synthetic biology, discourse on informed consent for clinical application is bound to occur. This is important while contemplating therapeutic intervention that could affect future generations, and raises an issue of the rights of unborn individuals. The complexity of genetic modification has imparted a sense of responsibility to the scientists, as they need to navigate the balancing act of coupling safeguards and ethical control into research practices. Although new solutions like CRISPR/Cas9 have immense potential for health advances, the risk of misuse or unintended harm underscores the necessity of ethical guidelines that guarantee human health and environmental health (Kaupbayeva et al., 2024). Finally, the regulatory environment for synthetic biology remains underdeveloped, contributing to the necessity of ethical guidelines. The rapid pace at which technological progress is made in synthetic cells requires policymakers to develop strong regulations that are capable of harmonizing innovation and ethics. Policymakers and scientists ought to collaborate to develop clear-cut policies that dictate the application, dispersal, and tracing of genetically modified organisms. As science moves forward, ongoing dialogue among stakeholders like ethicists, scientists, and the public will be critical to settle questions of equity, access, and the ethical sensibility of applications in synthetic biology. Otherwise, the potential promise of synthetic cells and minimal genomes would be undermined by ethical failure and public distrust, severely hindering future progress in the promising area of research (Zhou et al., 2024).

8.6.1. Potential Risks of Synthetic Biology

The rapid advancement in synthetic biology raises significant questions regarding potential risks associated with its applications, particularly with genetically modified microorganisms (GEMs). While the potential for innovation is expansive, the unintended consequences of releasing such organisms into natural systems cannot be dismissed. The success of biocontainment techniques is vital to prevent accidental or deliberate escape, since GEMs possess the potential to interact with and destabilize present ecological processes. Although present data suggest that such organisms will not compete in natural environments, the uncertainty surrounding synthetic biology renders them a built-in threat. Thus, strategies such as the design of novel genetic circuits and synthetic auxotrophies are essential to improve these biocontainment strategies to the extent that synthetic cells would not pose a threat to biodiversity and human health (Csibra et al., 2016). Besides, the philosophical underpinning of synthetic biology's goals does carry some element of overreaching with regard to the ambition to control and remake life itself. While scientists are predisposed to hail the precision and predictability of the achievements in synthetic biology, skeptics question whether such is overly sanguine or even deceptive. The DARPA Biodesign initiative is the prime example for this tension, considering that its ambitions in ending randomness in evolutionary processes may feel simultaneously visionary and visionary. This epistemic opportunism highlights a split between synthetic biologists, ethicists, and critical activists in the shared vision for the future relevance of this field. Without ethical foundations, research in synthetic biology may be grounded in techno-utopian ideals that do not accurately reflect the realities and complexities entailed by its practical usage (Bensaude-Vincent et al., 2013). Moreover, the relationship between technological advances and potential danger necessitates ongoing discussion and regulation in the field of synthetic biology. As researchers are exploring producing synthetic cells and stripped-down genomes, understanding the implications of these achievements is important to informed policy and governance. Regulatory frameworks must adapt to the specific challenges that synthetic biology presents, rather than merely containment of GEMs, but rather the ethical, social, and environmental aspects of their construction and deployment. Publicity and participation in discussions are imperative, as transparency and accountability can reduce the risks associated with synthetic biology. In acknowledging these multifaceted challenges, stakeholders are able to craft an equitable strategy that unlocks the maximum potential of synthetic biology while reducing the risks associated.

8.6.2. Ethical Implications of Creating Life

Synthetic cell and minimal genome creation have raised far-reaching ethical discussions regarding making and designing life. As scientists work on constructing minimal living systems, the ethical implications of creating life become more abstract. Of one thing, there is very little doubt. One of the biggest concerns involves the potential unintended consequences of creating organisms that are superficially indistinguishable from natural ones. The ethical concerns extend beyond laboratory environments, challenging researchers to identify and address the dangers of synthetic biology. Moreover, human attitudes toward life would potentially shift, as distinctions between natural and artificially created organisms crumble, prompting us to rethink our picture of living systems. Hence, the ethical standards for synthetic biology practice must be strict, addressing these nuances as well as making public engagement and understanding possible. The meeting of systems biology and ethics is particularly pertinent in the context of synthetic cell research. Systems biology, including assembling molecular data into a complete grasp of life processes, contains a preoccupation with the dynamicity of living systems. This strategy is in direct contrast to static conceptions of genomes and underlines the necessity to identify life as a complex, interdependent process. Ethical critique of systems biology has been severely inadequate so far, yet there is a need to consider the socioethical issues arising with its implementation in synthetic biology and minimal genomes. Concerns about commercialization, *in silico* research, and the changing cultural meanings of life are some of the growing issues requiring ethical exploration. Their resolution will not only serve to defend the integrity of scientific research but also add depth to the discussion about ethical procedures in life sciences (Calvert et al., 2013). Moreover, the metaphoric discourse of synthetic biology research reveals deep underlying philosophical undertones in how life and identity are perceived. Creating a synthetic cell, typically defined through metaphors like the barcode or "book" of life, raises images of life as one that can be decoded and reconstructed. These perceptions shape ethical perceptions of commodification and ownership of living organisms. Looking through the archetypal window of the mandala, which signifies wholeness and unity, synthesizing life is more than a scientific process but an involved exploration of what it takes to make life. Such an integrative approach involves the convergence of various voices and forms of know-how, affirming that the creation of synthetic cells is as much about ethics and philosophy as it is science (Zwart et al., 2018). Therefore, engaging with these metaphors is important to grasp the broader consequences of synthetic biology for our future understanding of life itself.

8.6.3. Regulatory Frameworks and Public Perception

The regulatory frameworks for synthetic biology, like synthetic cells and minimal genomes, are too intricate and multilayered. With the rapid rate of advancement in bioengineering, the government of Europe is severely tested to respond to ethical, legal, and social issues that accompany these sciences. The STOA project report identifies the controversy that continues with bioengineering and its further intersection with technology and biology. This overlap calls for a set of real-time regulations that are not just wide-ranging but also adaptable to unforeseen developments. In addition, existing frameworks need to account for public opinion, which can significantly affect the adoption and success of real-world synthetic biology technologies. Ensuring real-time transparency in regulatory systems and encouraging citizen participation is essential to building confidence and mitigating concerns about real-time safety and ethical considerations of synthetic cells (N/A, 2011). Public perception of synthetic biology is the force behind the formulation of policies and regulatory measures. As technologies like synthetic cells and minimal genomes continue to develop, public sagacity will be needed in order to shape facilitating regulatory frameworks. Public perception is typically shaped by ethical concerns related to genetic engineering, environmental risks, and threats of biosecurity.

Consequently, policymakers must prioritize engaging with diverse stakeholder groups, including scientists, ethicists, and the general public, to foster an inclusive dialogue. The incorporation of public concerns into regulatory frameworks can not only enhance the legitimacy of the regulations but also facilitate more effective oversight. For instance, increasing acceptability of plant-based protein production technologies, supported by advancements in CRISPR/Cas9 technology, illustrates how public trust can be created through transparent communication of benefits and risks (Evju et al., 2023). Eventually, the regulatory scheme-public attitude dance will set the tone for synthetic biology's future. Policymakers must balance facilitating innovation and public safety with ethical concerns. As the debate over synthetic biology unfolds, continuous assessment of regulatory frameworks will be vital for reacting to technological advances and changing public opinion. This involves a proactive policy, where regulation is not a reaction but informed by science and public opinion as they currently exist. By aligning regulatory strategies with societal values, the acceptance of synthetic cells and minimal genomes can be enhanced, paving the way for future breakthroughs while simultaneously addressing public concerns (Evju et al., 2023; N/A, 2011).

SUMMARY

In examining the realm of synthetic cells and minimal genomes, it becomes evident that these innovations hold significant promise for advancing our understanding of biological systems. With the use of the simplicity of minimal genomes, researchers are able to define basic genetic building blocks necessary for life, thereby providing an understanding of the cellular function components. Synthetic cell construction also encourages exploration of novel areas of application in biotechnology, such as drug delivery, biosensing, and environmental cleanup. This intersection of minimal genome research and synthetic biology not only enables the examination of life's building blocks but also the design of biological systems tailored to specific needs, with potential applications across a variety of industries and therapeutic methods. Lastly, the development of synthetic cells allows for rigorous testing of biological hypotheses in a controlled setting, a form of experimentation that may be prohibited by traditional models. With the assistance of methods such as genome editing and gene synthesis, researchers can tune protein interactions and pathways at the genetic level. For instance, using methods of targeted sequencing, it has been possible to identify disease trajectories in pediatric cohorts, as reflected by examining circulating tumor DNA in young high-risk patients for mature B-cell non-Hodgkin lymphoma (Werner et al., 2025). These advances underpin the realization that synthetic biology is more than the creation of life from non-biological materials but a capacity to enhance our knowledge of and ability to manipulate extant biological systems. Moreover, the challenging puzzle of defining protein-protein interactions lies at the heart of a molecular definition of protein function in synthetic systems. The yeast two-hybrid approach, commonly employed to screen for such interactions, is a model for the analysis of complex biochemical networks (Tulum, 2025). In minimal genomes, detecting and characterizing these interactions may lead to breakthroughs in basic and applied biological research. The study of protein interactions can allow researchers to continue developing more sophisticated synthetic cells with designer functions. This all-encompassing strategy encourages the possibility of cutting-edge applications across disciplines, further establishing the position of synthetic biology as a revolutionary field with the potential to redefine life sciences strategy.

Investigation into synthetic cells and minimal genomes is a revolutionary edge to the discipline of synthetic biology, highlighting the necessity for an engineering-based strategy toward biological systems. This area of research seeks to construct increasingly advanced biological entities with specific functions according to the principles of standardization, abstraction, and deliberate design. As artificial gene networks, de novo DNA synthesis, and protein engineering advance, researchers are capable of executing engineered biological functions such as toggle switching and oscillation. This engineering perspective not only allows the assembly of pharmaceuticals but also allows the disassembly and reassembly of functional modules in proteins as well, which further allows for the possibility of biotechnological innovation. Therefore, the strategies of synthetic biology are essentially revolutionizing our understanding of biological function and design, allowing new applications and the improvement of existing biological information (Heinemann et al., 2006). Central to understanding synthetic cells is the study of minimal genomes—life forms with the least amount of genes required to live autonomously. But as researchers pursue, it has been found that genome-reduced bacteria experience growth constraints under normal laboratory settings, exhibiting novel phenotypic effects. This is evidence of the intricacies of genome function and the advanced interactions of biological mechanisms, as the removal of certain genes acts to destabilize metabolic functions. The application of adaptive laboratory evolution (ALE) has proven to be a principal method in re-optimizing such strains, revealing the necessity for dynamic adaptation and genetic reassessment. The findings point towards the limitations in organismal behavior prediction being more apt to be linked with variation in a comprehensive understanding of systemic interaction rather than in distinct functions of missing genes, thereby establishing the role of systems biology in research (Cho et al., 2019). The results from experiments with genome-scaled organisms have profound effects on both fundamental and applied studies in synthetic biology. Specifically, the adaptive laboratory evolution metabolic rewiring emphasizes that growth performance can be significantly improved by understanding the broader metabolic network in action. Mutations within the RNA polymerase's promoter binding, as seen in evolved cells, facilitate not only increased translation of high-copy number mRNAs but also the possibility of more extensive regulatory regimes controlling cellular resources. Multiomic analysis also clarifies complex interactions by measuring changes at the transcriptomic and translatomic levels and correlating them with growth patterns. These findings provide the gateway to new synthetic biology technologies, making future efforts more likely to predict and manage the behavior of these engineered systems, enhancing the utility and efficiency of synthetic organisms across a broad spectrum of scientific and medical uses (Cho et al., 2019).

REFERENCES

1. Abbas, Z., Abbas, Z., Afzal, A., Bin Muhammad Ishaq, M., et al. (2024). Commands of synthetic biology to modernize and re-design the biological systems. https://core.ac.uk/download/616936549.pdf (accessed on 3rd October 2025).
2. Ali, A., Ham, J. D., Zairi, F., Patel, V., Tarannum, M., Dinh, K., Abdulhamid, Y. Z., et al. (2024). Abstract 5231: NK and T cells armed with EVE16, a novel synthetic immune receptor, mediate robust anti-tumor responses through dual direct cytotoxicity and ADCC with minimal trogocytosis and exhaustion. *Cancer Research.* https://www.semanticscholar.org/paper/577ea77e63d7ba1f897bc39190d60edc00217345 (accessed on 3rd October 2025).
3. Arkin, A. P., Cate, J. H. D., DeLoache, W., Dueber, J., et al. (2014). Selection of chromosomal DNA libraries using a multiplex CRISPR system. https://core.ac.uk/download/323081404.pdf (accessed on 3rd October 2025).
4. Bensaude-Vincent, B. (2013). *Between the Possible and the Actual: Philosophical Perspectives on the Design of Synthetic Organisms.*
5. Berg, G., Rybakova, D., Fischer, D., Cernava, T., Champomier Vergès, M. C., Charles, T. C., Chen, X., et al. (2020). Microbiome definition re-visited: Old concepts and new challenges. *Microbiome, 8.* https://doi.org/10.1186/s40168-020-00875-0 (accessed on 3rd October 2025).
6. Bernstein, D., Dewhirst, F., & Segre, D. (2018). Quantifying biosynthetic network robustness across the human oral microbiome. https://core.ac.uk/download/286677178.pdf (accessed on 3rd October 2025).
7. Boel, A., Callewaert, B., Coucke, P., De Paepe, A., et al. (2016). BATCH-GE: Batch analysis of next-generation sequencing data for genome editing assessment. https://core.ac.uk/download/74705686.pdf (accessed on 3rd October 2025).
8. Boyen, C., Børresen, T., Dobson, A., Höfle, M., et al. (2025). Marine biotechnology: A new vision and strategy for Europe. https://core.ac.uk/download/pdf/29241634.pdf (accessed on 3rd October 2025).
9. Bulcha, J., Wang, Y., Ma, H., Tai, P. W. L., & Gao, G. (2021). Viral vector platforms within the gene therapy landscape. *Signal Transduction and Targeted Therapy, 6.* https://doi.org/10.1038/s41392-021-00487-6 (accessed on 3rd October 2025).
10. Calvert, J., Dupré, J., & O'Malley, M. A. (2013). The study of sociometrical issues in systems biology. https://core.ac.uk/download/12825890.pdf (accessed on 3rd October 2025).

11. Carrera, J., & Jaramillo, A. (2013). Automated design of bacterial genome sequences. https://core.ac.uk/download/19553375.pdf (accessed on 3rd October 2025).
12. Chaneton, B., Gottlieb, E., Ruppin, E., Yizhak, K., Kiuru, M., Li, L., Lobo, C., et al. (2015). Modeling cancer metabolism on a genome scale. https://core.ac.uk/download/42363494.pdf (accessed on 3rd October 2025).
13. Cho, B. K., Cho, S., Choe, D., Hwang, S., et al. (2019). Adaptive laboratory evolution of a genome-reduced *Escherichia coli*. https://core.ac.uk/download/323065557.pdf (accessed on 3rd October 2025).
14. Csibra, E., Gianni, E., Kruger, A., Pinheiro, V. B., et al. (2016). Synthetic biology approaches to biological containment: Pre-emptively tackling potential risks. https://core.ac.uk/download/74205606.pdf (accessed on 3rd October 2025).
15. Danchin, A., Boel, G., Danot, O., de Lorenzo, V., & Danchin, A. (2019). Omnipresent Maxwell's demons orchestrate information management in living cells. https://core.ac.uk/download/187718269.pdf (accessed on 3rd October 2025).
16. Deplazes-Zemp, A. (2018). The conception of life in synthetic biology. https://core.ac.uk/download/159152194.pdf (accessed on 3rd October 2025).
17. Ding, Y., Tan, C., & Wu, F. (2014). Synthetic biology: A bridge between artificial and natural cells. https://core.ac.uk/download/323068426.pdf (accessed on 3rd October 2025).
18. Evju, E. (2023). General aspects and applications of plant genome editing: Advancements in recombinant protein production and viral vaccine efficiency through CRISPR/Cas9-edited *N. benthamiana*. https://core.ac.uk/download/591078709.pdf (accessed on 3rd October 2025).
19. Franco, B., Padilla-Vaca, F., & Anaya-Velázquez, F. (2015). Synthetic biology: Novel approaches for microbiology. https://core.ac.uk/download/159084117.pdf (accessed on 3rd October 2025).
20. Freemont, P., Kelwick, R., MacDonald, J. T., Webb, A., et al. (2014). Developments in the tools and methodologies of synthetic biology. https://core.ac.uk/download/76995845.pdf (accessed on 3rd October 2025).
21. Froese, T., Ikegami, T., & Virgo, N. (2013). Motility at the origin of life: Its characterization and a model. http://arxiv.org/abs/1311.2531 (accessed on 3rd October 2025).
22. Heinemann, M., & Panke, S. (2006). Synthetic biology—putting engineering into biology. https://core.ac.uk/download/pdf/12944386.pdf

(accessed on 3rd October 2025).

23. Hürtgen, D. (2018). Towards synthetic life: Establishing a minimal segrosome for the rational design of biomimetic systems. https://core.ac.uk/download/287653452.pdf (accessed on 3rd October 2025).
24. Kaupbayeva, B., Tsoy, A., Safarova (Yantsen), Y., Nurmagambetova, A., Murata, H., Matyjaszewski, K., & Askarova, S. N. (2024). Unlocking genome editing: Advances and obstacles in CRISPR/Cas delivery technologies. *Journal of Functional Biomaterials, 15.* https://www.semanticscholar.org/paper/92f48b7ca74d07164220798fdaf5162c1c81d881 (accessed on 3rd October 2025).
25. Muzzi, J. C. D., Magno, J. M., Cardoso, M. A., de Moura, J. F., Castro, M. A. A., & Figueiredo, B. (2021). Adrenocortical carcinoma steroid profiles: *In silico* pan-cancer analysis of TCGA data uncovers immunotherapy targets for potential improved outcomes. *Frontiers in Endocrinology, 12.* https://www.semanticscholar.org/paper/fd98c064dce3379d0833b43eadbff334a49f434d (accessed on 3rd October 2025).
26. N/A. (2011). *European Governance Challenges in Bio-Engineering: Making Perfect Life—Bio-Engineering (in) the 21st Century: Final Report.* https://core.ac.uk/download/577016149.pdf (accessed on 3rd October 2025).
27. Roguev, A., Byrne, A. B., Rizzitello, A. E., Tong, A. H., Tong, A. H., Typas, A., Wanner, B. L., et al. (2008). High-throughput, quantitative analyses of genetic interactions in *E. coli.* https://core.ac.uk/download/323074731.pdf (accessed on 3rd October 2025).
28. Rothschild, L. J., Averesch, N. J. H., Strychalski, E. A., Moser, F., Glass, J. I., Cruz Perez, R., Yekinni, I. O., et al. (2024). Building synthetic cells—From the technology infrastructure to cellular entities. *ACS Synthetic Biology, 13,* 974–997. https://www.semanticscholar.org/paper/9a15197f7a705e006ce0ba8de25e5617b02cdea6 (accessed on 3rd October 2025).
29. Todd, K. (2019). The promising viral threat to bacterial resistance: The uncertain patentability of phage therapeutics and the necessity of alternative incentives. https://core.ac.uk/download/213020895.pdf (accessed on 3rd October 2025).
30. Tulum, I. (2025). Analysis of *Mycoplasma pneumoniae* protein-protein interactions by yeast two-hybrid method. *Osmaniye Korkut Ata Üniversitesi Fen Bilimleri Enstitüsü Dergisi.* https://www.semanticscholar.org/paper/7b1f2e4f79665c30080567fc7c31f8796ea8c859 (accessed on 3rd October 2025).

31. Wedmark, Y. K., Vik, J., & Øyås, O. (2024). A hierarchy of metabolite exchanges in metabolic models of microbial species and communities. *PLOS Computational Biology, 20.* https://www.semanticscholar.org/paper/0b0bd8de61c2e875beddfb5cfda4d025b2fe9f80 (accessed on 3rd October 2025).

32. Werner, J., Damm-Welk, C., Rohde, M., Hundsdörfer, P., Vieth, S., Karow, A., Barnbrock, A., et al. (2025). Circulating tumor DNA in pediatric mature B-cell non-Hodgkin lymphoma for genotyping and minimal disease monitoring. *Pediatric Blood & Cancer.* https://www.semanticscholar.org/paper/1a19f242cdadd316eb2cae21c2f26420dc491092 (accessed on 3rd October 2025).

33. Zhou, L., & Simonian, A. L. (2024). CRISPR/Cas technology: The unique synthetic biology genome-editing tool shifting the paradigm in viral diagnostics, defense, and therapeutics. *Annual Review of Biomedical Engineering.* https://www.semanticscholar.org/paper/81c848b55e5a024831c98659110a55e887255df2 (accessed on 3rd October 2025).

34. Zwart, H. (2018). Scientific iconoclasm and active imagination: Synthetic cells as techno-scientific mandalas. https://core.ac.uk/download/186329944.pdf (accessed on 3rd October 2025).

CHAPTER 9

PROTEIN ENGINEERING

Contents

INTRODUCTION TO PROTEIN ENGINEERING

Protein engineering is a meeting of biological sciences and engineering principles aiming to create, construct, and alter proteins to embody desired functionalities. Essentially, the field of protein engineering extends our understanding of protein structures and dynamic properties. The establishment of molecular biology techniques, such as recombinant DNA technology, enables scientists to regulate the genetic code that dictates the synthesis of proteins. This manipulation enables the creation of novel proteins with enhanced properties or entirely novel functions. Protein engineering is ultimately not just academic curiosity but also central to biotechnology and pharmaceuticals, the basis of new therapies and diagnostic tools that can have a significant impact on public health. The combination of multidisciplinary approaches lies at the center of protein engineering as it entails transferring knowledge from several scientific fields, including bioinformatics, materials science, and nanotechnology. The interconnection offered across these fields ensures effective protein synthesis and characterization, especially for drug delivery systems and targeted therapies. For example, proteins engineered to encapsulate therapeutic drugs can improve their stability and delivery efficiency, which corresponds to better treatment effectiveness. The recent focus on nanomedicine is a testament to the revolutionary aspect of these engineered proteins.

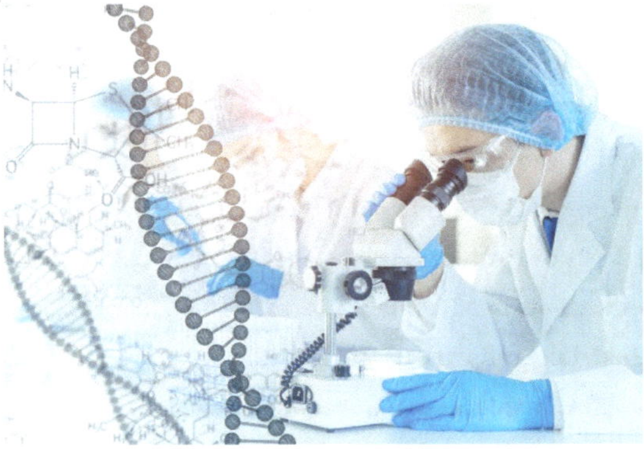

Figure 9.1. Scientists employ protein engineering techniques to design and create proteins with useful properties for medicinal and industrial applications.

Source: https://cdn.the-scientist.com/assets/articleNo/72286/iImg/54648/brushup-geneticengineering-1800x1260-l.webp.

Through facilitating drug delivery and improving the specificity of therapeutic action, this union of protein engineering and nanotechnology is a testament to the promise that can come from cross-disciplinary solutions to complicated medical problems (Anurogo et al., 2017). An interesting aspect of protein engineering includes the application of nature-inspired processes using new methods, such as bio-imprinting. This method entails chemically modifying proteins to change their structure and enhance functionality, enabling higher expression levels and improved storage stability. As more demand is created by industries for efficient ways to produce sustainable proteins, such methods gain importance. The focus on renewable energy resources and the need for sustainable approaches can be met through the development of bioengineered proteins that serve such purposes (Khan and Akhtar, 2011). On this note, protein engineering is not merely the study and modification of proteins for existing applications, but also the design of a sustainable future in which biomolecules play a central role in solving humanity's problems (Figure 9.1).

Definition and Significance of Protein Engineering

Protein engineering is the definition encompassing the design, manipulation, and optimization of proteins to obtain desired functionalities and features. This cross-disciplinary field combines elements of molecular biology, biochemistry, and genetic engineering in a way that allows for the alteration of protein structures at the molecular level. Scientists use various approaches, including site-directed mutagenesis and directed evolution, to create proteins with higher stability, catalytic activity, or specificity. The potential uses of protein engineering are diverse, ranging from the production of biofuels and drugs to the construction of therapeutic agents that target specific diseases. By altering the amino acid composition of proteins, researchers can investigate how certain modifications alter the function and behavior of proteins, and thus expand the level of biochemical studies. Not only for its applications but also for its biomedical and biotechnology consequences, the significance of protein engineering exists. Engineered enzymes, for instance, significantly enhance industrial processes by streamlining them to be more efficient and less invasive to the environment. In addition, therapeutic proteins such as antibodies and hormones are assisted by engineering methods that maximize their activity while reducing side effects. Genetically engineered organisms such as AquAdvantage Salmon's approval show the regulatory issues that accompany protein engineering uses. As quoted, the FDA's treatment of genetically engineered food under the New Animal Drug Application provisions raises transparency and accountability issues in risk assessment. This shows the tightrope that must be walked between innovation and safety in creating engineered products that can modify ecosystems and human health (McEvilly et al., 2013). Furthermore, protein engineering research is ongoing, with new technologies on the horizon that will unlock higher levels of knowledge and ability. New developments in synthetic biology and computational models enable more accurate changes, enabling the tailoring of proteins to carry out new functions previously out of reach. The ongoing research in this field, along with the collaborations between industries and universities, puts forward the requirement for ongoing research and development in protein engineering. As new techniques are emerging, there is a requirement for viewing their efficacy and safety as a whole. So, the applicability of protein engineering is not limited to its success on a scientific scale; it also encourages constant debate over its ethical implications and the scientists' responsibility in maintaining public trust through transparency in their endeavors and their applications (Eller et al., 2025) (Figure 9.2).

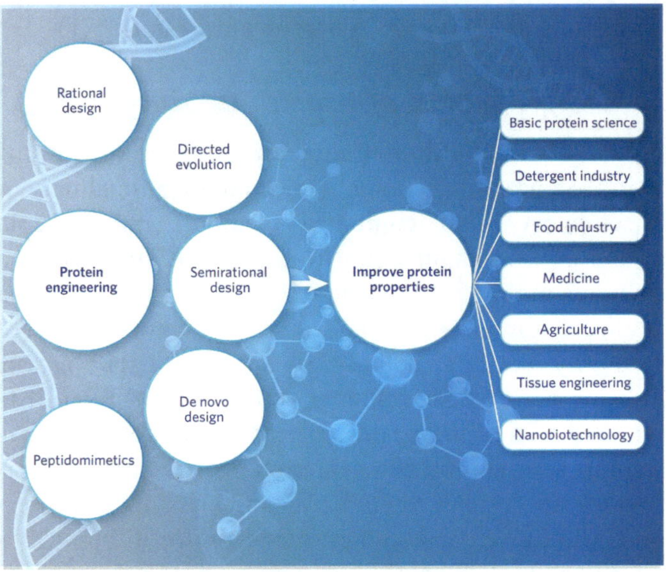

Figure 9.2. Protein engineering strategies such as rational design, directed evolution, semirational design, de novo design, and peptidomimetics help scientists improve protein properties for a range of applications.

Source: https://cdn.the-scientist.com/assets/articleNo/72286/iImg/54645/72867-ts-brushup-proteinengineering-3p-l.webp.

Historical Context and Development of the Field

The development of protein engineering owes much to the deep-seated innovations in molecular biology and biochemistry that occurred over the course of the late 20th century. Emerging from basic genetic research, the field transitioned to revolutionary advancements with the introduction of recombinant DNA technology in the 1970s. Through this technology, researchers could manipulate genetic material, enabling the possibility of creating target proteins in host organisms, paving the way for future innovations. The development of techniques such as the polymerase chain reaction (PCR) further accelerated protein engineering by enabling the amplification and modification of genes. And so, with the potential to engineer and produce proteins with functions aligned to meet particular requirements came the fusion of biochemistry and biotechnology into a cohesive science. And so this historically evolved factor has been pivotal to shaping the approaches and applications encountered in current protein engineering. The search for understanding protein structure and function has further driven protein engineering research. By the 1980s and 1990s, methods of structural biology such as X-ray crystallography and nuclear magnetic resonance spectroscopy (NMR) allowed it to be more straightforward to comprehend how protein conformation makes biological activity. This led to the development of directed evolution techniques by which researchers could mimic natural selection in the laboratory to produce proteins with particular attributes. Protein engineering thus developed as an area of keen interest not only for research scientists but also for industries that wanted to exploit the new opportunities offered by engineered proteins in therapeutics, agriculture, and biodegradable materials. The overlap of scientific interest and practical application is an indication of how earlier comprehension of structure-function relationships in proteins continues to guide the methods employed today in protein engineering. In addition, the social effects of protein engineering have become increasingly clear as the field continues to develop. The possibility of engineering proteins for medical use, for instance, provokes essential questions in the areas of ethics, safety, and public acceptance. Highlighted by lessons of history, the creation of similar interdisciplinary sciences emphasizes the necessity of considering social concerns along with scientific progress. The field of artificial intelligence, for example, has brought risk and responsibility forums into the picture as it is integrated into daily life (Ahuja et al., 2013). The protein engineering industry, thus evolving in tandem with societal demands, must consider all these ethical issues. An understanding of the historical underpinnings and ongoing developments will lend perspective to a greater appreciation of how protein engineering can be utilized responsibly to address global concerns while advancing scientific knowledge and practice.

Overview of Applications in Biotechnology and Medicine

Protein engineering, in the rapidly evolving fields of biotechnology and medicine, is an important stimulus for therapeutic applications and diagnostic development. By the promise to design protein structures, researchers are able to design more specific and stable enzymes and antibodies. These engineered proteins form the foundation of targeted therapeutics for diseases such as cancer, where monoclonal antibodies are now an important reagent in immunotherapy. Finally, protein engineering enables the development of biosensors to monitor in real-time biological processes, enhancing the specificity of medical diagnostics. Overall, the medical application of protein engineering not only promises new treatments but also improves upon the efficiency and precision of current diagnostic techniques, representing a compelling unification of biotechnology and the practice of medicine. Therapeutic uses aside, biotechnology has an even wider list of innovations, particularly in agriculture and environmental sectors. The integration of genetic engineering technologies like CRISPR and recombinant DNA has transformed crops by introducing nutritional traits and resistance to disease. These innovations are essential in dealing with food security challenges on a global scale. Furthermore, as highlighted in recent reviews, immobilization of bacterial cells and enzymes, as well as the production of new vaccines, showcases the viability of biotechnology in addressing both agricultural and health problems. The research into these new tools, such as epigenomics and bioinformatics, not only answers questions about fundamental biological processes but also opens up new niches in the biotechnology sector, towards sustainable agriculture and improved public provision of health (Baianu et al., 2004). The policy impacts of biotechnology are extremely broad in terms of both policy and market forces since new technologies affect both industry expectations and policy control. For instance, analysis of current activity in the biotechnology industry suggests urgent trends and problems to resolve. It is crucial that there be successful communication between scientists and policymakers for effective regulation that promotes innovation but solves issues of safety and effectiveness. A Center for Space and Advanced Technology report illustrates the importance of an awareness of industry views of novel biotechnological applications, portraying the need for responsive policies to support development and prevent market contraction. By linking research advances to pragmatic policy changes, the biotechnology sector can further apply protein engineering to achieve dramatic advances in medicine and agriculture, ultimately enhancing human health and environmental sustainability (Johnston et al., 2025).

9.1. FUNDAMENTAL CONCEPTS OF PROTEIN STRUCTURE

Fundamentals of protein structure form the core of being able to comprehend their function and, therefore, to the practice of protein engineering. Proteins are composed of long chains of amino acids that fold into their distinct three-dimensional conformations, necessary for the performance of their biological function. Folding is influenced by many factors, including chemical environment, temperature, and the presence of other molecules. Protein folding, stability, and interaction principles direct methods of designing proteins for a given functionality. Designing such proteins typically requires a thorough understanding of primary, secondary, tertiary, and quaternary structures because each level of organization is accountable for the overall behavior of the protein in biological functions. Studying these levels enables targeted alterations to enhance solubility or reduce aggregation, which is essential for the development of therapeutic proteins in drug discovery. Proteins are prone to aggregate under certain conditions, and this has been a significant challenge in protein research and engineering. Computational tool design, such as Aggrescan4D (A4D), has emerged as a major tactic for predicting protein aggregation tendency based on structural data alongside pH influence.

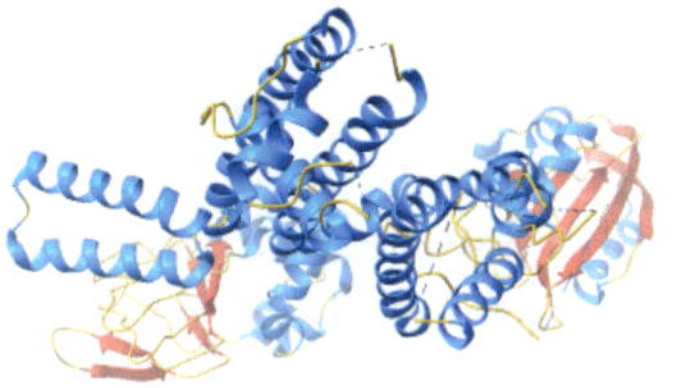

Figure 9.3. Protein structure.

Source: https://3d.nih.gov/api/submissions/14656/runs/657531d9-632d-4454-bbc6-d12757af548e/output-files/326330.

A4D is a sophisticated tool that analyzes the protein structure flexibility, which is directly associated with their tendency to aggregate under varied conditions. By efficient identification of probable sites of aggregation, protein designers can now make informed decisions about mutating specific residues to enhance the solubility of proteins and therefore overcome crucial issues in biopharmaceuticals. Because protein aggregation can cause efficacy delay as well as increase immunogenicity in clinical use, technologies like A4D offer valuable innovations towards attempting to maximize protein function and design in therapeutic use, melding cutting-edge computational power with traditional engineering methodologies. Also, the development of knowledge in protein structure is important for the advancement of protein design with the latest approaches in bioinformatics. Recent studies aim to merge structural and sequence-based features to generate numerical representations of protein characteristics with functional information. With machine learning techniques, particularly with the information from the Protein Data Bank (PDB), more accurate classification and protein function prediction can be done. Historically, sequence has been the focus; with the use of three-dimensional structural information, it gives a more complete picture on which to base analysis. An example is the development of contact map-based embeddings that show how geometrical relationships within protein structures can enhance supervised learning performance in function prediction. These novel methods focus on the highest convergence of protein structure and engineering so that subsequent studies can overcome the challenges presented in protein function analysis and design toward the creation of innovations in therapeutic applications (Figure 9.3).

9.1.1. Levels of Protein Structure: Primary, Secondary, Tertiary, and Quaternary

Being aware of the protein structure levels is crucial when it comes to protein engineering since it tells us how proteins acquire their functional structures and functions within biological systems. The primary structure is a straight chain of amino acids linked by peptide bonds that not only defines the protein itself but also its potential for additional structural constructions. The composition of the amino acids inherently affects the characteristics of proteins since changes in this pattern will lead to substantial functional changes. This fundamental level is essential for the following folding processes that define secondary, tertiary, and quaternary structures. By modifying the primary sequence using techniques such as site-directed mutagenesis, protein engineers can rationally analyze how specific modifications influence higher-order structures and, ultimately, the protein's function. Protein secondary structure is a description of the local folding patterns due to hydrogen bonding between the backbone atoms of the polypeptide. The alpha helices and beta sheets are common and contribute to the stability and integrity of the protein. This is a significant level of structure as it lays the foundation for more complex folding in the tertiary structure, where the overall three-dimensional conformation is established. The forces that drive the formation of tertiary structure include ionic interactions, hydrophobic interactions, and Van der Waals forces, which collectively stabilize the final confirmation of proteins. Understanding the principles of secondary structure is essential in protein engineering, as changes at this level can greatly alter a protein's stability and functionality within biological systems, thereby informing the creation of proteins with novel properties. Finally, the quaternary structure is the assembly of more than one polypeptide chain into a single functioning unit, and this represents protein interaction complexities. This level of structure is most relevant to multi-subunit proteins, where interactions between different polypeptides can lead to cooperative or allosteric behavior. These kinds of interactions often dictate a protein's functionality and are contingent on each subunit's primary, secondary, and tertiary structures. Quaternary structure importance in protein design cannot be overstated, being a reflection of the dynamic function of proteins in various biological processes. Based on an understanding of the laws of energy landscapes and structural motifs, it is possible to design proteins that utilize iterated units or structural mosaics, thereby creating new functionalities in biomedicine and biotechnology. Indeed, structurally similar proteins would possess analogous biological functions, suggesting that their levels of comprehension provide essential data concerning their engineering and application (Othman et al., 2010) (Figure 9.4).

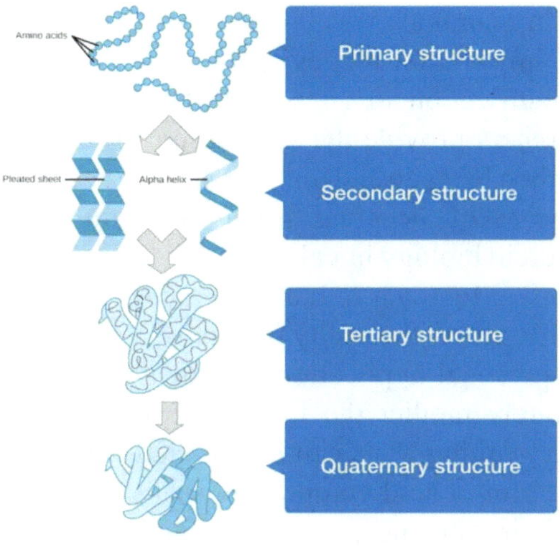

Figure 9.4. Levels of protein structure.

Source: https://geneticeducation.co.in/wp-content/uploads/2018/06/protein-structure.004-3623930124-1529393620113.jpeg.

9.1.2. Importance of Amino Acid Sequences in Protein Function

Amino acid sequences and protein function have a complicated relationship, which plays a central role in protein engineering. Amino acids, the building blocks of proteins, determine protein structure and consequently their diverse range of functions in biological systems. Amino acid sequence determines the folding into a three-dimensional shape of a protein, which is essential for stability and active sites of critical importance to its function. For example, proteins have specific sites where other molecules fit in and have interactions, and any mutation or modification of the amino acid sequence could have a significant impact on such interactions, leading to altered or compromised function. The importance of precise amino acid sequencing in achieving desired protein traits, thus contributing to the development of therapeutic and biotechnological applications, comes to the fore. New computational modeling advancements have allowed researchers to predict the functional consequences of amino acid sequence changes more adequately. The approach can be best exemplified by the study of the SIRT6 protein, where researchers employed structure prediction tools, i.e., RoseTTAFold, to contrast evolutionary conservation of amino acid sequences across species. Through the utilization of this analysis, a robust correlation between specific amino acid structures and maximum lifespan in mammals was created, underscoring that small changes might have critical impacts on longevity and protein function (Zhang et al., 2024). Through identification of sites critical to function through sequence optimization, researchers provide the foundation for novel interventions with the potential to lengthen lifespan and improve health outcomes. Accordingly, the interplay of amino acid sequences and structural dynamics not only enriches our understanding of protein biology but also consolidates the promise of designing better protein variants. Moreover, novel approaches to protein engineering, as in the case of ProMEP, affirm the utility of employing amino acid sequence data for the purpose of practical applications. Utilizing a deep learning platform, ProMEP can accurately predict the impact of a given mutation on protein function so that gene-editing tools can be targeted and improved. By learning how changes in the amino acid compositions of enzymes such as TnpB and TadA can improve their efficiency, engineers can design improved variants that greatly surpass their wild-type equivalents (Cheng et al., 2024). This ability to predict and manage mutational impacts emphasizes the pivotal role of amino acid sequences in the design of proteins for therapeutic use. With advancing science, the close interdependence of computational power and biological insight will undoubtedly change the pace of protein engineering, with profound implications for medicine and synthetic biology (Figure 9.5).

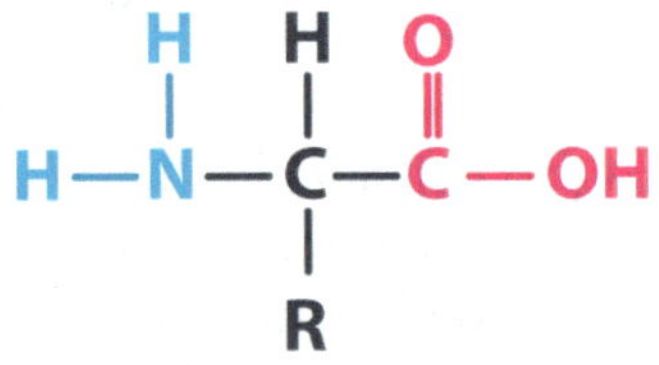

Figure 9.5. Amino acids formula structure.

Source: https://cdn1.byjus.com/wp-content/uploads/2017/11/amino-acid-formula.png.

9.1.3. Techniques for Studying Protein Structure (e.g., X-Ray Crystallography and Nuclear Magnetic Resonance (NMR))

The study of protein structures is a foundation for the branch of protein engineering, directed toward protein manipulation for various aims in medicine andbiotechnology. Severalmoderntechniques, particularly X-ray crystallography and NMR spectroscopy, are primary techniques for the determination of the three-dimensional structure of proteins. X-ray crystallography, first used for proteins in the 1950s, provides high-resolution structural data through analysis of the diffraction patterns of X-rays passed through protein crystal specimens. The technique allows researchers to acquire detailed atomic-level data regarding the conformation of proteins and how it correlates with their function and interaction. Moreover, the latest focus on polysterism, as mentioned in Fraser et al. (2010), underscores the importance of understanding conformational heterogeneity, which, in turn, shapes our view of protein dynamics and their roles in life across different environments. Unlike X-ray crystallography, NMR spectroscopy provides an exclusive method through studying proteins in solution, thereby preserving their native surroundings.

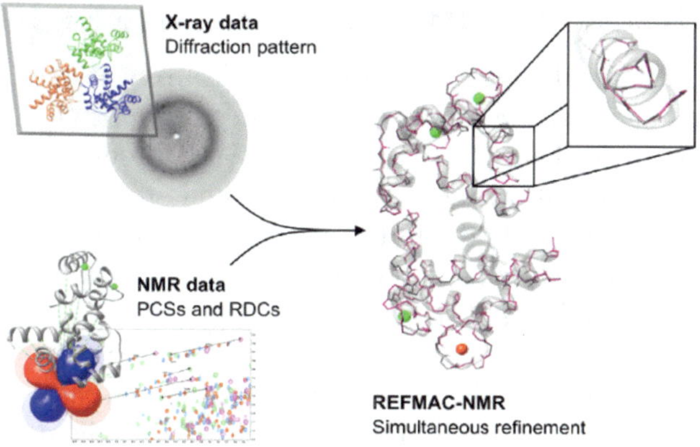

Figure 9.6. Information used in REFMAC-NMR structure refinement.

Source: https://www.creative-biostructure.com/images/Combining-X-ray-Crystallography-and-solution-NMR-data.png.

This technique is particularly helpful for the analysis of floppy regions of proteins and transient interactions that are lost in crystal forms. NMR provides a window into protein folding-unfolding dynamics and ligand-induced conformational adjustments. The availability of advanced NMR techniques allows researchers to interrogate protein interactions at the atomic level, which is critical for the rational design of protein engineering. Furthermore, application of NMR spectroscopy can complement X-ray crystallography to detect more regarding conformational substates with essential functions in proteins, again emphasizing the complex methodologies available to us in structural biology. With the ongoing development of protein engineering, the importance of accurate determination of protein structures has emerged more sharply into view, creating controversy over intellectual property implications of so doing. With the advancement in structural methods, the patentability of protein three-dimensional structures has been questioned, as seen in Quarmby et al. (2003). This is important when looking at the application of structural information to help protect protein function and design innovations. The intersection of patent law and structural biology highlights the rising recognition of how a better, more refined understanding of protein conformation can not only allow for the engineering of new proteins but also influence the biopharmaceutical industry as well. Developing further our techniques of inquiry into the structure of the protein allows us to open up new possibilities for advancements in both scientific progress and exploitable technology (Figure 9.6).

9.2. METHODS OF PROTEIN ENGINEERING

One of the fundamental aspects of protein engineering is the application of site-directed mutagenesis, a time-honored but fundamental approach. The method is a technique where intentional mutants are introduced at specific sites within a protein's gene, allowing researchers to study the relationship between amino acid composition and protein function. By making systematic sequence modifications, researchers can ascertain the effect of such modifications on the protein's stability, activity, and structure. The application of biocomputational tools enhances this process, as it assists in the prediction of the effects of the mutations based on the protein folds and structures presently available. The combination of empirical analysis and rational design has the potential to bring forth information that can galvanize significant advancements in therapeutic applications. However, a strong understanding of molecular biology, even in its subtle aspects, including the "intricacies of protein interactions" and the "principles governing protein functionality," is demanded by site-directed mutagenesis competence, testified to by the new character of protein changes described by Khan and Akhtar (2011). In contrast to the precision of site-directed mutagenesis, non-rational approaches like directed evolution exploit the effectiveness of natural selection in generating protein diversity. This strategy involves random gene mutation and screening of vast libraries of resulting proteins to identify variants with desired characteristics. Gene shuffling and error-prone PCR are among the methods used to produce libraries with broad mutation spectra. Through cycles of mutation and selection repeated over time, scientists can refine protein characteristics, such as stability, enzymatic activity, or binding affinity. This approach allows functional landscapes to be investigated beyond the boundaries of rational design, allowing for serendipitous discovery of protein function. The addition of advanced technologies enables the identification of optimal candidates, highlighting the use of directed evolution to develop proteins for specific applications, thereby improving the discipline of protein engineering, as outlined in Khan and Akhtar (2011). The intersection of protein engineering and therapeutic delivery technologies is an exciting new horizon for biomedical science. New methods are being developed to improve drug delivery systems, particularly those with proteins and genes as therapeutic agents. For instance, the adoption of the utilization of nanoscale delivery vehicles that are naturally derived or synthetically made has been innovative in improving the pharmacokinetics and biodistribution of therapeutic proteins. Physical property characterization of these nanomaterials is of significant relevance, as this dictates their performance within biological systems, addressing issues of stability and agglomeration difficulties. With therapeutic delivery technology advancing, they offer considerable potential for driving entrepreneurship and innovation in the biopharmaceutical sector. Enabling through rigorous research and analysis, these advancements not only provide greater therapeutic effects but also create the significance of protein engineering in addressing complex medical needs, as discussed in the growing interest in this area (Figure 9.7).

Figure 9.7. Scientists employ protein engineering techniques to design and create proteins with useful properties for medicinal and industrial applications.

Source: https://cdn.the-scientist.com/assets/articleNo/72286/aImg/54650/brushup-geneticengineering-1800x1260-l.jpg

9.2.1. Site-Directed Mutagenesis and Its Applications

Site-directed mutagenesis is the peak of protein engineering, and it offers a flexible array of tools for scientists to redesign individual amino acid sequences in proteins. Site-directed mutagenesis permits the introduction of designed mutations, allowing scientists to investigate the function and structure of proteins with monumental specificity. By altering single amino acids, scientists can explore the effect of these alterations on a protein's stability, activity, and interaction with other molecules. The resulting insights have wide-ranging applications in drug design, therapeutic protein production, and enzyme engineering. As mentioned, mutagenesis approaches involving multidisciplinary methodologies are needed to realize the potential of protein engineering by integrating traditional methodologies with modern computational predictions to achieve sought-after functional properties in proteins (Khan and Akhtar, 2011). Moreover, developments in recent years have blended site-directed mutagenesis with machine learning, streamlining the process with data-driven sophistication.

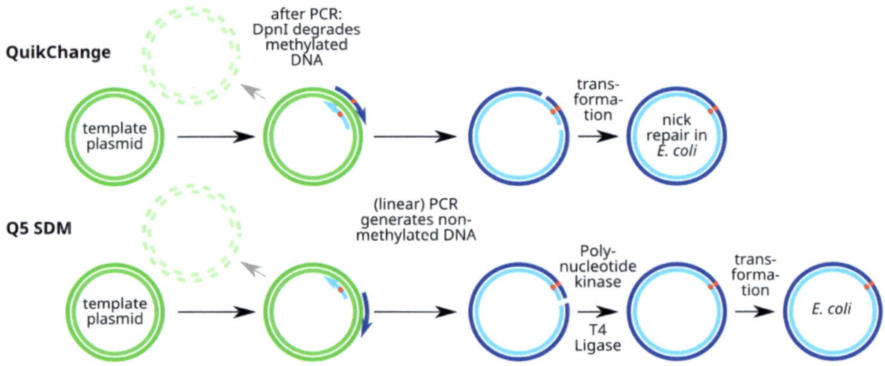

Figure 9.8. Site directed mutatgenesis.

Source: https://upload.wikimedia.org/wikipedia/commons/e/ed/Site_directed_mutatgenesis.svg.

This integration facilitates researchers to predict the effect of specific mutations on protein functions ahead of extensive biological expertise. Machine learning algorithms handle enormous sets of protein sequences and their corresponding activities, providing insights into which mutations will potentially optimize desired attributes. By utilizing these ML-assisted approaches, researchers can make the directed evolution process more efficient by focusing their experimental effort on the most promising candidates, thereby saving substantial time and lab resources (Arnold et al., 2019). Besides insight into protein dynamics in even deeper levels, the utilization of machine learning also results in an acceleration of the engineering rate of proteins with higher or novel functions. In conclusion, site-directed mutagenesis represents a central approach in protein engineering, which opens doors to advancements in all fields of science. With its ability to allow for the specific alteration of proteins, our strategy towards understanding biological systems and constructing therapeutic compounds has been revolutionized. This technique becomes even more powerful with the incorporation of machine learning, advancing the area towards faster discovery of protein functions and optimization. As researchers continue to take advantage of these advances, the horizon of protein engineering is bound to expand, and novel biomolecules with certain functions will be identified that can revolutionize medicine, industrial processes, and numerous other domains. The future is promising for this field, as it increasingly combines traditional lab techniques with cutting-edge computational methods, further establishing the boundaries of what can be achieved in protein design (Arnold et al., 2019; Khan and Akhtar, 2011) (Figure 9.8).

9.2.2. Directed Evolution Techniques for Protein Optimization

Directed evolution strategies have transformed protein engineering science by simulating the natural evolutionary process to improve protein functions. With applications like random mutagenesis, recombination, and selection, researchers are able to generate variant proteins that are screened for desired characteristics, such as increased stability or enhanced catalytic activity. This iterative process leads to the identification of variants with improved properties in a significantly reduced timeframe and cost compared to traditional methods of protein design. Directed evolution is powerful as it can effectively sample vast sequence spaces and identify new functions that are inaccessible by rational design alone. Use of machine learning (ML) in directed evolution strategies is a significant achievement in protein improvement. ML can handle data obtained from pools of mutant proteins and infer weak relations between function and sequence, optimizing the rounds of evolution iteratively. Through the prediction of mutation probability for improved traits, ML-directed evolution accelerates the optimization process. It has been achievable for researchers to utilize this paradigm to design models that not only guide mutagenesis strategy but also indicate specific variants to test according to their anticipated performance. This novel framework illustrates the potential of ML to enhance protein engineering quality and efficiency and provide new avenues for functional discovery and utilization in biotechnology (Arnold et al., 2019). In addition, advances in multistate design algorithms have provided increased insight into engineered protein activity and stability. The majority of traditional models of computation rely on single stable conformational assumptions and do not account for proteins' dynamic nature that exists in multiple forms. With the use of multistate design approaches, researchers have demonstrated that the incorporation of diverse structural data into optimization processes can lead to variants with significantly improved thermodynamic stability. This approach allows for a more extensive examination of protein function under various conditions and can result in the development of stable proteins with broad utility. Interestingly, studies have shown that designs are not required to precisely fit experimentally derived numbers to be deemed successful, suggesting the potential for the development of better protein design approaches in the future (Nisthal et al., 2010) (Figure 9.9).

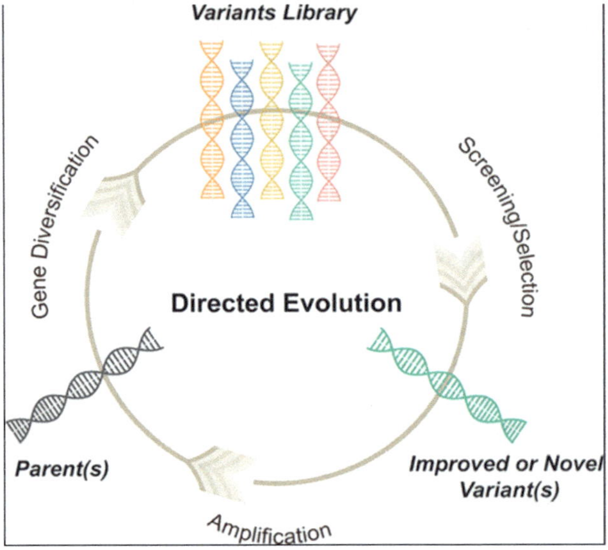

Figure 9.9. Directed evolution.

Source: https://pubs.acs.org/cms/10.1021/acs.chemrev.1c00260/asset/images/medium/cr1c00260_0037.gif.

9.2.3. Computational Methods in Protein Design

The integration of computational methods in protein design represents a significant advancement in the field of protein engineering, offering a systematic approach to understanding and manipulating protein structures and functions. By employing simulations and algorithms, researchers can predict how specific amino acid sequences will fold into three-dimensional structures, paving the way for the design of proteins that meet particular functional requirements. This theoretical basis enables exploration of novel protein structures, particularly in cases where conventional lab methods are lacking. Although many successes have been realized using computational design, the practicality of methods is frequently overwhelmed by the success of directed evolution, which has a tendency to yield more stable proteins within the laboratory. However, these improvements in computational techniques still direct and supplement our understanding of protein surfaces and their numerous applications in medicine and biotechnology (Mayo et al., 2015). As challenges in protein design become increasingly complex, the need for new methodologies has led to the emergence of a hybrid approach that combines computer-aided design with experimental methods such as directed evolution. An excellent example can be observed in the development of a red fluorescent protein (RFP), where researchers designed a hybrid construct that made it easier to control its oligomerization states. By successfully separating the complexity of surface design while simultaneously stabilizing the core protein, the research yielded a library of variants with excellent solubility and function.

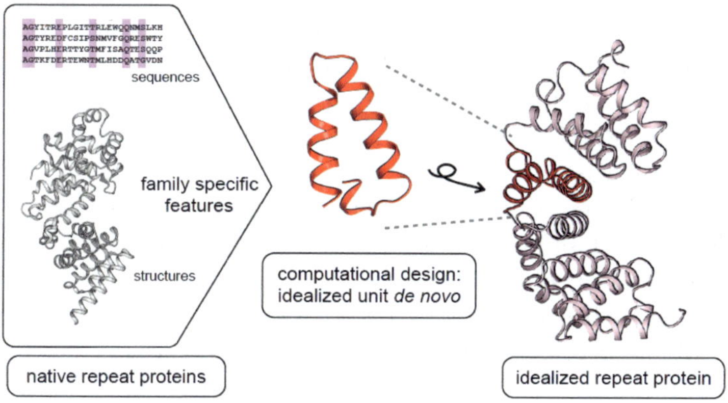

Figure 9.10. A general computational approach for repeat protein design.

Source: https://www.bakerlab.org/wp-content/uploads/2015/12/Fabio_repeat.png.

Therefore, this method emphasizes the necessity to maximize both computational models and experimental protocols in an attempt to discover desired engineering results. The interaction between computational design and empirical verification ultimately reveals significant information on the process of protein engineering, and thus validates that computational methods should not be used alone but in conjunction with experimental testing so as to gain maximum utilization of their contribution to protein design (Mayo et al., 2015). Taking its strength from computational methods, the focus on de novo protein design has now become a revolutionary trend in protein engineering. Here, researchers aim to create totally new proteins with desired functions based on rational design and computational modeling strategies. A demonstration of the potential indicated by this method lies in the iterative build-up of the Octarellins, highlighting successful structural characterization and bacterial expression. While the initial generations suffered from poor solubility and stability, taking a directed evolution approach significantly enhanced the quality of the produced proteins. This illustration not only shows the merit of combining computational design with laboratory-driven improvements but also the ongoing progress of the technologies as complementary drivers of protein engineering. The iterative loop between design and empirical research finally carries the potential for advancing knowledge of protein functionalities and is thus critical for future progress within the field (Figure 9.10).

9.3. APPLICATIONS IN MEDICINE

Protein engineering's medical applications are varied and evolving continuously, largely driven by advances in our understanding of protein functionality. One such important area is the development of biopharmaceuticals, in which proteins are engineered to function as therapeutic drugs. For instance, monoclonal antibodies have been designed to bind to certain antigens associated with a variety of diseases, such as cancer and autoimmune disorders. These modifications enhance efficacy, reduce side effects, and tailor treatment to the specific needs of an individual patient, thereby giving rise to personalized medicine. The ability to regulate protein functions and interactions by using techniques such as directed evolution and rational design has facilitated easier production of more effective therapeutic agents, illustrating the pivotal role of protein engineering in modern medicine (Krishna et al., 2015). In addition, the integration of systems biology approaches enhances protein engineering's value in a clinical setting. Several computational tools and databases have been created to make complex biological systems understandable, linking genetic data to functional protein outcomes. For example, novel platforms created by the Indiana Center for Systems Biology and Personalized Medicine have advanced remarkably in predictive models of cancer treatment. They enable scientists to graphically forecast interactions within biological pathways and, in the process, identify potential therapeutic targets. Such systems biology strategies are complemented by protein engineering activities, integrating experimental and computational data to ultimately enhance the design of novel therapies that are effective and targeted (Chen et al., 2011). Moreover, the use of protein engineering in regenerative medicine cannot be overstated. Gene editing and the development of tissue-engineered constructs rely significantly on engineered proteins for stimulating cell growth and differentiation. For example, scaffolds that consist of engineered proteins can serve as a biodegradable template for tissue development that facilitates healing processes in a variety of clinical applications. These innovations are a testament to the capability of protein engineering not only to generate therapeutic proteins but also to revolutionize approaches in tissue repair and organ regeneration. Since technologies get better and are integrated with newer technologies, the applications of protein engineering to impact regenerative medicine will expand further, making treatment outcomes and patient care better (Krishna et al., 2015) (Figure 9.11).

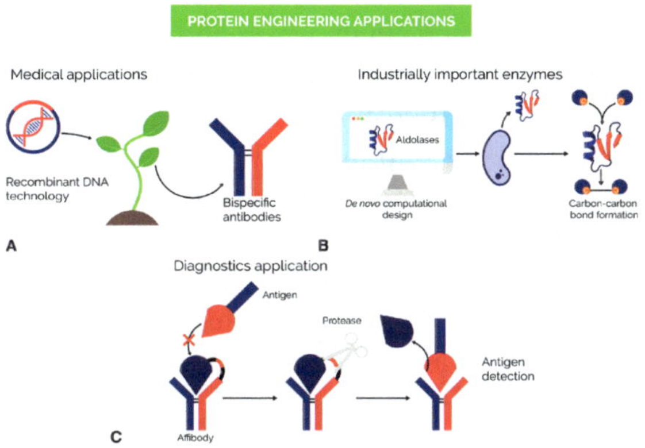

Figure 9.11. Protein engineering application. (A) Medical applications. Bispecific antibodies are a widespread medical application of protein engineering. This application is based on the physical fusion of two antibodies and the implementation of recombinant DNA technology to produce in plants. (B) Industrially essential enzymes. Protein engineering can create selective chemical modifications in enzymes with industrial applications. (C) Diagnostics application. The affibodies are very useful in the diagnostics field. This antibody is created using a rational design to modify the original antibody.

Source: https://www.researchgate.net/publication/373215903/figure/fig3/AS:11431281251869175@1718397472269/Protein-engineering-application-A-Medical-applications-Bispecific-antibodies-are-a.png.

9.3.1. Development of Therapeutic Proteins and Enzymes

The development of therapeutic proteins and enzymes is the core function of protein engineering because there is a need to treat various health ailments at the molecular level. Therapies are mostly in the form of biologically active proteins that have the capacity to correct or limit the severity of diseases. One of the main emphases is on glycoproteins, which are proteins that contain carbohydrate groups attached by glycosylation. This post-translational processing influences protein folding, stability, and biological activity, and thus its importance in the efficacy of therapeutic proteins. Sophisticated methods of protein engineering, such as site-directed mutagenesis and high-throughput screening, can be employed to engineer glycoproteins with better characteristics. It is thus important to understand the complexities of glycosylation pathways, particularly the N-linked glycosylation pathways, in an effort to rationally design therapeutic proteins with desirable properties (Helenius et al., 2009). Another notable advance in the therapeutic field pertains to the engineering of enzymes specifically for therapeutic use against lysosomal storage diseases (LSDs). Enzyme replacement therapy aims to replace the defective or deficient enzymes that are responsible for making the toxic substrates build up within lysosomes. One big step in this direction has been the construction of genetically engineered exosomes that deliver lysosomal enzymes, such as β-glucocerebrosidase (GBA) in Gaucher disease.

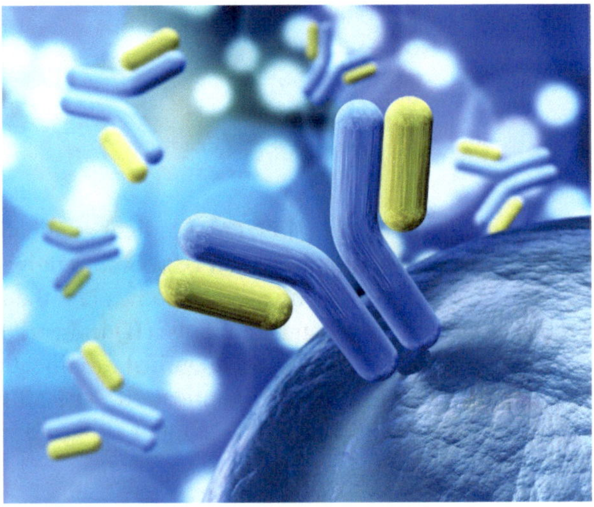

Figure 9.12. Therapeutic proteins.

Source: https://www.susupport.com/uploads/media/default/05/2485-therapeutic-proteins.webp?v=2-0.

Using a genetic engineering method that targets exosomes for delivery to cells' endocytic compartments, researchers have enhanced the therapeutic activity of these enzymes. Efficient delivery of some enzymes into exosomes and efficient integration allows their targeting at the precise location where their activity is needed, which holds the promise of strengthening the efficacy of treatment in central nervous system disorders (Brown et al., 2019). Continuous advancements in protein engineering technology are the key to the creation of therapeutic proteins and enzymes, which hold vast possibilities for personalized medicine. With advancing knowledge of protein structure and function, the ability to design personalized therapeutics with defined pharmacokinetics and biological activity enters the realm of possibility. Methods like CRISPR gene editing and artificial intelligence-guided design are enabling more intelligent therapeutic interventions, such as the ability to design customized proteins tailored to individual patient needs. Furthermore, the implications of these advances extend beyond treatment since they open new avenues for more effective management programs in chronic diseases. Therefore, the design of therapeutic proteins and enzymes not only reflects the significance of protein engineering but also raises basic questions regarding ethics and access to such sophisticated medical therapies (Figure 9.12).

9.3.2. Role in Vaccine Design and Production

The role of protein engineering in vaccine manufacturing and design is most prominent, particularly with the growing demand for rapid responses to novel infectious diseases. By manipulating protein structure, scientists can enhance the immunogenicity of candidate vaccines to induce more potent immune responses. This is typically achieved by epitope identification and modification—specific regions on antigens recognized by the immune system—with consequent increased ability to elicit humoral and cell-mediated immunity. Additionally, advances in recombinant DNA technology enable the production of vaccines from expression systems that yield enormous quantities of target proteins, a requirement in curtailing public health crises rapidly. The use of these techniques illustrates the innovative role of protein engineering in modern vaccine technology to make vaccine candidates safe and efficacious for human use. In vaccine production, a critical factor is the optimization of the process scheme to reduce costs and promote environmental stewardship. Here, multicriteria optimization methods, e.g., Multicriteria Genetic Algorithm (MUGA) design, have been found effective in optimizing various production factors. For instance, through the use of stochastic algorithms combined with discrete event simulations, researchers can design multiproduct batch plants specifically designed towards producing recombinant proteins for vaccine production. This is particularly relevant where businesses are attempting to support fluctuating needs while minimizing the environmental impact of manufacturing processes. The use of such advanced methodologies not only allows for improved operational effectiveness but also safeguards against potential delays in vaccine delivery, thereby linking efficient protein engineering with environmentally friendly manufacturing practices (Azzaro-Pantel et al., 2007). Aside from that, resolving the uncertainty typically involved in demand prediction is also necessary to enable the optimization of vaccine production flexibility. Through the application of fuzzy logic concepts in the design phases of batch plants, scholars can create dynamic models with the ability to deal with fuzzy variables such as fluctuating demand and available resources. This approach facilitates the concurrent optimization of net present value, management of production schedules, and overall production flexibility. As such, stakeholders are now able to negotiate the complexities of vaccine development scenarios more effectively, leading to better decision-making. With the implementation of these cutting-edge methodologies, the utilization of protein engineering not only infiltrates ideation and formulation of vaccines but also strategic planning and optimization required to manufacture such vital products on a timely and efficient scale (Aguilar-Lasserre et al., 2008).

9.3.3. Protein Engineering in Gene Therapy

The application of protein engineering to gene therapy is a milestone in the management of genetic diseases. By manipulating proteins at the molecular level, scientists can make them more effective and specific, thereby improving therapeutic response. With this tailored approach, scientists are able to create proteins that can selectively target sick cells without affecting healthy tissue. These advancements are particularly pertinent to the delivery of genetic material to target cells. The development of nanoscale delivery carriers, as referenced in current literature, enables the delivery of such therapeutic agents with improved efficiency and specificity, which identifies the importance of probing protein interactions within the delivery vehicles (Savas et al., 2019). As the technologies advance, they have the ability to transform gene therapy from experimental treatments to widely accepted clinical therapies. One anticipated use of protein engineering in gene therapy is the development of tailored gene editing tools, such as CRISPR-Cas9 systems. The protein-engineered molecular scissors have the ability to offer accurate editing of the genome, and thus the correction of faulty genes causing inherited diseases. The ability to improve these proteins—through successive rounds of engineering—allows for greater precision and less off-target effect in gene therapy treatments. Moreover, advances in the knowledge of protein function in biological systems have added to their application in clinical use. For instance, recent studies indicate that microbubbles are effective delivery vehicles for CRISPR reagents to enhance the specificity and efficiency of gene editing (Browning et al., 2016). This blend of protein engineering and gene therapy not only suggests therapeutic strategies but also promises novel paradigms for treatment. The focus on protein engineering extends beyond facilitating therapeutic delivery systems; it also includes creating proteins that can induce target physiological effects in patients. Engineered proteins can be optimized to be stable and exhibit long-term bioactivity, thereby offering therapeutic effects over extended periods of time. Moreover, with the advent of machine learning and computational models, researchers can more accurately model protein structure and interactions. The confluence of these advances in technology makes efficient design of better-targeted therapies possible, which accelerates the development process and permits more rapid clinical translation of novel treatments. Since science continues to grow, the field of protein engineering and gene therapy will be the answer to curing complicated diseases not curable as of today, and hence it is an area of high interest for future research and development.

9.4. INDUSTRIAL APPLICATIONS

Industrial applications of protein engineering are becoming increasingly pivotal because the proteins can be designed specifically for application. One of the major applications is the production of biopharmaceuticals, where engineered proteins are designed to enhance therapeutic efficacy or reduce side effects. For instance, yeast systems such as *Pichia pastoris* have been extensively employed since they can be cultivated to cell densities and secrete a lot of recombinant protein. These systems employ strong promoters for stringently regulated expression levels and allow intracellular and secreted protein production. However, complexity comes when expressing complex proteins, particularly hetero-oligomeric or degradation-sensitive proteins. Ongoing efforts to enhance protein yields through designed protein engineering demonstrate the extensive impact of such developments on the pharmaceutical sector, broadening the scope and viability of biopharmaceutical formulations (Pichler et al., 2014). Furthermore, protein engineering-led innovations in functionalized materials hold significant promise for various industrial applications. One of these breakthroughs is the engineering of bacterial polyhydroxybutyrate (PHB) inclusions as a universal platform for applications ranging from vaccines to enzyme immobilization. Through the application of the bacterium *Bacillus megaterium*, researchers can isolate PHB inclusions that have the desired protein characteristics. By genetic modification of the organism to eliminate undesired traits, such as lipopolysaccharide secretion, the safety and efficiency of the generated material are enhanced. Mutagenesis of the PHB synthase enzyme has enabled the introduction of IgG binding domains to allow for efficient purification of antibodies from these engineered beads, furthering the applied utility of this technology to biopharmaceuticals and diagnostics (Grage et al., 2017). More generally across industrial applications, advances in protein engineering not only drive efficiency of production but also encourage sustainability in biotechnology. The ability to engineer novel proteins that catalyze reactions or improve product yields is significant for developing greener production processes. For example, engineered enzymes can be employed to degrade waste products or create biofuels using renewable feedstocks, lessening the reliance on fossil fuels. Protein engineering to optimize enzyme function at maximum performance is a quintessential example of biotechnology applying to the solution of environmental issues in response to economic requirements. The two-sided benefit emphasizes the revolutionary application of protein engineering for furthering industrial development and ecological stewardship, eventually becoming an integral aspect of modern manufacturing (Grage et al., 2017; Pichler et al., 2014) (Figure 9.13).

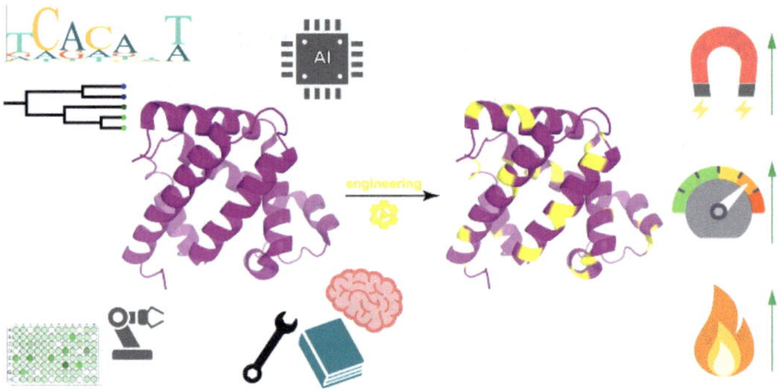

Figure 9.13. Protein engineering for industrial biocatalysis.

Source: https://pub.mdpi-res.com/catalysts/catalysts-15-00147/article_deploy/html/images/catalysts-15-00147-ag.png?1738811616.

9.4.1. Use of Engineered Proteins in Biocatalysis

The application of engineered proteins in biocatalysis depicts the revolutionary aspect of protein engineering in the scope of biochemical reactions. Altering protein structures through approaches such as rational design and directed evolution enables researchers to engineer biocatalysts not just to enhance the efficiency of reaction but also to maximize substrate specificity. These engineered forms tend to have novel properties that render them appropriate for use in industry, such as pharmaceuticals, where efficiency and accuracy are of the utmost importance. For example, circular permutation methods enable scientists to resequence amino acids, greatly modifying the protein's function without changing its master form. This new approach assists in forming robust and effective biocatalysts, thus showcasing the great capabilities of engineered proteins in tackling contemporary biochemical issues (Das and Ainavarapu, 2024). Evolution has occurred only in the field, leading to the evolution of enzymes that work exceptionally well in biocatalytic applications. A case in point is the bacterial SDR ketoreductase enzyme from *Thermus caliditerrae*, characterized for its ability to produce optically pure alcohols and ketones. The biooptimization of the enzyme with the aid of bioinformatics tools, such as the BioMatchMaker® platform, is a paradigm case of the integration of computational approaches into protein engineering. By the identification of wild-type sequences of potential desired chemical transformations, researchers can create more efficient and selective biocatalysts. Exploring the structure of the Tcalid SDR enzyme also provides further details on substrate interaction and stability, which are of utmost concern in the development of high-performance biocatalysts for use across various industries (Sosa et al., 2024). Furthermore, the use of engineered proteins as biocatalysts is inherently linked to their stability and working lifespan under the reaction conditions. The activity of engineered enzymes can largely be increased when they maintain their structurally intact conformation at elevated temperatures or in extreme pH levels. The investigation of protein stability, as in the instance of the aforementioned Tcalid SDR, has elucidated its good reactivity even after being heat-treated, demonstrating the potential of these biocatalysts in real-world applications. The knowledge thus gained by investigating protein folding, thermodynamic processes, and catalysis helps understand how efficient biocatalysts can be designed, not only functionally optimal but also environmentally responsive across various industrial environments. As protein engineering has progressed, the possibilities for creating more efficient as well as sustainable biochemical processes are enormous, providing scope for ground-breaking discoveries in biocatalysis (Das and Ainavarapu, 2024) (Figure 9.14).

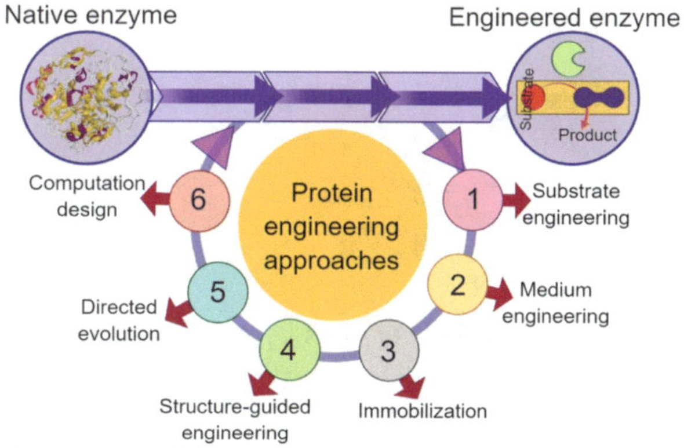

Figure 9.14. Protein engineering approaches to develop multipurpose biocatalyst.

Source: https://www.researchgate.net/publication/333196712/figure/fig1/AS:941530328092679@1601489731077/Protein-engineering-approaches-to-develop-multipurpose-biocatalyst.png.

9.4.2. Role in the Production of Biofuels

Protein engineering's contribution to the production of biofuels is increasingly important, particularly in the production of microbial fuel cells (MFCs). These cells use particular strains of electrogenic bacteria to convert biochemical energy obtained from organic substrates to electrical energy and thus represent a feasible technology for the production of waste-to-energy. By harnessing naturally occurring microbial activity, researchers aim to optimize the efficiency and sustainability of biofuel production. Engaging engineered proteins in MFCs has the capability to optimize electron transfer processes, thereby enhancing the performance and yield of biofuel production systems. Microbial metabolic research at the molecular level directs protein modification processes that enable engineered strains of bacteria to be able to cultivate on various organic waste substrates, thus producing renewable energy options in the process while addressing the issue of wastewater treatment (Duţeanu et al., 2012).

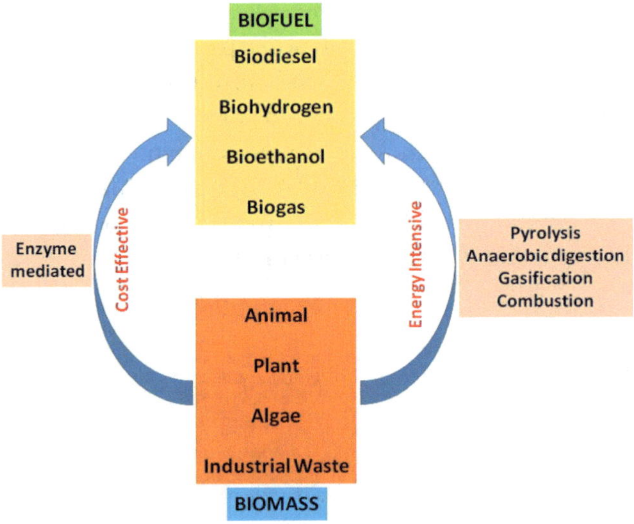

Figure 9.15. Production of biofuels options by contribution of effective and suitable enzymes.

Source: https://ars.els-cdn.com/content/image/1-s2.0-S2589299122000167-ga1_lrg.jpg.

Enzyme-based fuel cells also rely on protein engineering, in which immobilized enzymes catalyze reactions that generate electrical energy. Enzyme engineering for enhanced stability, efficiency, and substrate specificity is crucial for the creation of more efficient biofuel systems. Advanced methodologies such as directed evolution and site-directed mutagenesis allow researchers to design enzyme variants that are robust to severe operating conditions often experienced in biofuel processes. This engineered performance enhancement has extensive implications for cost savings in biofuels production as well as technological scalability. As enzyme-based fuel cells continue to develop, they will not only fuel portable devices but also offer green energy solutions in different environments, altering the way biofuels can become part of our energy profile (Duţeanu et al., 2012). The wider implications of protein engineering in biofuel production agree with the findings outlined in reports by bodies like the International Energy Agency (IEA) that offer a broad overview of the production of bioenergy and utilization of biomass. These reports emphasize the role of biorefining as well as the conversion of biomass to useful biofuels, and they all point to the need for new technologies that will make the processes more efficient and environmentally safer. As protein engineering continues to evolve, its integration into biofuel production systems could yield cleaner energy production methods that align with global energy policy and environmental goals. The hope for protein-engineered solutions provides a route to renewable, cleaner energy options that have the potential to significantly reduce reliance on fossil fuels while maximizing resource recovery from organic waste streams (Figure 9.15).

9.4.3. Applications in Food Technology and Safety

The use of protein engineering in food technology and safety is revolutionizing food product development and testing. One of the most critical applications is the creation of gene-manipulated organisms, such as the AquAdvantage Salmon. This new salmon, which has survived a stringent premarket approval process at the FDA, is a perfect example of the way protein engineering can cut through animal growth rates for food uses. It is controversial; however, the regulatory frameworks surrounding such technologies have been attacked for their openness and for posing risks to health. The manner in which the FDA has handled genetically altered animals under provisions of the New Animal Drug Application has further reemphasized the necessity for even more sweeping safety and effectiveness testing. Resolution of such is necessary, as it mitigates against the benefits of engineered protein to take over potential health and environmental implications (McEvilly et al., 2013). In addition to animal-source food, protein engineering is also applied in enhancing the quality of food at various stages of the food supply chain. New technologies such as Multiplexing Electric Field Sensors (MEFS) have also been identified for their capability to check food safety and quality non-destructively. These sensors offer a very valuable advantage in food processing by enabling the detection of foreign matter, and also the measurement of some parameters regarding quality, like fat content in meat. The low cost and programmability of MEFS offer an additional attraction towards large-scale implementation of this technology in food safety applications. With increasingly strict food safety regulations, the implementation of next-gen technologies can help reinforce consumer confidence and improve food quality evaluation. In this regard, advancements in protein engineering transcend production innovations, significantly lending themselves to food safety activities throughout the industry (Capdeboscq et al., 2018). Furthermore, applications of protein engineering in food technology continue to advance, once again illustrating the harmony between innovation and food safety regulations. One specific focus is the ethical issue with genetically modified organisms (GMOs) from a public acceptance and attitude point of view. Although opportunities exist to improve food production and nutritional value through engineered proteins, public concern often stems from a lack of understanding of the science. It is essential to enhance communication among regulators, scientists, and consumers to foster trust and facilitate acceptance of new technology. Additionally, ongoing research plays a vital role in the determination of long-term impacts of engineered proteins on human well-being and environmental integrity. Finally, safe utilization of protein engineering in food technology relies on attaining innovation without compromising rigorous safety assessments in order to offer a safe and healthy food supply to future generations (McEvilly et al., 2013).

9.5. ETHICAL CONSIDERATIONS

The field of protein engineering has ushered in extraordinary advancements in biotechnology and medicine, but has associated enormous ethical concerns that must be accorded careful attention. One of the primary concerns is the potential for unforeseen consequences of modifying biological systems. The complexity and unpredictability that are inherent in living things have the potential to produce unforeseen ecological or health effects. For instance, depositing designer proteins in natural habitats is a threat of disrupting well-established biological equilibria, with potentially harmful effects on biodiversity. Secondly, the human health aspects, particularly concerning harmful immune responses or pathogen evolution, demand an ethical procedure that prioritizes caution. With expanding research in protein engineering, it is important to offset the potential benefit against the potential risks, and ethical principles should be the basis of decision-making. A further ethical consideration of protein engineering is the application of synthetic biology technology, particularly therapeutic applications such as xenotransplantation. The ethical consideration of access to organs from outside of human society creates the intricate complexity of the interrelationship between scientific innovation and moral principles. Though xenotransplantation promises to ease the shortage of human organs for transplant, it invites concerns regarding animal rights, species integrity, and zoonotic disease transmission. As this frontier is explored, bioethicists underscore the need to develop exhaustive guidelines that can address these challenges. The scientific community needs to interact with various stakeholders, integrating cultural, ethical, and scientific considerations into a balanced discussion. Strong comprehension of the ethical ramifications of such applications is a necessity to build public trust and enable responsible scientific progress (Reiss et al., 2000). Furthermore, the design cycle that underlies synthetic biology itself brings about ethical matters that require resolving in detail. The cycle involves specifying, modeling, analyzing, and testing for engineered biological systems, which, while promising, opens up the door towards ethical ambiguity. The logical engineering of biologically-derived systems and components should carry some vision about their long-term impact, whether expected or unexpected, or their potential to be misused. With the emergence of new techniques and tools, transparency and access questions come even more into prominence, particularly to provide equitable access to biotechnological advances. Ensuring an equitable distribution of the benefits of protein engineering can prevent ethical issues regarding disproportionate health results among populations. Therefore, as the science evolves, it is necessary that ethical attention within the scientific circle ensures that the development and utilization of engineered proteins occur in an environment centered on society's well-being (Freemont et al., 2014).

9.5.1. Potential Risks and Benefits of Protein Engineering

The protein engineering field is one where there is a complicated relationship between risks and benefits that must be approached with care. On the other hand, the possible benefits come in the form of curative treatments for diseases that were hitherto untreatable, such as autoimmune and cancer diseases. For instance, engineered proteins and antibodies can be created to stimulate stronger immunity against some pathogens or tumors, leading to more effective treatments. However, this effective technology also carries built-in risks of unintended consequences. For example, the manufacturing of GMOs is a significant concern for the environment and public health, particularly due to the unclear interactions within ecosystems. The Precautionary Principle (PP) recommends that where there is a lack of certain scientific knowledge, such as releasing GMOs, precautionary actions must be used to prevent potential global harm (Bar-Yam et al., 2014). Hence, rigorous assessment and careful planning are called for in resolving these matters in protein engineering. Additionally, breakthroughs in protein engineering techniques, such as CRISPR and exosome therapy, have brought promise for their therapeutic applications. Exosomes, the naturally occurring nanovesicles, also have promising potential as drug delivery systems, allowing non-invasive approaches to treat rheumatoid arthritis (Curley et al., 2017). The non-immunogenic nature of exosomes further contributes to their appeal, reducing the possibility of unwanted immune responses compared with conventional therapies. While the potential of such novel treatments to improve patient outcomes is dramatic, they also introduce new risks regarding their long-term safety and effectiveness across different biological systems. The potential for these engineered proteins and treatments to behave unexpectedly within complex biological systems needs to be evaluated with caution. Thus, further research must take into account not only the possible uses but also the ethical and safety concerns that come with these scientific advances. In the end, the debate about the possible advantages and disadvantages of protein engineering shows the need for multidisciplinary collaboration in the field. Scientists, ethicists, and regulatory officials must engage in extensive dialogue to develop guidelines that are safety-oriented without stifling innovation. The engineered protein design and deployment should involve considerations like the PP, fostering cautious experimentation along with the potential for breakthroughs. The next area of research should focus on designing methods for the prediction and prevention of undesirable effects of protein engineering. By developing a system that identifies and controls risks while concurrently promoting widespread therapeutic benefits, the scientific community can realize the full potential of protein engineering, ultimately fostering medical and environmental applications that advance human health and ecosystem integrity.

9.5.2. Regulatory Frameworks Governing Protein Engineering

Legal frameworks controlling protein engineering are complex and multidisciplinary, reflecting the uses of engineered proteins in biotechnology, pharmaceuticals, and agriculture across many multifaceted dimensions. These are prepared to deliver safety, efficacy, and ethical guarantees while encouraging innovation. In regions such as the European Union, technology development and commercialization of technologies such as plant molecular farming (PMF) and new plant breeding techniques (NPBTs) are dominated by regulatory policies. Regulatory issues in the EU have been referred to for years by stakeholders as the primary barrier to the growth of PMF and NPBT programs, due mainly to stringent safety tests and popular resistance to genetic modification. Comprehending the intricacy of such laws will enable researchers to better avoid the pitfalls and possibilities of the ever-changing protein engineering environment. And with the advancement of the protein engineering field comes the emergence of biosimilars, which are transforming regulatory thinking. As increasing numbers of biologic products reach the end of their patents, the production of biosimilars is a huge prospect for both established players and new entrants to the market. Regulators must adjust existing regulations to deal with the unique issues posed by biosimilars so that such drugs are safe and effective as far as their respective reference biologics. India and China, as well as other countries, are rapidly enhancing their biosimilar manufacturing capacities, with questions about the adequacy of current regulatory procedures in multiple jurisdictions. This evolution necessitates global dialogue on best practices and standards that can meet rising levels of complexity associated with biosimilars in a way that serves the interests of patients in different healthcare systems. Furthermore, the interaction between regulatory strategies and public attitude will have a critical role to play in the future of protein engineering. The communication strategies employed by scientists and companies can significantly influence social acceptance of GMOs and engineered proteins. By way of example, transparent depiction of benefits accruable to NPBTs in PMF can shift public opinion towards a favorable direction concerning these technologies, thereby allowing for concomitant regulatory approval. It is essential that stakeholders be able to communicate adequately with media and communities to de-mystify the protein engineering science and convey its possible advantages. Building open and inclusive discussions of the effects of biotechnology can produce a good regulatory environment that favors innovation and addresses people's concerns, paving the way for the acceptance of protein engineering into common utilization.

9.5.3. Public Perception and Societal Implications

Public opinion has a significant influence on the advancement and acceptability of protein engineering technologies, particularly genetic modification. Through the past, resistance to GMOs, with the European Union as an important exception, has shaped a regulatory system that centers on safeguarding the public and the environment. This opposition reached a peak in a de facto moratorium on commercializing GM crops in the late 1990s, demonstrating the significant influence that public opinion could have on policy. In efforts at rebuilding consumer confidence, regulation was imposed through precautionary principles and heightened transparency programs to provide information to the public about GM food products. Nevertheless, in spite of such efforts to catch public attention and divert safety concerns, the impact of public opinion continues to be a central challenge in cultivating broad-based acceptance of biotech advances such as protein engineering (Waele et al., 2006). The idea of plant molecular farming (PMF) is illustrative of the intricacies revolving around public image when it comes to protein engineering. As an emergent pathway to high-added-value recombinant protein manufacture, PMF has been subject to skepticism as to its feasibility and safety, contributed to by the current EU regulatory climate. Stakeholders on PMF projects have identified public perception of biotechnology as a significant barrier to PMF development. Qualitative interviews find that among the key issues is that individuals are worried about public perception of the risk of genetic modification, and this causes resistance to embracing new technologies such as new plant breeding approaches (NPBTs). To bypass these barriers, stakeholders find that effective communication of the benefits and purpose of PMF is necessary in order to achieve social acceptance and render the technology scalable (Drake et al., 2020). Further, the media are also critical in shaping public opinion and the social implications of protein engineering. Depictions of biotechnology's scientific advances have far-reaching impacts on public awareness and public attitudes, tilting towards acceptance or rejection in the majority of instances. The science-media interface highlights the importance of accurate and positive reports that convey the benefits of protein engineering, particularly improving health and sustainability. In addition, the efforts that aim to amplify public input through open debate can help improve the level of trust in science. The interaction cycle between public perception and technology advancement underscores the importance of adopting a strategy that incorporates societal concerns, ethics, and scientific innovation to ensure protein engineering evolves in an acceptable social and acceptable manner (Drake et al., 2020; Waele et al., 2006).

9.6. FUTURE DIRECTIONS IN PROTEIN ENGINEERING

The future of protein engineering is in revolutionary developments, particularly with the integration of artificial intelligence and machine learning methods. These technologies are rapidly transforming the discipline by allowing scientists to filter through vast databases at unprecedented speed. Protein engineers use these algorithms to discover patterns related to protein stability, solubility, and overall function. The ability to predict substrate specificity and enhance rational design methods will drive the production of tailored biocatalysts, which are essential in applications in biotechnology and pharmaceutical utilization. Through existing publications, the discipline is operating under many limitations in current procedures, but advancements in machine learning methods offer the primary potential directions for resolving these issues and pushing the field forward (Damborský Jiří et al., 2020). This influx of computational power is the catalyst for a shift towards more efficient, data-driven approaches to pursuing novel protein solutions. Coupled with machine learning, the continued exploration of biomaterials and scaffold design is also an integral part of protein engineering's future evolution. By incorporating cross-disciplinary scientific knowledge from fields like materials science, cellular biology, and tissue engineering, scientists are pioneering the development of biocompatible protein-based materials. Interdisciplinary methodologies have already shown promise for application in regenerative medicine technologies, particularly in the development of tissue-mimicking materials for dental and oral health (Amrollahi et al., 2016). The merging of these fields will not only improve the mechanical properties of engineered proteins but also the functional performance thereof. As the understanding of protein interactions and material properties continues to advance, future research will likely be focused on optimizing the interface between engineered proteins and biological systems, thus optimizing therapeutic effectiveness and duration. The discipline of protein engineering also looks to synthetic biology's horizon, for the combination of protein engineering with genome editing reagents can revolutionize the field. Reagents like CRISPR/Cas9 offer unparalleled precision in editing genetic information, enabling the design of site-specific proteins with enhanced or novel functions. This collaboration between protein engineering and synthetic biology facilitates the development of extremely selective enzymes, more productive metabolic pathways, as well as new protein constructs that can be utilized for numerous industrial and therapeutic purposes. With further research, it is probable that the engineered proteins will be essential for green solutions and applications to health, energy, and conservation challenges. The new innovation in these interrelated sciences has revealed an intensified opportunity for scientists to rethink traditional protein engineering techniques and embrace innovation with far-reaching implications.

9.6.1. Advances in Synthetic Biology and Its Impact

Recent advances in synthetic biology have greatly transformed the science of protein engineering to enable researchers to manipulate biological systems at scales never thought possible. With the exploitation of synthetic biology, researchers can effectively design and construct new proteins with desired functionalities, providing long-term scope for therapeutic, industrial, and environmental applications. With the integration of new technologies, such as the de novo synthesis of biological systems and the design of genetic circuits, one not only facilitates the engineering of proteins but also enables the discovery of new avenues to synthesize complex biomolecules economically. With the growth of synthetic biology, it also stimulates greater understanding of biological processes and encourages the development of hybrid proteins that combine aspects from various organisms to enhance their utility in various industries, including medicine and biotechnology. One of the major characteristics of synthetic biology's contribution to protein engineering is that it is capable of generating unprecedented diversity in engineered proteins. Techniques such as epPCR and site-directed mutagenesis enable the creation of many genetic variations that expand the functional scope of proteins. It is through such molecular diversity that proteins can be designed to target specific therapeutic targets, with drug development and disease therapy breakthrough potential. Further, the integration of synthetic biology with computational methods enhances the capability of researchers to predict the structure, behavior, and stability of proteins, thus revolutionizing the realm of protein engineering. With more studies into and development of such multidisciplinary techniques, the therapeutic potential of engineered proteins is poised to address complex health conditions, thus demonstrating the significance of advances in synthetic biology (Battisti et al., 2018). The ethical implications of protein engineering and synthetic biology must be carefully considered as advancements unfold. Advances in the area must balance the advantages against the risks, particularly regarding environmental impacts and biosecurity. As synthetic biology continues to offer innovative solutions, it also raises questions of accessibility to its advantages and the consequences of dual-use technologies. Ongoing public discourse is required to develop ethical norms that inform research and application in synthetic biology. By fostering exchange among scientists, ethicists, and regulators, society can ensure that protein engineering remains a tool for good, ultimately resulting in sustainable development and the appropriate use of biotechnology gains sensitive to human interests (Battisti et al., 2018).

9.6.2. Emerging Technologies in Protein Design

In recent years, new technologies have transformed protein design into a new field with significant breakthrough applications in academic and industrial research. Synthetic biology, machine learning, and high-throughput screening have played a key role in designing proteins with enhanced functionalities and specificities. For example, protein structures and functions can be predicted using machine learning algorithms trained on large datasets, thereby making the design easier. Furthermore, high-throughput screening techniques enable the rapid examination of thousands of candidate protein variants at increased speeds, accelerating the discovery of those with the ability to mediate desired biological activities. The fast prototyping not only indicates the advancement in protein engineering but also promises a future for exploiting these advances for drug discovery, industrial biotechnology, and environmental remediation. Synthetic biology is at the forefront of these emerging technologies, with the rational design and engineering of biological systems being emphasized. Progress in the area is towards the creation of an overarching field of knowledge grounded in the range of sciences while applying engineering methods to address the inherent complexity of biological systems. Among the advances made possible by synthetic biology is the design of standard biological parts that can be combined to generate new proteins and pathways. Because biological systems are inherently variable, designing by a systematic cycle of specification, modeling, analysis, assembly, and testing considerably enhances the likelihood of success. The systematic method allows scientists to repeatedly refine their designs, ultimately leading to stable biological systems that can be used reliably for specific purposes (Freemont et al., 2014). Additionally, not all new protein design technologies pose problems. The use of these sophisticated methods always poses questions regarding safety, quality control, and environmental impact. As nonthermal food technologies demonstrate, while they offer considerable advantages in terms of food safety and quality improvements, they can introduce unforeseen quality-related issues that should be thoroughly investigated (Djekic et al., 2018). As such, it becomes essential for protein engineering scientists to not only innovate but also investigate the macro-level implications of their technology innovations. By utilizing environmental impact assessments and putting forward holistic approaches, researchers are able to better understand the complexity of protein design, ensuring the benefits of novel technologies are maximized with minimized risk.

9.6.3. Potential for Personalized Medicine Through Protein Engineering

The rapidly emerging discipline of protein engineering has tremendous potential for driving personalized medicine. By engineering proteins to the particular demands of particular patients, researchers are inventing novel therapeutic strategies that have the potential to drastically improve the outcome of treatments. This potential is grounded on the ability to precisely regulate protein structure and function in order to enable the design of therapeutic agents that modulate targeted molecular pathways responsible for various diseases. Therefore, protein engineering not only facilitates a more accurate comprehension of the underlying biology but also allows proteins to be formulated to tailor to the genetic profile of individual patient populations, ultimately leading to better and customized treatment regimens. The translation of these engineered proteins to the clinical field can potentially usher in a new era of therapeutics based on patient-specific parameters and thus emphasize the revolutionary power of protein engineering in medicine. Another potential meeting of protein engineering and nanomedicine underscores the promise of personalized therapeutic modalities. Nanoinformatics, being relatively new, plays a crucial role in bridging the advances in nanomaterials to their use in medicine. By directing the development of tailored therapeutic drug delivery systems and diagnostic devices, nanoinformatics leverages the unique characteristics of engineered proteins at the nanoscale. By customizing these materials to a patient's individual profile, researchers can enhance not only the safety but also the efficacy of therapeutics, particularly in the use for diseases such as cancer and regenerative medicine. Moreover, the convergence of the principles of protein engineering with nanoinformatics can unlock unprecedented opportunities in drug development towards new directions that not only improve the precision of drug targeting but also enable improved insight into personalized therapeutic strategies (Anguita et al., 2012). The arrival of human-induced pluripotent stem cells (iPSCs) as a convenient platform for personalized medicine is but one illustration of the potential of protein engineering to revolutionize treatment approaches. By the creation of patient-specific cells, iPSCs allow researchers to examine the unique response to treatment strategies of an individual patient, something that will become increasingly significant as precision medicine receives more widespread acceptance. With advancements in differentiation protocols and phenotyping strategies, these cells are able to be personalized to mimic a patient's unique physiological conditions and thus provide unparalleled promise for drug discovery and testing. This not only points to the shift away from traditional animal models that do not take into account interspecies variations, but also enables the tailoring of drugs to match an individual's customized genetic profile. Consequently, the confluence of protein engineering and iPSC technology represents a milestone towards more effective and specific medical interventions, whereby the discipline advances significantly (Chandy et al., 2020).

SUMMARY

In the discipline of protein engineering, the culmination of research offers an appreciation of the complexity that comes with biological systems, culminating in new applications in various industries. As we consolidate observations across multiple experiments, we notice that advances in theoretical comprehension and hands-on practices have significantly impacted the production of recombinant proteins, enzymes, and other biological constructs. Such developments harmonize with those of synthetic biology, which aims for the creation of new functions through an engineering approach to biological synthesis and design. With its emphasis on purpose-orientation and deeper insight into the underlying scientific principle, protein engineering has undoubtedly tremendous potential for even more targeted and efficient outcomes in the years to come. Furthermore, integration with new methodologies, such as those identified in recent studies, is poised to enhance both the scalability and efficiency of protein expression systems. For instance, the application of plant cell packs, as documented in Arfi et al. (2019), is a key milestone in establishing high-throughput screening methodologies for protein production. This new system not only enables rapid protein expression across numerous plant species but also aligns with the fundamental paradigms of metabolic engineering and synthetic biology. Hence, they bring into focus the necessity to enhance methods in which proteins can be strongly expressed and resources conserved as well as time, which will drive the discipline towards sustainable strategies in biomanufacturing in the long term. Last but not least, protein engineering innovations bring to light the ways of leveraging advanced biological systems in useful application-specific design. The convergence of various disciplines, ranging from protein design to DNA synthesis, shows a tangible mechanism of breaking barriers in pharmaceutical manufacturing and beyond. As emerged by Heinemann et al. (2006), engineering-based strategies applied at the DNA and protein levels provide a rich soil for the development of high-level biological constructs with the potential to perform desired functions. This convergence of fields marks a revolutionary era in biotechnology, with interdisciplinary research driving the boundaries of innovation in protein engineering and synthetic biology, promising a world full of new possibilities and discoveries that can enhance human health and sustainable development.

Protein engineering is a groundbreaking achievement in biotechnology, and it has become possible to engineer proteins with tailored functionalities for varied purposes. Areas of focus commonly discussed during this discussion include the techniques used to modify proteins, such as site-directed mutagenesis and recombinant DNA technology. The methodologies allow researchers to introduce specific amino acid modifications that enhance targeted properties, such as stability, activity, or specificity. For example, the physical adsorption or covalent immobilization of proteins onto carbon nanotubes illustrates the ability of engineering techniques to provide stable and reusable protein conjugates that preserve structure and function under harsh conditions (Asuri et al., 2009). This interdisciplinarity of protein design emphasizes the possibility of protein engineering as the method of choice for addressing actual practical problems in numerous fields of science, including pharmaceuticals, agriculture, and environmental science. Furthermore, the integration of proteins into scaffolding structures forms the basis for advances in regenerative medicine. A recent investigation has investigated the successful combination of recombinant human bone morphogenetic protein-2 (rhBMP-2) with various carriers in vertebral models of reconstruction. Results showed the two rhBMP-2 formulations were 100% fusion when they were combined with novel carrier systems, such as compression-resistant matrices (CRM) and absorbable collagen sponges (ACS) mixed with ceramic granules (Mobasser et al., 2016). These findings highlight the groundbreaking impact of protein engineering towards stimulating biological mechanisms of repair with possible morbidity elimination associated with traditional autograft methods. The implications of this kind of research extend beyond spinal surgery, with potential applications in bone tissue engineering and the development of targeted therapies. Overall, the consideration of protein engineering is a fast-evolving and dynamic field that is characterized by new techniques and promising potential. The methods utilized in protein engineering and the mechanisms by which they transform their environment play a critical role in the discovery of new therapeutic uses, as well as in optimizing existing technologies. Protein immobilization on nanomaterials and the effectiveness of rhBMP-2 formulations for fusion capture the magnitude of success in protein engineering thus far (Asuri et al., 2009; Mobasser et al., 2016). As research improves, the intersection of disciplines such as materials science and molecular biology will yield increasingly sophisticated techniques for protein design, ultimately pushing the boundaries of what is possible in medicine and beyond.

REFERENCES

1. Abbas, Z., Abbas, Z., Afzal, A., Bin Muhammad Ishaq, M., et al. (2024). Commands of synthetic biology to modernize and re-design the biological systems. https://core.ac.uk/download/616936549.pdf (accessed on 3rd October 2025).
2. Aguilar-Lasserre, A., Azzaro-Pantel, C., Dietz, A., Domenech, S., et al. (2008). A fuzzy Mult objective algorithm for multiproduct batch plant: Application to protein production. https://core.ac.uk/download/12039646.pdf (accessed on 3rd October 2025).
3. Ahuja, P. (2013). Man, and machine: Questions of risk, trust and accountability in today's AI technology. http://arxiv.org/abs/1307.7127 (accessed on 3rd October 2025).
4. Ali, S., Chourasia, P., & Patterson, M. (2023). When protein structure embedding meets large language models. *Genes, 15.* https://www.semanticscholar.org/paper/ee58c3e0214c7f5d90e127c64c46abd3b3b0ad5b (accessed on 3rd October 2025).
5. Amrollahi, P., Seifi, A., Shah, B., Tayebi, L., et al. (2016). Recent advancements in regenerative dentistry: A review. https://core.ac.uk/download/213080485.pdf (accessed on 3rd October 2025).
6. Anguita, A., Pazos, A., Geissbuhler, A., Smith, B., Kim, B. Y., Kulikowski, C., Rosse, C., et al. (2012). Nano informatics: Developing new computing applications for nanomedicine. https://core.ac.uk/download/28969094.pdf (accessed on 3rd October 2025).
7. Annevelink, E., & van Ree, R. (2010). Country report The Netherlands 2010: IEA Bioenergy Task 42. https://core.ac.uk/download/pdf/29240740.pdf (accessed on 3rd October 2025).
8. Anurogo, D. (D.), Ikrar, T. (T.), & Parikesit, A. A. (A.) (2017). Bio nanomedicine: A "panacea" in medicine? https://media.neliti.com/media/publications/179998-EN-none.pdf (accessed on 3rd October 2025).
9. Arfi, M., Bendandi, M., Buyel, J. F., Buyel, J. J., et al. (2019). Plant cell packs: A scalable platform for recombinant protein production and metabolic engineering. https://core.ac.uk/download/187423824.pdf (accessed on 3rd October 2025).
10. Arnold, F. H., Wu, Z., & Yang, K. K. (2019). Machine learning-guided directed evolution for protein engineering. https://core.ac.uk/download/222931970.pdf (accessed on 3rd October 2025).
11. Asuri, P., Bale, S. S., Banerjee, A., Dordick, J. S., et al. (2009). Preparation

and characterization of protein-nanotube conjugates. https://core.ac.uk/download/215441831.pdf (accessed on 3rd October 2025).

12. Azzaro-Pantel, C., Dietz, A., Domenech, S., & Pibouleau, L. (2007). Eco-design of batch processes: Optimal design strategies for economic and ecological bioprocesses. https://core.ac.uk/download/12039698.pdf (accessed on 3rd October 2025).
13. Baianu, I. C., Lin, H. C., Lozano, P., Prisecaru, V., et al. (2004). Novel techniques and their applications to health foods, agricultural and medical biotechnology: Functional genomics and basic epigenetic controls in plant and animal cells. https://core.ac.uk/download/pdf/85962.pdf (accessed on 3rd October 2025).
14. Bar-Yam, Y., Douady, R., Norman, J., Read, S., et al. (2014). The precautionary principle (with application to the genetic modification of organisms). http://arxiv.org/abs/1410.5787 (accessed on 3rd October 2025).
15. Battisti, A., & Persiani, S. (2018). Frontiers of adaptive design, synthetic biology and growing skins for ephemeral hybrid structures. https://core.ac.uk/download/188829284.pdf (accessed on 3rd October 2025).
16. Brown, A., Do, M. A., Levy, D., Lu, Y., et al. (2019). Targeted delivery of lysosomal enzymes to the endocytic compartment in human cells using engineered extracellular vesicles. https://core.ac.uk/download/286362291.pdf (accessed on 3rd October 2025).
17. Browning, R. J., de Jong, N., Eckersley, R. J., Luan, Y., et al. (2016). Characterization of contrast agent microbubbles for ultrasound imaging and therapy research. https://core.ac.uk/download/46558242.pdf (accessed on 3rd October 2025).
18. Buscetta, M., Figueroa Yévenes, M., Lejeune, A., Martial, J., et al. (2012). *In silico* and *in vivo* combinatorial design of Octarellin VI, an artificial protein modeled on the (B/A)8 fold. http://orbi.ulg.ac.be/bitstream/2268/160463/1/OCTA%20VI%20poster.pdf (accessed on 3rd October 2025).
19. Capdeboscq, Y., Rittscher, A. E., Rittscher, J., Sulaimalebbe, A., et al. (2018). Foreign object detection and quantification of fat content using a novel multiplexing electric field sensor. http://arxiv.org/abs/1806.08596 (accessed on 3rd October 2025).
20. Chandy, M., Paik, D. T., & Wu, J. C. (2020). Patient and disease-specific induced pluripotent stem cells for discovery of personalized cardiovascular drugs and therapeutics. https://core.ac.uk/download/323062894.pdf (accessed on 3rd October 2025).

21. Chen, J. Y., & Wu, X. (2011). Predictive and personalized medicine with systems biology solutions. https://core.ac.uk/download/46964267.pdf (accessed on 3rd October 2025).

22. Cheng, P., Mao, C., Tang, J., Yang, S., Cheng, Y., Wang, W., Gu, Q., et al. (2024). Zero-shot prediction of mutation effects with multimodal deep representation learning guides protein engineering. *Cell Research, 34,* 630–647. https://www.semanticscholar.org/paper/292506d08d2c4c68eaabe24945e6a2890833375d (accessed on 3rd October 2025).

23. Curley, K., & Duong, N. (2017). Exosome membrane bound TNF-receptor for the treatment of rheumatoid arthritis. https://core.ac.uk/download/84118681.pdf (accessed on 3rd October 2025).

24. Damborský, J., Mazurenko, S., & Prokop, Z. (2020). Machine learning in enzyme engineering. https://core.ac.uk/download/392357028.pdf (accessed on 3rd October 2025).

25. Das, D., & Ainavarapu, S. R. K. (2024). Protein engineering using circular permutation–structure, function, stability, and applications. *The FEBS Journal, 291.* https://www.semanticscholar.org/paper/d0406e4d7e18af713d7536780c612fa76df97bbe (accessed on 3rd October 2025).

26. De Waele, D., Devos, Y., Reheul, D., Van Speybroeck, L., et al. (2006). The interplay between societal concerns and the regulatory frame on GM crops in the European Union. https://core.ac.uk/download/55749078.pdf (accessed on 3rd October 2025).

27. Djekic, I., Donsi, F., Paniwnyk, L., Režek Jambrak, A., et al. (2018). Three pillars of novel nonthermal food technologies: Food safety, quality, and environment. https://core.ac.uk/download/304335323.pdf (accessed on 3rd October 2025).

28. Drake, P. M. W., Fuller, S. S., Hobbs, M., Ma, J., et al. (2020). Shotguns vs lasers: Identifying barriers and facilitators to scaling-up plant molecular farming for high-value health products. https://core.ac.uk/download/288436395.pdf (accessed on 3rd October 2025).

29. Duţeanu, N. M., Erable, B., Ghangrekar, M. M., Scott, K., et al. (2012). Biological and microbial fuel cells. https://core.ac.uk/download/16754572.pdf (accessed on 3rd October 2025).

30. Eller, D. G. (2025). Biotechnology—Diversification, and the food industry. https://core.ac.uk/download/pdf/6988570.pdf (accessed on 3rd October 2025).

31. Espada, R., Ferreiro, D. U., Parra, R. G., Sippl, M. J., et al. (2013).

Detecting repetitions and periodicities in proteins by tiling the structural space. https://core.ac.uk/download/19042832.pdf (accessed on 3rd October 2025).

32. Fraser, J. S., & Jackson, C. J. (2010). Mining electron density for functionally relevant protein polyesters in crystal structures. https://core.ac.uk/download/323074355.pdf (accessed on 3rd October 2025).

33. Freemont, P., Kelwick, R., MacDonald, J. T., Webb, A., et al. (2014). Developments in the tools and methodologies of synthetic biology. https://core.ac.uk/download/76995845.pdf (accessed on 3rd October 2025).

34. Grage, K., McDermott, P., & Rehm, B. H. A. (2017). Engineering *Bacillus megaterium* for production of functional intracellular materials. https://core.ac.uk/download/157569015.pdf (accessed on 3rd October 2025).

35. Heinemann, M., & Panke, S. (2006). Synthetic biology—putting engineering into biology. https://core.ac.uk/download/pdf/12944386.pdf (accessed on 3rd October 2025).

36. Helenius, A., Ishii, A., Lal, A., Varki, A., Barabási, A. L., Feist, A. M., Sburlati, A. R., et al. (2009). Centralized modularity of N-linked glycosylation pathways in mammalian cells. https://core.ac.uk/download/pdf/8332683.pdf (accessed on 3rd October 2025).

37. Johnston, R. S., Norton, D. J., & Baldwin, T. (2025). Space and biotechnology: An industry profile. https://core.ac.uk/download/pdf/42821064.pdf (accessed on 3rd October 2025).

38. Khan, I., & Akhtar, M. W. (2011). Different approaches for protein engineering in industrial biotechnology. https://core.ac.uk/download/pdf/289804.pdf (accessed on 3rd October 2025).

39. Khan, I., & Akhtar, M. W. (2011). Molecular bio-imprinting of biocatalysts. https://core.ac.uk/download/pdf/289836.pdf (accessed on 3rd October 2025).

40. Krishna, V., Poelwijk, F. J., & Ranganathan, R. (2015). The context-dependence of mutations: A linkage of formalisms. http://arxiv.org/abs/1502.00726 (accessed on 3rd October 2025).

41. Laubichler, M., Loettgers, A., & Maienschein, J. (2008). How can history of science matter to scientists? https://core.ac.uk/download/4879106.pdf (accessed on 3rd October 2025).

42. Mayo, S. L., Moore, M. M., Mou, Y., Wannier, T., et al. (2015). Computational design of the β-sheet surface of a red fluorescent protein allows control of protein oligomerization. https://core.ac.uk/download/33121124.pdf (accessed on 3rd October 2025).

43. McEvilly, M. P. (2013). Lack of transparency in the premarket approval process for AquAdvantage salmon. https://core.ac.uk/download/62559747.pdf (accessed on 3rd October 2025).
44. Mobasser, J. P., Schwartz, D. G., Toth, J. M., Williams, K., et al. (2016). Instrumented lumbar corpectomy and spinal reconstruction comparing rhBMP-2/compression-resistant matrix, rhBMP-2/absorbable collagen sponge/ceramic granules mixture, and autograft in two different devices: A study in sheep. https://core.ac.uk/download/213087657.pdf (accessed on 3rd October 2025).
45. Nisthal, A., Allen, B. D., Allen, C., Ambroggio, X., Arnold, F. H., Aucamp, J. P., et al. (2010). Experimental library screening demonstrates the successful application of computational protein design to large structural ensembles. https://core.ac.uk/download/4887227.pdf (accessed on 3rd October 2025).
46. Othman, F. (2010). Bipartite graph algorithm with reference frame representation for protein tertiary structure matching. https://core.ac.uk/download/32599925.pdf (accessed on 3rd October 2025).
47. Pichler, H., Schwab, H., Hirz, M., & Ahmad, M. (2014). Protein expression in *Pichia pastoris*: Recent achievements and perspectives for heterologous protein production. https://core.ac.uk/download/pdf/81755868.pdf (accessed on 3rd October 2025).
48. Quarmby, B. (2003). 3D molecular structures: Patentable subject matter under 35 U.S.C. §101? https://core.ac.uk/download/62559157.pdf (accessed on 3rd October 2025).
49. Reiss, M. (2000). The ethics of xenotransplantation. https://core.ac.uk/download/pdf/111612.pdf (accessed on 3rd October 2025).
50. Savas, P. E. (2019). Therapeutic delivery technology and its economic impact. https://core.ac.uk/download/213463611.pdf (accessed on 3rd October 2025).
51. Sosa, Y., Kapur, B., Hurtak, J., Kingsley, L. J., Wu, H., Gruber, S., Nar, H., et al. (2024). *In silico* enzyme screening identifies an SDR ketoreductase from *Thermus caliditerrae* as an attractive biocatalyst and promising candidate for protein engineering. *Frontiers in Chemical Biology.* https://www.semanticscholar.org/paper/d54228b7da507aa6dcc547ca41b20c12c0fe2ec1 (accessed on 3rd October 2025).
52. Zalewski, M., Iglesias, V., Bárcenas, O., Ventura, S., & Kmiecik, S. (2024). Aggrescan4D: A comprehensive tool for pH-dependent analysis and engineering of protein aggregation propensity. *Protein Science: A*

Publication of the Protein Society, 33. https://www.semanticscholar.org/paper/909d44ceb9e99f0b7189bd5203e1e3aa7b255c26 (accessed on 3rd October 2025).

53. Zhang, Y., & Ruan, Z. (2024). Optimizing human SIRT6 protein with deep learning of 3D structures based on maximum lifespan. *Highlights in Science, Engineering and Technology.* https://www.semanticscholar.org/paper/c5094d9353f596a007a3a0724508ee760fce2a80 (accessed on 3rd October 2025).

CHAPTER 10

CHEMICAL BIOLOGY IN SYNTHETIC SYSTEMS

Contents

INTRODUCTION TO CHEMICAL BIOLOGY

The field of chemical biology is an important multidisciplinary area that bridges the gap between chemistry and biology and assists in the progression of understanding molecular mechanisms behind biological processes. By applying chemical concepts, researchers can manipulate biological systems to examine interactions among biomolecules and their roles in biological function. The importance of this science has been reinforced by breakthroughs in synthetic systems, in that the integration of synthetic biology makes it possible to design and build new biological parts and systems. By applying chemical methods along with biological insights, scientists can create innovative solutions for complex biological challenges, from drug discovery to disease modeling, thereby illustrating the applicability of chemical biology to contemporary scientific investigation. As synthetic biology advances, so does the linkage between synthetic and chemical biology, particularly in the building of synthetic systems to mimic or replicate natural biological mechanisms. The precision of synthetic techniques, as demonstrated in the construction of dendrimers, enables researchers to engineer molecular devices that perform specific functions within biological systems. Dendrimers with their branched molecules illustrate the power of controlled chemical synthesis to result in more cell-compatible and bioavailable compounds and therefore excellent candidates for drug delivery applications (Prakash et al., 2025).

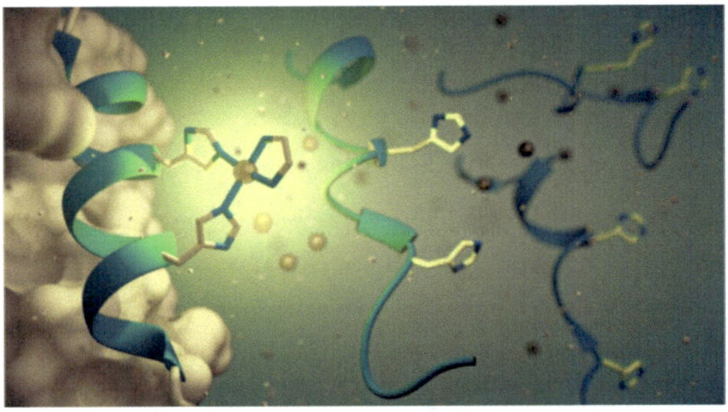

Figure 10.1. Understanding chemical biology

Source: https://media.springernature.com/w580h326/nature-cms/uploads/collections/COMMSCHEM-22-0080B-6476b17265d63f5ec8714a98c057f658.png.

Research and application of these functionalities highlight the potential that chemical biology holds in discovering synthetic systems that will address essential biomedical challenges. This convergence illustrates the potential of synthetic biology, unlocking new therapeutic paradigms and greater insight into complex biological systems. Continued technological advances in chemical biology not only shed new light on biochemistry but also indicate the growing use of bioengineering paradigms in exploring living systems. Synthetic biology applies principles of engineering to facilitate the design of genetic and metabolic circuits towards the stimulation of innovation in various applications, including environmental sustainability and pharmaceuticals (Hanczyc, 2020). The potential for the construction of living organisms as green technologies offers alternative methods to traditional industrial processes towards optimizing our capacity to solve global challenges. By situating chemical biology within synthetic systems, researchers can leverage the strengths of molecular engineering to create life-like systems with desired properties. Through such an interdisciplinary strategy, not only does this lead to fantastic advances in biological science, but it also paves the way towards sustainable and efficient solutions in various areas, eventually exhibiting the great contribution of chemical biology within synthetic systems (Figure 10.1).

Definition and Scope of Chemical Biology

The definition and scope of chemical biology have evolved significantly, emerging as a core discipline that encompasses chemistry, biology, and related fields. At the core of its premise is the application of chemical tools and methodology to explore and manipulate biological systems. This interdisciplinary approach enables researchers to analyze complex biological processes at the molecular scale, enhancing our understanding of cellular mechanisms and biomolecule interactions. By using chemical approaches, such as small molecules, genetic tools, and bioconjugation methods, researchers can probe the function of proteins, nucleic acids, and other biomolecules. The scope of chemical biology is not only confined to foundational research but extends to applied applications in drug discovery, synthetic biology, and biotechnology, thus connecting theoretical science with applied use in numerous domains. In synthetic systems, the importance of chemical biology is further increased. As synthetic biology aims to design and construct new biological parts, devices, and systems, detailed understanding of chemical interactions is essential. Chemical biology provides the basis for the knowledge required to develop synthetic organisms with particular functions, such as bioremediation or biofuel production. Researchers in this discipline use chemical concepts to design metabolic pathways and genetic circuits and facilitate the development of organisms with desired phenotypes. Notably, with rising automation and standardization in synthetic biology, the integration of chemical biology concepts will be central to enhancing productivity and efficiency, ultimately transforming the design of biological systems and their applications across various fields, including agriculture and environmental sustainability (Beal et al., 2020). The rise in overlap between chemical biology and other technologies further emphasizes the need for well-defined standards and shared practices. Forum and workshop debates, such as those led by the MicrobiomeSupport project, highlight the need to reach agreement on defining key terms and practices in the microbiome field, which is increasingly grounded in chemical biology approaches (Berg et al., 2020). Having a unified framework may lead to increased reproducibility and interoperability between research groups, which can lead to more efficient knowledge transfer from basic research to practical applications. As synthetic systems improve, investment in standardized protocols and interdisciplinary collaborations will be the driving force towards innovation in chemical biology that will enable researchers to tackle tough problems in health, agriculture, and environmental systems more effectively. This interdisciplinary synergy emphasizes the potential of chemical biology as a driver of follow-on biotechnological advancement (Figure 10.2).

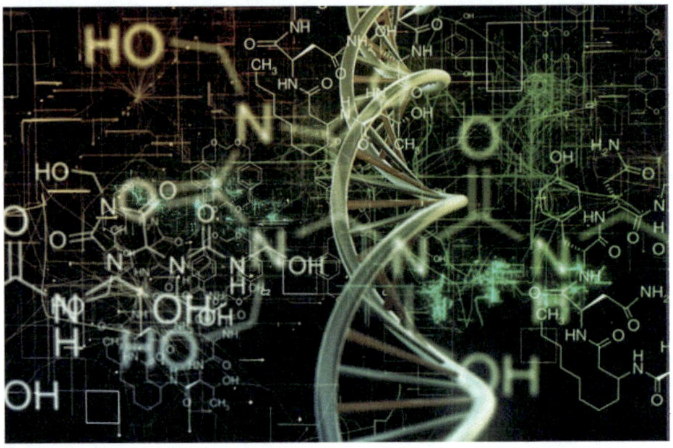

Figure 10.2. Chemical biology.

Source: https://education.scripps.edu/_files/images/graduate/doctoral-program/customizable-curriculum/600x400_chemical_biology_i.jpg.

Historical Context and Development

The historical development of chemical biology in synthetic systems traces back to the early 20th century, when researchers were beginning to understand the complexity of biochemical processes. This initial understanding laid the ground for the later union of chemistry and biology, which played a key role in the creation of synthetic compounds used in various aspects. One of the significant developments in this period was the discovery of synthetic polymers, which illustrated how chemical biology could be utilized in a bid to create molecules that were replicating nature molecules. Early investigation of synthetic systems not only revolutionized material sciences but also generated excitement for prospecting the applications in medicine and agriculture. When the investigation mounted, it became apparent that synthetic biology not only enhanced what already exists but also led to new things entirely that combined biological components with synthetic components, taking the subject far ahead. Over the past few years, the interface between synthetic systems and microbial technology has undergone a revolutionary change, marking a turning point in the history of chemical biology. Microbes have been harnessed as biocatalysts and integrated into synthetic pathways to produce sustainable, environmentally friendly products. Microbial cell factories embody a paradigm shift toward renewable biomaterials, which have far-reaching implications for industrial applications. Through the optimization of these biological systems, researchers have had an industry-wide effect on sectors such as pharmaceuticals and biofuels. Specifically, the ability of microorganisms to transform renewable resources into valuable chemicals not only makes cleaner production a reality but also positions biological systems as a central component of sustainable development initiatives. This evolution brings into focus the increasing reliance on chemical biology as a paradigm for the solution of contemporary global challenges, expanding the frontiers of synthetic systems to address environmental concerns. Furthermore, biopesticides from plants have come onto center stage as a green solution to chemical pesticides, accentuating the utilitarian evolution of chemical biology in synthetic systems. Historical records indicate that traditional agriculture practices have always involved botanical elements; the institutionalization of such practices as contemporary biopesticide manufacturing is new. Research into various plant extracts for use as biopesticides indicates the advantages of using naturally derived solutions that contribute to food security and environmental health. This renewed interest is not only a return to traditional practices but is heavily rooted in the developments of technology in extraction and formulation processes, such as nanoencapsulation, that enhance the efficacy and shelf life of these biopesticides. As indicated in recent reviews, despite lingering issues with production costs and regulatory regimes, the complementarity of chemical biology and synthetic systems has revolutionary potential in future agricultural uses, reflecting the ongoing evolution of this historical context (Shaizee et al., 2025). Moreover, integrating multi-omics technologies and synthetic biology approaches promises a new frontier for biopesticide research, reiterating the role of chemical biology in creating innovative, eco-friendly products (Qumsani, 2024).

Importance in Modern Scientific Research

Synthetic systems and chemical biology have emerged as the focus point of recent scientific research, presenting new frontiers for innovation across disciplines. These man-made systems allow researchers to mimic and manipulate biological processes, which allows them to gain a better understanding of molecular processes at work. The possibilities are endless, ranging from the development of new medicines to the design of organisms that yield valuable biofuels. By the use of advances in genetic and molecular engineering, chemical biology facilitates the synthesis of smart biomaterials that are capable of dynamically reacting to external signals. This adaptability is vital in optimizing therapeutic interventions and optimizing the efficacy of bioengineering solutions. Research, hence, not only responds to theoretical matters but also evokes pragmatic solutions for acute worldwide challenges, capitalizing on the significance of chemical biology in achieving a sustainable world. Chemical biology research in synthetic systems emphasizes the importance of cross-disciplinary interaction between chemistry, biology, and materials science. Such an interdisciplinary context encourages innovative solutions and a holistic understanding of complex biological processes. For example, finding aromatic compounds for agriculture shows how chemical biology can impact environmentally friendly practices. These compounds enhance plant growth and insect resistance, creating a convergence of natural and synthetic methodologies to improve food security. Integration of technologies such as genetic engineering and nanoencapsulation increases the therapeutic value of such bioactive compounds, demonstrating the effectiveness of combinatorial strategies. As scientists continue to investigate the intricate coexistence of synthetic systems with nature, the research will equip future generations with the ability to find solutions to environmental and health problems. Ultimately, the meaning of chemical biology in artificial systems extends beyond the laboratory and holds potential for real-world applications that can transform industries and improve human well-being. Advances in tissue engineering and drug delivery systems, for example, highlight the requirement for targeted, accurate therapies that have minimal side effects and maximal efficacy. With chemically responsive matter and stimulus-responsive systems, researchers can design advanced therapeutic strategies and diagnostic tools and open the door to personalized medicine. This focus on precision-based therapies is accompanied by a drive to bring scientific inquiry closer to practical outcomes. Moreover, as sustainability comes to reclaim the science discourse, the role of chemical biology in enabling environmentally sustainable processes cannot be overemphasized. With the application of advances in synthetic biology, researchers are capable of bridging the chasm between technological utilization and biological insight, ultimately resulting in an era of increased health, sustainability, and innovation in modern scientific research.

10.1. PRINCIPLES OF SYNTHETIC BIOLOGY

Synthetic biology relies on the principle of the generation and assembly of novel biological systems from the deconstruction of current cellular functions. This engineering approach employs deoxyribonucleic acid (DNA) as the fundamental building block, allowing for the assembly of biologically active components, or bioparts. The bioparts are simply assembled into devices, which are then assembled into more advanced biological systems. Computer science is a key component throughout the Design-Build-Test-Learn cycle, ranging from mathematical modeling to employing robots in assembly processes. These strategies contribute significantly to the efficiency of synthetic biology research, allowing researchers to design and build quickly and iteratively. Importantly, the principles of standardization, modularity, and abstraction are central to this field, facilitating collaborative research across scientists globally and fostering a rich information-exchange culture and collective advancement in biological engineering (Garner, 2021). Moreover, advances in sequencing and synthesis have advanced synthetic biology, bringing together knowledge from systems biology with creative engineering. While systems biology investigates biological interactions across various scales, from molecular pathways to whole organisms, synthetic biology aims to engineer new systems that are able to perform specified functions. Such a differentiation is crucial because it positions synthetic biology as a transformative force that will address complex healthcare and environmental sustainability issues. The applications of synthetic biology run the gamut from optimizing diagnostic capability to delivering new therapies that improve patient outcomes. In addition, by the design of organisms that can detect environmental pollutants or enhance sustainable agriculture, synthetic biology is an extremely critical tool in redefining interactions between biology, technology, and environmental integrity (Garner, 2021). New findings also illustrate the extensive applications of synthetic biology in medicine. For instance, engineered microbes can be tailored to express some therapeutic proteins with improved pharmacological properties. One such instance is the creation of a semi-synthetic organism capable of producing a library of personalized cytokines that interact selectively with immune receptors, thereby provoking enhanced immune responses with no side effects typical of conventional therapeutics. An example of such a case study is the cytokine IL-2, whose manipulations have created variants like THOR-707, which selectively stimulate good immune cells without increasing regulatory populations. Such recent advances have rendered the tremendous potential of synthetic biology to not only direct drug discovery but also revolutionize treatment modalities of oncology and other clinical disorders, showcasing its tremendous capability to enhance health outcomes.

10.1.1. Overview of Synthetic Biology Concepts

Synthetic biology is a multidisciplinary discipline that integrates biology along with engineering principles to specifically design and construct novel biological components, devices, and systems. By Relying on the consensus of these principles, scientists can play around with genetic materials, create organisms, and come up with new applications in medicine, agriculture, and bioenergy, among others. One of the drivers of synthetic biology is the ability to engineer microbial systems for specific objectives. This engineering involves the accurate design of genetic circuits that control cell behavior so that engineered organisms can react favorably to environmental stimuli or generate therapeutic products. The overall goal is to establish a toolbox of standardized biological building blocks that can be put together systematically so that efficiency and reproducibility in biological engineering operations are maximized.

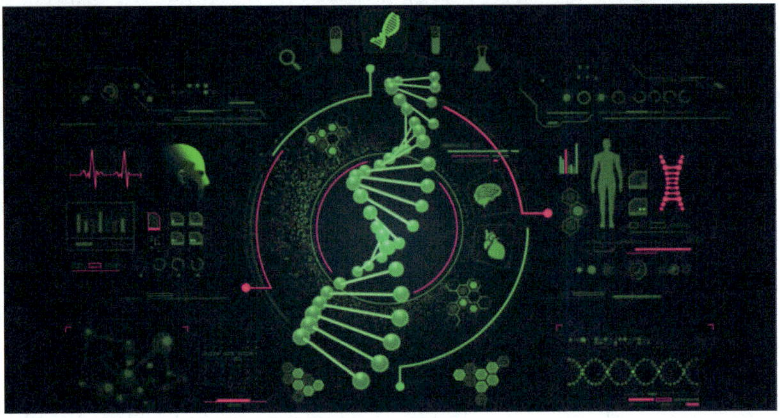

Figure 10.3. Overview of synthetic biology.

Source: https://docmode.org/wp-content/uploads/2023/03/2020-top-synthetic-biology-articles_03-730x390-1.jpg.

One of the biggest applications of synthetic biology is the drug delivery system, where cells are genetically engineered to transport biological drugs more precisely. Recent studies have shown that the application of synthetic biology to develop manipulable and intelligent drug delivery systems has revolutionized therapeutic protocols, especially for diseases such as cancer. Synthetic biology strategies enable cell behavior to be engineered for greater specificity and lower side effects, an important advance over traditional delivery systems. Further, with the integration of synthetic biology and nanotechnology, scientists are able to design multifunctional nanomaterials that can deliver medicines with high specificity and targeted localizations in biological tissues. This convergence is evidence of the innovative potential of synthetic biology in addressing pressing problems in drug delivery and developing advanced therapeutic modalities that can respond to the dynamic nature of diseases (Yi et al., 2025). Advances in synthetic biology have also been transformed by methods such as photoclick chemistry, which can simplify the operations of biological systems through precisely controlled chemical reactions. Photoclick chemistry offers the virtue of temporal control and high selectivity, thus making it an appropriate tool in bioconjugation and modular synthesis. Modern developments, including the PANAC photoclick reaction, showcase the generality of primary amines as reactive handles and expand applications in synthetic biology. This programmability allows scientists to develop complex biological constructs and build new platforms for applications such as the proteolysis-targeting chimera (PROTAC) library setup. Furthermore, the ability to probe interactions and dynamics of biomacromolecules in living systems through endogenous bioconjugation is an example of the intersection of chemical biology and synthetic systems, which provides new avenues to therapeutic discovery and development (Hu et al., 2025) (Figure 10.3).

10.1.2. Key Techniques and Methodologies

Chemical biology, particularly in synthetic systems, has developed immensely through the application of new techniques and methodologies. The most significant methodology is the use of synthetic nucleic acids, which enables researchers to enhance existing biological processes and design novel catalysts. Nucleic acids such as RNA and DNA have been extensively engineered to create synthetic DNAzymes and ribozymes whose functions go beyond the storage of genetic information to encompass catalytic and recognition functions. Recent scholarly research outlines the synthesis of xeno-nucleic acids (XNAs), which incorporate nonnatural backbones that significantly alter the properties of natural nucleic acids. These advances have allowed not only the creation of novel protein and peptide constructs, but also the exploration of artificial biopolymers that can perform specific biochemical functions, thereby providing insights into future synthetic biology strategies (Kuwahara et al., 2013). Another new advancement in synthetic systems is the application of high-throughput selection techniques such as phage display and mRNA display. These technologies allow for the swift identification and optimization of proteins and peptides with desirable characteristics and subsequently genotype to phenotype. One such technology is phage display, which allows for screening of vast libraries of peptides for favored interactions to ultimately lead to novel drugs or biomolecular reagents. Additionally, *in vitro* compartmentalization (IVC) provides a very efficient means of evolving proteins by encapsulating genetic material within small droplets, allowing the direct relationship between genotype and fitness. This has been useful in creating functional transfer RNAs and peptide mimics, thereby expanding the toolset for synthetic biologists to design new molecular systems (Zagnoni, 2012). Moreover, developments in microfluidic systems have transformed the ability to study cellular processes by building artificial cellular environments. The lab-on-a-chip technologies provide for the precise control of picoliter volumes of fluids to shape an environment favorable for high-throughput experimentation. By offering a method to create artificial lipid bilayers that closely mimic the cellular membrane, researchers are able to study the interaction of small molecules and membrane proteins in an artificial environment. This convergence of biological and engineering concepts has opened up new horizons for biosensing and pharmaceutical creation, allowing molecular interactions to be responded to in rapid real-time. The advancements in microfluidic technologies highlight the coupling of synthetic systems with chemical biology, pointing to an existence where biological response to different stimuli on a personalized level could be possible (Zagnoni, 2012).

10.1.3. Applications in Various Fields

The chemical biology forefront of cutting-edge synthetic systems has opened up avenues to numerous applications in various areas, most prominently environmental biotechnology and medicine. In environmental applications, synthetic microbial communities are engineered to enable the enhancement of bioremediation against environmental contaminants, e.g., heavy metals and persistent organic pollutants. Applying artificial community selection, researchers can exploit various microbial characteristics that optimize degradation pathways and increase the rate of biodegradation, which remain important challenges in pollution management. Emphasized in recent studies, such dynamic microbial systems are more promising compared to monocultures due to their enhanced stability and productivity, with a better breakdown of poisons (Borchert et al., 2021). Accordingly, chemical biology not only advances theoretical understanding but also provides actual solutions to pressing environmental problems. Further, the integration of chemical biology into synthetic systems is transforming the biomedical industry, particularly in cancer treatment. New strategies aimed at the molecular mechanisms of chromatin remodeling, for instance, the mSWI/SNF complex, illustrate the promise of the design of medicines aimed at particular genetic mutations in most cancers. The evidence indicates that over 20% of cancers harbor mutations in subunits of the mSWI/SNF complex, suggesting the potential for synthetic lethality strategies (Centore et al., 2020). This therapeutic potential results from the ability to selectively disrupt errant complex activities, thus paving new avenues to precision medicine. This confluence of synthetic biology and cancerology highlights the necessity for sustained investigation into principles of chemical biology to allow for the design of new treatment modalities for countering the complexity of cancer. Besides applications in the environment and biomedicine, chemical biology in synthetic systems makes a substantial contribution to agriculture and food production. The ability to engineer microbial consortia to obtain biopesticides and biofertilizers is great to enhance agricultural yields while minimizing reliance on synthetic chemicals. Such microbial systems can enhance nutrient uptake and improve plant health through synergistic interactions in soil. Furthermore, synthetic biology utilizes the tools of gene editing and metabolic engineering to breed crops that are more resistant to pests and diseases or possess superior nutritional quality. This new application not only addresses food security problems but also fosters sustainable approaches within agricultural systems. By employing chemical biology knowledge, these synthetic systems illustrate the potential to enhance the efficiency of food production and ecological stability (Borchert et al., 2021).

10.2. CHEMICAL BIOLOGY TECHNIQUES

Chemical biology as a discipline is essentially characterized by its hybridity, wherein biology and chemistry methodologies are fused to explain and regulate intricate biological processes. The pivotal feature of this science is the use of synthetic tools that enable the development and modification of biological molecules. Whereas biochemistry tends to specialize in the dynamics and interactions of biomolecules, chemical biology progresses a step further towards synthesizing these biomolecules based on the principles of synthetic organic chemistry. This way, scientists are capable of generating new tools and molecules replicating or altering biological processes, hence capable of bridging the gap between laboratory experiments and actual applications for medicine and biotechnology (Zhang et al., 2024). It allows researchers to create artificial systems that can be used in many applications, ranging from the discovery of drugs to the manufacture of new materials. Single-Particle Analysis (SPA) employed in cryo-electron microscopy is one of the most significant techniques in the field of chemical biology. This method has revolutionized structural biology by enabling high-resolution structural analysis of biological macromolecules and their complexes without the need for crystal formation, a common limiting factor in structural studies. Recommended also, SPA can be applied to artificial chemical systems with discrete clusters such as metal-organic frameworks and supramolecular compounds (Zhang et al., 2024). The ability to investigate such complex structures not only provides information on synthetic materials but also on the behavior of reaction intermediates. Since synthetic systems tend to possess diverse structures, the technique is helpful in ascertaining functionality and optimizing the synthesis process. Furthermore, the growth of chemical biology tools inspires the creation of new methods to explore and modulate cellular processes at a molecular level. Techniques such as CRISPR-Cas9 gene editing, high-throughput screening, and chemical probes enable researchers to study biological pathways and interactions with unprecedented precision. Chemical probes, for instance, can be designed to selectively bind to specific biomolecules so they may be examined across various contexts within a cell. This capability supports synthetic biology through enabling the design of entire pathways or systems, which in turn can result in response-specific or novel functionalities (Natarajan, 2021). Taken together, these approaches reflect the dynamic evolution of chemical biology, underscoring its capability to enhance our knowledge of both synthetic and biological systems, and ultimately culminating in revolutionary breakthroughs in science and technology.

10.2.1. Molecular Cloning and Gene Editing

Gene editing and molecular cloning are among the most important advances in the field of chemical biology, whereby genetic material can be manipulated for numerous applications. The development of tools like recombinant DNA technology allows researchers to isolate, replicate, and even alter genes across various organisms. Molecular cloning is a fundamental resource for gene isolation, providing the first step in preparing to design genetically engineered organisms. In the context of artificial systems, these techniques make it possible to design and construct novel genetic circuits, with potential for innovation in applications such as metabolic engineering and synthetic biology. This basic research in gene cloning serves as a stepping stone to more target-based editing technologies, ultimately leading to the exploration of gene function and interaction in biological systems (Li et al., 2020). Gene editing tools, in particular, engineered nucleases-based tools, have transformed the landscape of genetic studies. The introduction of CRISPR/Cas9 and the more mature tools, such as zinc-finger nucleases (ZFNs) and transcription activator-like effector nucleases (TALENs), has provided researchers with unprecedented accuracy for targeting specific genomic locations. These technologies not only facilitate the development of genetically modified organisms but also enhance our understanding of how genes predispose an individual to various diseases (Li et al., 2020). For example, the ability to make precise edits in eukaryotic genomes has facilitated the development of cellular and animal models that accurately mirror human pathological processes. Therefore, the integration of molecular cloning and such gene editing technologies highlights their synergistic roles toward transmuting basic science and applied research, as well as ultimately transmuting into new therapeutic strategies. Beyond their implications in research, molecular cloning and gene editing technologies are also finding applications in biotechnology and medicine. Methods such as CRISPR/Cas9 have been developed for application in various fields, such as agriculture, where they are used to engineer disease-resistant crops, and medicine, where they promise potential for genome therapy. Specifically, gene targeting of the causative genes of parasites and vectors using CRISPR/Cas9 underscores the relevance of these tools in infectious disease management. In particular, efforts are underway to optimize these systems for large-scale use and to make them more accurate and efficient. As researchers persist in overcoming the hurdles around these technologies, gene editing and molecular cloning will significantly enhance our ability to design and engineer biological systems for medicine and green solutions, and further solidify their role in the area of chemical biology in synthetic systems.

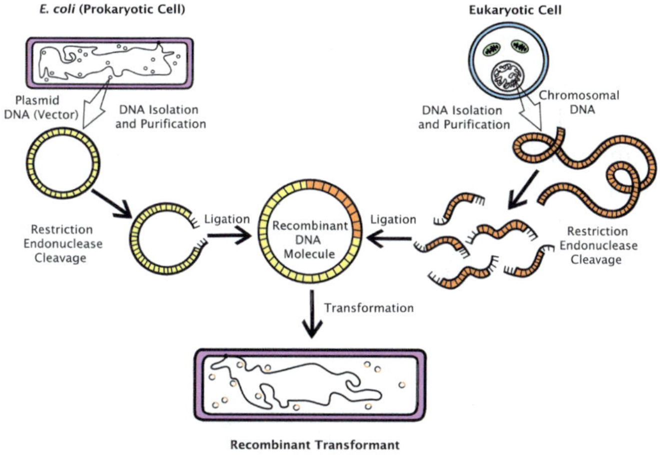

Figure 10.4. The overview of the procedure of gene cloning.

Source: https://static.cambridge.org/binary/version/id/urn:cambridge.org:id:binary:20201222024324969-0688:S2058850X20000053:S2058850X20000053_fig1.png?pub-status=live

10.2.2. High-Throughput Screening (HTS) Methods

HTS technologies revolutionized the landscape of chemical biology to enable scientists to efficiently screen vast libraries of compounds for biological activity. The techniques leverage automation, miniaturization, and advanced data analysis to rapidly analyze multiple assays simultaneously. HTS is, therefore a crucial element in drug development and discovery because it allows scientists to find promising candidates with the capacity to modulate biological pathways. In synthetic systems, HTS technologies not only boost the discovery rate but also provide valuable information on the potency and mechanism of action of lead compounds, paving the way for subsequent optimization and validation. Therefore, the integration of these screening technologies marks a significant step forward in chemical biology, facilitating a more rational strategy to unravel complicated biological interactions and drug targets. Furthermore, the application of HTS in synthetic biology is important in the functionalization of biomolecules as well as in the construction of novel synthetic systems. For example, advances in combinatorial chemistry have led to the synthesis of large libraries of complex compounds, which are amenable to screening for preferred bioactivity to identify lead candidates quickly. The use of high-throughput technologies has also streamlined the investigation of metabolic pathways, where researchers can identify the roles of several metabolites in biological processes. One good case is the metabolomic profiling of 3,5-DCAA-treated barley seedlings, which is a compound that triggers defense processes. Using sophisticated screening techniques, researchers can detect hidden patterns and treatment-associated alterations in metabolite profiles, thereby advancing the field of plant immunity and enabling the development of new crop defense and improvement strategies (Djande et al., 2025).

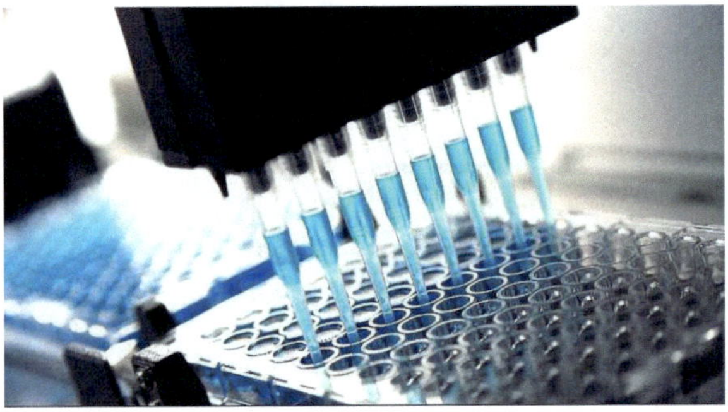

Figure 10.5. High throughput screening (HTS).

Source: https://assets.technologynetworks.com/production/dynamic/images/content/288926/high-throughput-screening-accelerating-drug-discovery-efforts-288926-960x540.jpg?cb=11727967.

Apart from applications in drug discovery and synthetic biology, HTS methods are being developed to incorporate new chemical methodologies such as photoclick chemistry. This novel approach has a number of significant advantages, including exquisite control over reaction conditions and high selectivity, which are essential for the investigation of complex biological systems. Development of procedures like the light-induced Primary Amine and o-Nitrobenzyl Alcohol Cyclization (PANAC) is a significant advance in photoclick chemistry that enables multifaceted use in modular synthesis and bioconjugation. By leveraging the potential of HTS, researchers have been able to harness these novel chemical reactions to accelerate the discovery of bioactive compounds, build DNA-encoded libraries, and obtain more comprehensive profiling of biomolecular interactions within living organisms. Overall, the combination of HTS with sophisticated chemical techniques opens up new platforms in chemical biology toward groundbreaking advancements in drug discovery, bioconjugation, and synthetic systems (Hu et al., 2025) (Figure 10.5).

10.2.3. Metabolomics and Proteomics Approaches

Metabolomics and proteomics are two of the most important approaches in chemical biology that expand our understanding of complex biological systems. Metabolomics is a holistic analysis of the metabolic profiles, owing to both genetic and environmental influences on biological organisms. With techniques such as mass spectrometry and nuclear magnetic resonance, researchers can identify and quantify small molecules or metabolites in biological samples. This data can be useful in the knowledge of metabolic pathways and can help researchers observe how organisms respond to diverse stimuli such as environmental stressors or chemical treatments. One such example is observed in the study where the metabolomic profile of 3,5-Dichloroanthranilic acid-treated barley seedlings was investigated, and impressive defense mechanisms were shown to be triggered via altered metabolic pathways (Djande et al., 2025). At the same time, proteomics seeks to gain information regarding protein dynamics and functions within biological systems. It employs various techniques such as two-dimensional gel electrophoresis, mass spectrometry, and protein microarrays to detect, characterize, and quantify proteins and their expressions across different conditions. Proteomic conclusions uncover not only the abundance and modification of proteins but also protein interactions and functional capabilities in metabolic networks. This is particularly relevant in the area of targeted disease therapies, where information regarding protein interactions can be utilized to create novel therapeutic approaches. One such example is cereblon as a drug target, illustrating how proteomic strategies can define protein interactions and provide data that is accountable for drug efficacy, as with thalidomide and lenalidomide (Xu et al., 2025). The union of metabolomics and proteomics with synthetic biology is yet another new paradigm in chemical biology. By examining the interaction between protein functions and metabolic pathways, researchers can design biological systems with higher function. This cooperation enables a better understanding of cellular processes and the design of more effective means of crop protection and agricultural sustainability, as indicated by the research on 3,5-DCAAs' action on barley (Djande et al., 2025). In addition, by employing ideas of chemical biology, including selective protein destruction, scientists are capable of managing metabolic pathways and protein expression patterns to achieve desired effects in synthetic systems. The convergence of the application of these methodologies not only drives fundamental research but also has practical applications in biotechnology and medicine, ultimately in the development of improved and enhanced biological systems.

10.3. ROLE OF ENZYMES IN SYNTHETIC SYSTEMS

Enzymes, as biocatalysts, play a catalytic function in the development of synthetic systems, demonstrating the merging of chemical biology and biomedical applications. Their specificity and efficiency allow complex biochemical conversion under mild conditions and thereby make them valuable tools in synthetic biology. For instance, the application of engineered enzymes allows selective control of biomolecules, which is paramount in synthesizing sophisticated synthetic pathways. The rapid advancement in enzyme engineering has also allowed the introduction of new catalytic functions and thereby extended the field of synthetic chemistry. By enhancing substrate specificity and reaction rates, enzymes enable the synthesis of compounds that would be challenging to make through conventional chemical routes, underscoring their significance in industrial and academic settings with interests in innovation and the efficiency of chemical reactions. The emerging field of photoclick chemistry also emphasizes the key position that enzymes occupy in synthetic systems. Recent advances have identified a novel strategy known as Primary Amine and o-Nitrobenzyl Alcohol Cyclization (PANAC), where the selectivity of primary amines is employed to enable specific conjugation in complex biological contexts (Hu et al., 2025). The method not only exhibits the selectivity of enzyme-catalyzed reactions but also has potential for modular synthesis and bioconjugation. With the introduction of enzymes to photoclick reactions, researchers are able to effectively synthesize many bioactive molecules and facilitate intricate molecular interactions within cells. In return, applying enzymes to synthetic systems not only enhances the tunability of biological functions but also accelerates the discovery of new therapeutic strategies, simplifying links between synthetic chemistry and valuable biomedical functions. Besides their applications in synthetic chemistry and photoclick strategies, enzymes have an important function in the development of intelligent biomaterials, such as hydrogels, that have both biological and technological elements for enhanced biomedical uses (Segneanu et al., 2025). Hydrogels can be engineered to adapt based on various stimuli of pH, temperature, and enzymatic action, and hence support tailored drug delivery systems and tissue engineering strategies. Enzymes invest such materials with dynamic functionality, enabling them to mimic biological processes and effectively respond to environmental cues. Such responsiveness optimizes the therapeutic effectiveness of hydrogels more efficiently, enabling targeted release profiles and increased bioactivity. Thus, the inclusion of enzymes in synthetic systems promises new materials that are functionally superior and able to address complex biological problems, marking a big stride in chemical biology.

10.3.1. Enzyme Engineering and Design

Enzyme engineering and design are a key interface between synthetic systems and chemical biology, in which principles of biochemistry are used to redesign or optimize enzyme function. Enzyme engineering has enabled the design of expert biocatalysts over the past few years, which have uses in a variety of industrial applications, including pharmaceutical and biofuel production. With techniques such as directed evolution and rational design, scientists can improve the performance, stability, and specificity of enzymes. Such modifications are particularly significant in synthetic biology, where engineered enzymes may be incorporated into larger synthetic pathways to achieve desired biochemical conversions. The ability to design enzymes to some specifications not only promotes efficiency in biocatalysis but also opens the door to cleaner processes, as part of global efforts towards greener chemistry and resource conservation. Recently, enzyme engineering has focused on leveraging the unique properties of novel materials and nanostructures to maximize enzyme performance. For instance, enzyme systems have been of great interest for incorporation of graphdiyne (GDY) due to its pore structure and reactivity, which can improve the stability and efficiency of enzymes in catalysis. Biomaterials composed of GDY can enable better immobilization of enzymes to create heterogeneous catalytic systems with high activity in a variety of applications, including drug delivery and molecular sensing. With scientists venturing to couple such materials with engineered enzymes, possibilities for applications in biomedical sciences become drastically increased. These advances represent a promising synergy of chemical biology and materials science and are central to the advancement of multifunctional systems that can transform several industries, notably in the guise of targeted therapeutic agents and diagnostics, as emphasized by recent reviews. In addition, biological whitening technology is one of the emerging trends in enzyme engineering, where sustainability is of greatest concern. The development of green agents, i.e., microbial enzymes and plant-based agents, demonstrates how enzyme design can address industrial challenges with minimal ecological impact. Biological whitening agents possess inhibitory activity on melanin biosynthesis and antioxidant activity, and are proof of the versatility of enzymes in cosmetic and textile industries. Inclusion of synthetic biology approaches enhances enzyme performance even better, so that it facilitates industry-specific solutions that strictly meet industry requirements. Even if problems like scale-up production and stability pose challenges, new approaches such as metabolic optimization and nano-encapsulation usher in new gates of applicability in daily situations. While more advanced research is conducted, enzymatic solutions highlight the enormous contribution enzyme engineering has made to synthetic systems, providing new ways towards sustainable industrial processes (Jia, 2025; Zhao et al., 2025) (Figure 10.6).

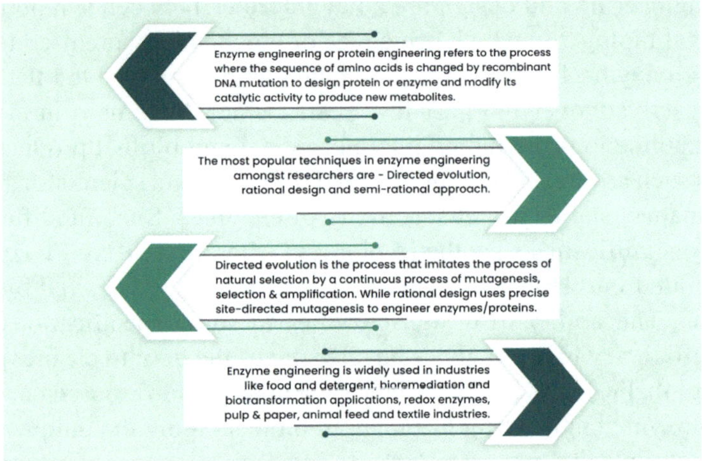

Figure 10.6. Enzyme engineering.

Source: https://infinitabiotech.com/wp-content/uploads/2022/02/enzyme-engineering-super-final.jpeg.

10.3.2. Applications of Enzymes in Synthetic Biology

Enzymes are critical components in synthetic biology, which dramatically improve the efficiency and versatility of biochemical processes. Among the prominent applications is the design of Type II polyketide synthases (PKS) to generate diverse polycyclic aromatic molecules relevant in the pharmaceutical industry. These enzymes allow one to build intricate molecular frameworks through the catalysis of polyketide core synthesis with alteration of chain length and modes of cyclization. The modular cassette of Type II PKS, consisting of a series of tailoring enzymes, is one such example. By enhancing these enzyme platforms, researchers hope to identify novel therapeutic compounds and biofuels, highlighting the importance of enzyme application in the design of synthetic biology projects and their impact on chemical diversity in synthetic systems (Rivers et al., 2024). Another promising aspect of enzyme application in synthetic biology involves the integration of artificial small molecules capable of furthering enzyme function. Recent advances in protein design have enabled one to apply such molecules, like metal cluster mimics and designer cofactors, that supplement traditional enzymic functions. Such technology allows for the enrichment of enzymic reactions as well as metabolic pathways and sets up a new platform for the design of synthetic biological systems. For instance, it is possible to incorporate non-natural amino acids and nucleotides into protein scaffolds to engineer new enzymes with distinct characteristics, thereby significantly expanding the limits of traditional biocatalysis. The ability for the design of enzymes to catalyze novel reactions is one of the areas where synthetic biology has revolutionized the biotechnology field, introducing new approaches to solutions in the pharmaceutical and biofuel industries (Liu et al., 2023). The applications of enzymes have implications in environmental sustainability, with engineered enzymes allowing for the practice of green chemistry. Enzymes provide a biocompatible method for chemical synthesis compared to conventional methods, with milder conditions and fewer byproducts being required. In the creation of bioremediation methods, enzymes are designed to degrade environmental pollutants, thereby aiding in ecological rehabilitation processes. The advances in synthetic biology enable the engineering of enzymes to specifically target and degrade toxic compounds, with a practical demonstration of enzyme engineering to address real-life problems. With continuous advancements in science, the union of synthetic biology and enzymology is poised to generate sustainable bioprocesses and minimize environmental impact, thus becoming a critical component of present-day chemical biology [cite29; cite30].

10.3.3. Case Studies of Successful Enzyme Applications

The past years have witnessed enormous progress in the area of chemical biology, utilizing enzymes in artificial systems. One of the striking examples is the engineering of enzymes for increasing the production of biofuels, particularly ethanol. Researchers have succeeded in optimizing the process of metabolism in microorganisms like *Saccharomyces cerevisiae* to increase their ability to utilize unfavorable sugars like xylose. Through the accurate engineering of the pivotal genes in the yeast's metabolic pathway, scientists have made significant improvements to ethanol production, demonstrating the biotechnological promise for renewable energy resources (Liu et al., 2023). This shows how enzyme engineering and metabolic regulation converge, as is the case with synthetic biology, solving the world's key problem of sustainable fuel production. Another notable case study is the use of designed enzymes for the biosynthesis of high-value drugs. By using non-native amino acids and designed cofactors in metabolic pathways, scientists have expanded the chemical range of compounds that can be biologically synthesized. For instance, the yeast system has been redesigned to produce dihydroartemisinic acid, a precursor of the antimalarial drug artemisinin. This approach leverages the dual strengths of in silico modeling and in vivo experimentation to identify non-trivial metabolic targets, thereby optimizing enzymatic steps for higher yields (Liu et al., 2023). Achievements in these uses illuminate the potential of enzyme engineering towards the production of sophisticated, therapeutically worthwhile molecules as well as the overall influence on drug development and public health. Moreover, the use of enzymes for environmental cleaning is also one sector that has been one of the success stories of synthetic biology. Design of microbial systems for the degradation of environmental toxins is particularly critical when addressing the issue of industrial waste. For instance, researchers are engineering synthetic microbial communities that employ enzymes that are capable of degrading toxic substances in contaminated water and soil. By using synthetic biology concepts, these consortia can be designed to optimize interactions among microbial populations, thereby improving the efficiency of degradation (Zhang et al., 2014). This demonstrates how the utilization of enzymes not only stimulates biotechnological advancements but also aids in addressing important ecological issues, demonstrating the value of multidisciplinary approaches in chemical biology and in synthetic systems in general.

10.4. CHEMICAL BIOLOGY IN DRUG DEVELOPMENT

Chemical biology has transformed drug development over the past few years, primarily through the mechanism of targeted covalent inhibitors. These inhibitors create stable and tight interactions with targets that enhance both efficacy and selectivity in treatment. For example, recent studies on RMS-07, a first-in-class covalent inhibitor of the protein kinase TTK, demonstrate significant potential for the therapeutic treatment of several distinct forms of solid tumors. Not only does the promise of this agent lie in its selectivity, but also in the strategic design of its binding kinetics, engineered to create an optimal therapeutic index. However, the problem persists regarding the synthetic accessibility of key reagents, such as [1-(ethoxycarbonyl)cyclopropyl] triphenylphosphonium tetrafluoroborate, applied during the synthesis of such new drug candidates. Limited access and complex synthesis pathways pose problems that necessitate innovative solutions in chemical biology to rationalize drug development processes (Rebhan et al., 2025). Moreover, the synthetic preparation of synthetic systems that can mimic biological processes is crucial during drug development.

Application of chemical systems designed to replicate natural signaling mechanisms has revealed promising avenues for new therapeutic methods. As an example, present studies have centered on minimalist photoresponsive molecule development that can be utilized as artificial photoreceptors. These molecules allow for insertion into lipid bilayers and utilize the power of light to modulate biological activity, illustrating the union of synthetic biology and therapy. By leveraging reversible photoisomerization, such artificial systems can be engineered to control enzymatic catalysis in a cellular environment. This new approach not only enhances understanding of natural biochemical processes but also has potential for future applications, such as intelligent drug delivery and targeted therapy, an important milestone in employing chemical biology for drug discovery (Yang et al., 2022).

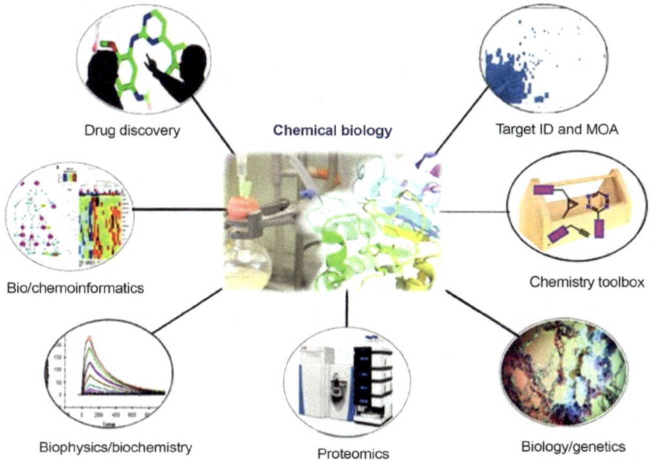

Figure 10.7. Chemical biology in drug discovery.

Source: https://ars.els-cdn.com/content/image/1-s2.0-S0065774317300179-f10-01-9780128130698.jpg.

The use of chemical biology in drug discovery is a paradigm shift towards a more mechanistic understanding of disease and targets. Utilizing synthesis strategies that mimic intricate biological processes, scientists are poised to create new therapies with improved therapeutic profiles. As illustrated by RMS-07 and photoresponsive systems, the ability to design compounds that interact precisely with biological pathways is necessary to advance targeted therapies. Such breakthroughs are a vision towards precision medicine, in which treatments are tailored to the molecular characteristics of certain diseases. Scalability and reproducibility in synthesis techniques, however, remain current concerns, such that more work and innovation are needed in the field of chemical biology. Lastly, the convergence of synthetic systems with drug discovery is poised to transform therapeutic paradigms, advancing more efficient and targeted medical therapy (Rebhan et al., 2025; Yang et al., 2022) (Figure 10.7).

10.4.1. Target identification and validation

Target identification and validation are significant processes in drug discovery and therapeutic development in the field of chemical biology. Target identification begins with the identification of potential biological targets, which are usually proteins or other biomolecules that participate in disease mechanisms. Effective target identification can employ both experimental and computational approaches, including genomics, proteomics, and high-throughput screening. Chemical probes, such as small molecules and biological assays, can be used to uncover crucial interactions in biological systems. Following identification of putative targets, the next step is validation, ensuring that they are useful and applicable in therapy. Validation typically encompasses rigorous screening to determine whether putative targets can be confidently modulated by compounds of interest, thereby validating their potential as viable drug leads, as discussed in the drug discovery paradigm (Riaz et al., 2020). After successful identification of biological targets, validation is a process that validates the functions of such targets in the desired biological processes. Validation not only establishes the biological significance of the target but also yields a clear mechanism of action for potential therapeutics. Techniques such as RNA interference, CRISPR/Cas9 gene editing, and the use of specific inhibitors can be utilized to define the function of the identified targets. In precision medicine, the role of target validation is even more pronounced, as the efficacy of targeted therapies often relies on a good understanding of the molecular basis associated with disease states. With the aid of sophisticated proteomics approaches, e.g., SWATH-MS-based methodologies, target interactions can be quantified more accurately and reproducibly, thereby unveiling critical information needed for effective drug design and discovery (Ludwig et al., 2018). Lastly, the interplay between target identification and validation is key to speeding up the process of drug discovery in chemical biology. With improved strategies, the ability to identify and validate drug targets at faster rates will certainly speed up the translation of scientific discovery to clinical application. This interplay between chemical biology and new technologies offers a fertile ground on which new therapy strategies are possible and can tackle unmet medical needs. By the methodical development of target identification and validation ideas, scientists can warp drug development's turns of fate to their advantage and deliver novel medications that are effective and optimized to the very particular demands of human health (Riaz et al., 2020) (Figure 10.8).

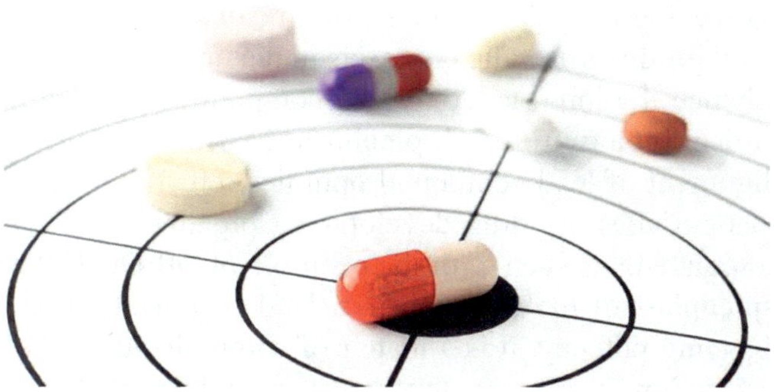

Figure 10.8. Target identification and validation in drug discovery.

Source: https://assets.technologynetworks.com/production/dynamic/images/content/312290/target-identification-validation-in-drug-discovery-312290-1280x720.webp?cb=12688410.

10.4.2. Lead Compound Optimization

Optimization of lead compounds is a critical component of drug discovery, particularly in the realm of chemical biology in artificial systems. Not only does this phase optimize the chemical structure of lead compounds, but also the potency, selectivity, and bioavailability of lead compounds. One great example of this can be observed with the optimization of the isocyanoterpene analogue, MED6-189. This compound has shown potent antimalarial activity against drug-sensitive and drug-resistant *Plasmodium falciparum* strains, showing its potential in combating malaria's resurgence despite the development of resistance to the existing treatments. By the use of a multi-faceted approach that entails phenotypic assays and chemical syntheses, researchers have optimized MED6-189s therapeutic profile, confirming its safe profile in humanized mouse models, ultimately addressing the urgent need for new antimalarial agents in the current health context (Chahine et al., 2024). Parallel to advances in antimalarial drugs, the optimization of lead molecules in the field of antiviral drugs has also been a major focus, especially amid recent viral outbreaks. Optimization of antiviral molecules typically exploits natural products due to the broad array of chemical structures and biological activities they possess. The case of the high-throughput screening platform designed for the screening of microbial natural product extracts against human coronaviruses, namely HCoV-OC43, highlights this novel approach. With the COVID-19 pandemic highlighting the pressing need for efficient antiviral therapies, these approaches have accelerated the discovery of promising new antiviral molecules. The ability for mechanistic screening at scale and delivering results within a matter of hours positions natural products as valuable assets in the ongoing battle against novel viral threats, hence the importance of lead compound optimization to remain ahead of chemical biology research pipelines (Martínez-Arribas et al., 2023). Moreover, alignment of lead compound optimization strategies across diverse disease models enriches the drug development pipeline such that results can be better extrapolated. As seen with the example of MED6-189 and its action against the apicoplast of malaria parasites, lead optimization is not merely a matter of enhancing potency. It is a matter of comprehending the compounds' mechanism of action, resistance mechanisms, and interactions in complex biological systems. These are supplemented by sophisticated approaches such as transcriptomic and metabolomic profiling, which give detailed insights into how molecules influence cellular pathways. A comprehensive approach to optimizing lead compounds thus not only accelerates the discovery of new therapeutics but also sheds light on pathway directions for new drug innovation. This future-oriented strategy is at the core of sustaining chemical biology breakthroughs that facilitate a robust reaction to public health challenges (Chahine et al., 2024) (Figure 10.9).

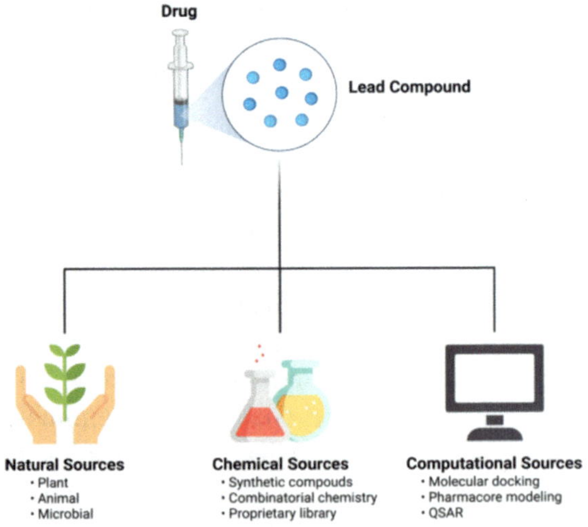

Figure 10.9. Sources of lead compounds.

Source: https://ars.els-cdn.com/content/image/3-s2.0-B9780323903004000896-f08-02-9780323903004.jpg.

10.4.3. Case Studies of Synthetic Drugs

Synthetic drugs are dominated by intriguing case studies that illustrate the tremendous impact of chemical biology on modern medicine. A classic example is the creation of synthetic opioids that are engineered to be potent pain relievers without the risk parameters inherent in natural opioids. The example of synthesizing drugs such as fentanyl, initially created as an anesthetic, shows how slight modifications in molecular structure can result in immense differences in potency and pharmacokinetics. This case is a classic example of the double-edged sword of synthetic drugs, where the therapeutic benefits of targeted therapy must be conscientiously compared against the risk of abuse and unwanted effects. Thus, this ongoing debate in terms of safety and effectiveness illustrates the need for high-capacity regulatory policies and safety assessments, a view similarly echoed in the discussion for synthetic biology applications (Firestone et al., 2024). Another significant case study is the process of creating fluorescent compounds as pharmaceuticals through synthesis, specifically their use in cancer diagnostics.

Fluorescent agents are now widely utilized in imaging technologies, offering real-time information on tumor dynamics and therapeutic outcomes. Not only are these compounds efficient in their native forms, but they can also be artificially modified to increase their biocompatibility and specificity. The potential to tailor these fluorescent molecules enables greater visualization of cellular functions during drug delivery, thereby optimizing therapeutic effectiveness. Knowledge of their unique chemical structures and photophysics is crucial to expanding their uses across various biomedical disciplines. These developments are indicative of the integration of chemical biology and synthetic systems to establish new solutions in the medical arena, putting fluorescent agents at the forefront of anti-cancer treatments (Joshi et al., 2024). Finally, case studies of drugs synthesized also move into the antibiotic development area, where drug-resistant microorganisms have precipitated an impending need for a firm response from the field of chemical biology. Synthesis of novel antibiotic compounds, such as synthetic derivatives of existing classes, is designed to outsmart resistance mechanisms most commonly employed by bacteria. By deploying high-throughput screening technologies and structure-activity relationship analysis, researchers can identify molecules that remain effective against resistant pathogens. It is through such proactive thinking that synthetic chemistry takes central stage in solving issues of global health. Moreover, integrating risk-based analyses and assurance processes can help ensure the safety and efficacy of synthetic antibiotics—all important puzzle pieces in the battle against antibiotic resistance (Firestone et al., 2024). These case studies demonstrate the revolutionary potential of synthetic drug development, revealing the pivotal convergence of chemistry, biology, and medicine in addressing complex health issues (Figure 10.10).

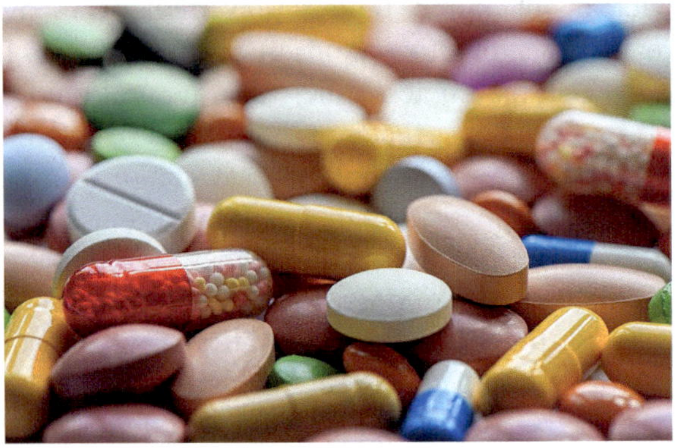

Figure 10.10. Synthetic drugs.

Source: https://www.zelusrecovery.com/wp-content/uploads/2023/11/What-Drugs-are-Classed-as-Synthetic-Drugs.jpg.

10.5. ETHICAL CONSIDERATIONS IN CHEMICAL BIOLOGY

The discipline of chemical biology in artificial systems certainly possesses transformative capability, but is not without ethical issues to carefully consider. The tension between the natural and the artificial strongly influences public understanding of such advances. The ongoing conflict between imitation and authenticity, identified by Orvell, serves as a compass in working through meaning within this novel context. The subtleties of separating natural phenomena from artificial constructs, as explained by Birnbacher, serve to demonstrate the contentiousness of the discussions, particularly in measuring society's impact of artificial systems. The intersection of ethics, technology, and biology raises critical issues regarding the verisimilitude of synthetic biology applications, which impact not only scientific culture but also the wider population and policymaking. Hence, the moral debate regarding chemical biology has ramifications that extend beyond the lab setting to question the way that innovations become accepted or rejected in society. Examining the moral responsibilities generated by advances within artificial systems demonstrates concerns to include human well-being and environmental impact. Growing reliance on metabolomic research is just one example that demonstrates the dual nature of innovation: its ability to extend our understanding of complex biological interactions and the potential to harm. As suggested by the review in metabolomics, though it increases our level of knowledge about microbial interactions and their influence on the health of humans, application of such technologies also creates ethical concerns regarding data purity, privacy, and unintended consequences of messing with biological processes (Muhamadali et al., 2023). Chemical biologists are required to maneuver these complexities to ensure scientific progress does not happen at the cost of ethics or compromise health outcomes. It is essential that there is a development of a moral framework that will guide research and implementation in this new science, with consideration of the potential benefits as well as the risks of negative effects of synthetic biological applications. As chemical biology advances as a science, its ethical implications on its synthetic systems become increasingly pertinent, necessitating interprofessional discourse and regulation. Since artificial intelligence and automation play such a pivotal role in this field, the issues of accountability and commodification of scientific development come into focus. AI-driven platforms, for example, are able to accelerate personalized medicine but also pose concerns regarding emotional dependence and commodification of care (Nairn et al., 2024). In addition, the intersection of engineered and natural systems raises ethical and philosophical issues with regard to ownership, sustainability, and access to biotechnological developments. There is a need for inclusive regulatory frameworks that incorporate diverse perspectives in responding to the complex moral context of synthetic biology. Fostering ethical public discourse will facilitate rational innovation in chemical biology, such that the advancement of synthetic systems is harmonious with societal standards and ecological health.

10.5.1. Biosecurity and Biosafety Issues

In the realm of synthetic biology, issues of biosecurity and biosafety are now the dominant themes, demanding extensive research into the risks posed by engineered biological systems. As the world in the field accelerates by the minute, the possibility of unintended consequences—such as the release of modified organisms into environments—gives rise to grave issues regarding governing such risks. It then becomes crucial to have rigorous biosafety measures that ensure containment and restrict the extent of harm possible. These not only serve to protect biodiversity but also the public health by preventing harmful pathogens from entering the general environment from the laboratory. The use of advanced technologies in synthetic biology increases the stakes in such challenges since complexity in engineered systems may generate unintended interactions in biological networks, which may require stringent biosecurity practices in research and application settings. Moreover, the regulation of synthetic biology must emphasize public involvement to impart transparency and confidence-building. Successful public engagement can address concerns about synthetic biology innovations, particularly misuse or accidents. Science Cafés then emerged as significant spaces for dialogue where community members and specialists could debate the safety, ecological, and ethical implications of synthetic biology in a casual and more accessible context. Throughout these encounters, people radiate enthusiasm along with fear regarding the technologies' benefits and drawbacks, which emphasizes the dual-use promise of synthetic biology. The trust of the public in researchers conducting such research is built upon proper regulatory frameworks being aligned with such developments. The requirement for sustained contact to bridge the knowledge gap and enhance community understanding of biosafety and biosecurity issues in synthetic biology is present, as per Science Café research (Navid et al., 2012).

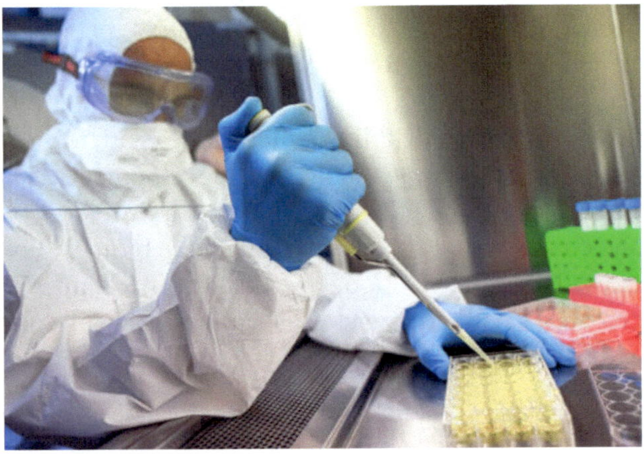

Figure 10.11. Biosecurity and biosafety.

Source: https://www.orfonline.org/public/uploads/posts/image/1723206649_Source%20Image%20-%20871x613.png.

In addition, as synthetic biology continues to progress, biosecurity research must follow the technologies. The use of artificial intelligence in this area also offers new ways to enhance biosafety and biosecurity measures, potentially enabling real-time monitoring of biological systems and predictive risk modeling. In the process, however, these technologies present ethical issues regarding data privacy as well as the responsibilities of the inventors and implementers. Therefore, interdisciplinarity is imperative in formulating policies not only for safety concerns but also incorporating ethical considerations that guide good research. During the investigation of artificial intelligence in synthetic biology, collective collaboration by researchers, policymakers, and society will be necessary to render these advancements beneficial while preventing likely harm (Amaan et al., 2024) (Figure 10.11).

10.5.2. Ethical Implications of Synthetic Organisms

The emergence of synthetic biology raises complex ethical implications that move beyond the biological to societal, legal, and moral dimensions. While scientists create synthetic organisms, they are confronted more and more with concerns regarding the potentiality of ecological disruption, unintended destruction, and the resulting liability of those who build such living things. Ethical issues should address the environmental impacts of released organisms, particularly in the natural biomes. This has created concerns over biosafety features, such as genetically embedded features, to prevent unintended proliferation or gene transfer. This calls to mind earlier calls for responsible innovation, as in keeping with models that suggest integrating public participation and ethical deliberation into biotechnological research and development (Kallergi et al., 2020). Confronting these ethical dilemmas is necessary in order to build public confidence and inform the appropriate development of synthetic biology. Coupled with environmental issues, synthetic organism manipulation calls into question autonomy and the very definition of life itself. The act of synthesizing life that holds a changed or completely synthetic genetic map calls us to question our philosophical conception of nature and existence. What are the rights, if any, that these bioengineered species ought to enjoy? More importantly, this supplements human debates regarding biotechnology and gene modification, placing synthetic organisms at the center of debates over natural and artificial life. The ethical systems informing these debates must balance technology with respect for biodiversity and existing ecosystems. In going into these interstices, the discussion of genetic protections must balance broader ethics and values of society on the basis of varying spheres of application, so to situate the discussion in a paradigm that is context-sensitive (Muratov et al., 2020). Finally, the implication is to equity and access factors because possible advantages to medicine, agriculture, and environmental sustainability through synthetic organisms have the ability to create inequalities as to who benefits from these innovations. Thus, it is essential that all individuals have a voice in how these technologies are used and created. The variations may not only affect access to benefits but also yield polarized views concerning the acceptability of synthetic organisms based on cultural and socioeconomic factors. Herein, the RRI principle entered the scene and advocated for proportional representation of stakeholders from mixed backgrounds in synthetic biology development decision-making. As technology advances, the management of these social challenges is a significant factor in promoting equitable access and in creating a shared vision of the consequences that accompany synthetic organisms (Kallergi et al., 2020).

10.5.3. Regulatory Frameworks and Guidelines

Regulatory frameworks and guidelines must be established in overseeing the rapidly evolving field of synthetic biology, particularly with regard to chemical systems. As biologists push the biological engineering field further each day, the need for a robust governance structure is greater than ever before. Current frameworks lag behind technologies that involve complex interactions between synthetic and biological materials. This is a regulatory gap that may lead to poor control, with potential misapplication or unforeseen consequences, such as environmental damage or biosecurity risks. Therefore, having comprehensive guidelines that can evolve to adapt to new technologies is crucial to realizing the full potential of synthetic biology without threatening public security and environmental integrity.

It is one of the major issues to incorporate emerging technologies like artificial intelligence (AI) into synthetic biology in the development of effective regulatory frameworks. Incorporating AI enhances the design and engineering functions, facilitating unprecedented efficiency in designing biological systems. Integration, as such, adds complexity regarding accountability and transparency in making decisions by means of AI. Existing laws must be developed to combat these challenges with provisions for human intervention and ethical needs within autonomous processes. Hardening laws with a focus on AI's dual-use nature can help instill a culture of responsibility among bioengineers so that the revolutionary potential of AI-enabled synthetic biology can be realized without sacrificing ethics and public trust (Vindman et al., 2024). Apart from theoretical issues, regulatory authorities need to urgently consider pragmatic approaches in the assessment of synthetic systems, especially via cutting-edge analytical methods. Cryo-electron microscopy and single-particle analysis (SPA) have proven to be valuable tools for determining the structures of biological macromolecules and synthetic chemical systems. While SPA has been effectively applied in the field of structural biology, its possible application in the study of synthetic intermediates remains underleveraged (Zhang et al., 2024). Policy frameworks need to facilitate research using these approaches by providing guidelines specifying best practices and safety measures in their use. Such practices would not only promote the growth of scientific knowledge but also ensure that the results achieved from synthetic systems are credible and ethical, paving the way to their safe integration into use within society.

10.6. FUTURE DIRECTIONS IN CHEMICAL BIOLOGY

The future of chemical biology will benefit from synthetic systems that replicate the metabolic capabilities of living organisms. These biological networks are not simply mimics of chemical plants; rather, they take simple input molecules to generate useful compounds through intricate metabolic pathways. By tapping into the natural efficiencies of microbial production, researchers aim to bridge the gap between metabolic engineering and traditional industrial methods. This area embraces advanced systems and control engineering principles to drive metabolic networks to optimal capacity, allowing for greater yields and new product attributes. With each improvement in analytical methods, the capability to optimize these systems increases, with potential to drive novel uses in fields from medicine to energy production. Therefore, the search to harvest both synthetic and natural processes will inform the next generation of chemical biology breakthroughs, rewriting our comprehension of biochemical processes and broadening the industrial horizon (Sarkar et al., 2019). As synthetic biology matures, it is emerging as a significant paradigm in chemical biology, where biological research and engineering design come together to solve global problems.

The double-edged character of synthetic biology—designing novel biological components and reengineering existing systems—opens up the possibility of engineering living systems that function like highly efficient biomachines. This capability holds significant promise in the applications of biomedicine, environmental remediation, and the production of renewable energy. Future endeavors will focus on the development of synthetic genomes and minimal cells, which could revolutionize our approaches to drug development and environmental cleanup. Although challenges such as public perception and ethical implications persist, research initiatives like the International Genetically Engineered Machine competition exemplify how collaborative, creative efforts can foster innovation in synthetic biology. This merging can lead to innovations that directly respond to societal needs, e.g., reductions in pollution and enhancements in health, pointing out the intrinsic merit of engineering living systems (Porcar et al., 2010). Chemical biology is in the future, standing at the threshold of a revolutionary era as it attempts to address worldwide challenges with synthetic systems. Interdisciplinary collaboration will be the main approach, bringing together knowledge from biology, engineering, and computation to produce innovative solutions. Efforts at standardizing biological parts and improving communication between scientists of multiple disciplines will push the development of effective synthetic biological tools. Furthermore, initiatives like TARPOL highlight that building a strong educational system will be central to developing the next generation of synthetic biologists. By fostering a culture of learning and innovation, Europe and other advanced regions can drive the pace of novel synthetic biology advances that facilitate proactive approaches to environmental challenges. All these structured initiatives will ultimately encourage scientific advancement and contribute to the development of a sustainable bio-economy beneficial to society at large (Porcar et al., 2010).

10.6.1. Emerging Technologies and Innovations

The marriage of chemical biology and synthetic systems has spurred the development of advanced technologies that have the promise of changing many fields, particularly medicine. Hydrogels are one of those advances that are versatile in nature as biomaterials, and they exhibit unique attributes suitable for a broad variety of applications, like drug delivery and tissue engineering. Given their water content and biocompatibility, hydrogels facilitate the creation of tissue-like environments where cell growth and functionality can be promoted. Interestingly, recent advances in polymer chemistry have allowed for strict control of the mechanical and biochemical properties of natural as well as man-made hydrogels to make them more efficient in biomedicine. As highlighted by recent reviews, responsive therapeutics as intelligent hydrogels that respond to physiological signals, are the cutting edge of next-generation therapeutics with the potential to revolutionize how patients are treated by providing adaptive medicine tailored to individual needs (Segneanu et al., 2025). Novel syn-bio technologies increasingly expand the scope of chemical biology, as evidenced by the emerging use of non-natural cofactors like nicotinamide mononucleotide (NMN) in enzymatic reactions. The unique regulation of NAD^+ metabolism by NMN has opened new avenues for biocatalysis, enabling more efficient synthesis of complex biomolecules. This achievement not only enhances metabolic engineering processes but also sidesteps the obstacles faced by conventional multi-enzyme cascades and biofuel production. The complementary application of AI-enzyme design with NMN-dependent mechanisms reflects how interdisciplinarity maximizes synthetic biology applications. In addition, as scientists aim to address pertinent gaps in the scalable synthesis of NMN and its integration into biotechnological applications, the ground-breaking potential of such innovations becomes that much more evident (Naaz et al., 2025). The ongoing progression of chemical biology and synthetic systems is marked by the dynamic interplay between the central scientific discovery and emerging technological advancements. As new materials and methods are developed, the possibilities for interdisciplinary partnerships increase. These partnerships can help overcome problems with scalability, regulatory acceptance, and therapeutic applications, eventually refining and enhancing the capabilities of synthetic systems. Moreover, the emphasis on developing smart materials, such as adaptive hydrogels, and on the judicious use of cofactors, such as NMN, aligns with the visions of precision medicine and green bioprocessing prevalent in the present times. The thrust towards adaptive, modular solutions places these advances at the forefront of scientific advances, with a future that enables biotechnological progress to address complex health issues in an efficient manner while maintaining environmental sustainability (Naaz et al., 2025; Segneanu et al., 2025).

10.6.2. Potential Impact on Healthcare and Industry

The union of chemical biology with synthetic systems has the potential to revolutionize both health care and industry by enabling bioprocess and product innovation. Specifically, advances in synthetic biology have enabled the engineering of microorganisms to make useful drugs and agrochemicals in an environmentally responsible way. For example, the application of microbial fermentations has become increasingly important for creating bio-based products that can replace petrochemicals in their applications and thereby address environmental challenges while meeting industry needs. Besides, industries are only beginning to realize the necessity of academic entrepreneurship and partnership to bring such breakthroughs to create new solutions that would be able to counter challenges like antibiotic resistance and sustainable food production effectively (Luo et al., 2023). Through highlighting such methodologies, a community can be formed that fosters revolutionary breakthroughs and enhances overall healthcare outcomes while, at the same time, mitigating important industrial challenges. In the medical space, synthetic systems based on chemical biology are breaking ground for more effective production of vital medications and treatments. With the capability of engineering microbial cell factories, drug makers have been able to reduce the cost of production and enhance the affordability of life-saving therapies. The potential to produce complex biologics like insulin and monoclonal antibodies through fermentation processes is the revolutionary impact of microbial technology on the pharmaceutical industry. In addition, there is a growing practice of using such innovations to make drug development processes simpler and lower the time-to-market. This shift to biotechnology not only widens the set of therapeutics and reduces environmental signatures but also enhances global public health outcomes through easier accessibility for life-saving drugs and interventions (Nielsen et al., 2022). In addition to its effect on medicine, the incorporation of chemical biology into artificial systems can enhance industries such as agriculture and energy. The potential to produce biodegradable products through engineered microbes is of special relevance in the context of increasing plastic waste and environmental issues. As the world moves towards bio-based alternatives, microbial technology has the potential to enable the manufacture of sustainable chemicals, biofuels, and food ingredients that meet evolving consumer needs for more sustainable products. Public pressure and regulatory drive towards more sustainable operations are also propelling the pace in such advancements. By promoting innovation in industrial biotechnology, synthetic systems are able to play an indispensable role in transforming towards a circular economy and ultimately towards contributing to society's wellbeing while also driving economic growth and competitiveness worldwide (Luo et al., 2023).

10.6.3. Challenges and Opportunities Ahead

As chemical biology in synthetic systems evolves, it has numerous challenges that must be met by innovative means. The major challenge is the deployment of multi-enzyme catalysis, which holds the promise to revolutionize the synthesis of complex chemicals but is plagued by issues with coordinating the simultaneous operation of multiple enzymes within one vessel. The complexity of operation calls for exploration into new approaches to optimizing the performance of enzymatic cascade reactions. One of the hopeful methods is enzyme co-immobilization, facilitating higher interaction and stability of enzymes during the reaction process. Overcoming these problems requires not just interdisciplinarity that brings together synthetic biology with bioengineering but also offers a chance for developing sophisticated biocatalytic systems to self-organize autonomously, paving the way for groundbreaking progress in industrial applications of chemical biology (Min-Seo et al., 2021). Furthermore, the pursuit of sustainability in chemical manufacturing presents both challenges and opportunities. Bioinspired use of new carbon feedstocks, particularly wastes, emphasizes the potential of synthetic biology to minimize the environmental impact of traditional chemical processes. Nevertheless, for an opportunity to avail themselves of these opportunities, scientists must identify microorganisms that are not merely metabolically diverse but also have engineering potential. Traditionally, this has been a significant problem, particularly in the context of requiring effective genetic manipulation techniques. New technologies, including rapid molecular cloning techniques that have made use of organisms such as *Acinetobacter baylyi* ADP1, indicate a welcome direction. These new approaches not only facilitate the engineering of microbes for more efficient bioprocessing of untapped resources like lignin but also symbolize the effective intersection of sustainability and synthetic biology in solving existing problems (Biggs et al., 2020). Overall, while the future of chemical biology in synthetic systems is fraught with challenges, the combined possibilities of innovation are no less intriguing. The creation of engineering approaches and the exploration of biological systems within nature, can facilitate tremendous improvements in industrial processes. Genetically programmable self-organizing biocatalytic materials, for instance, represent a horizon of opportunity, as they would reduce reliance on synthetic chemicals and enhance manufacturing efficiency. Meanwhile, better genetic tools and strategies for handling microorganisms underscore their value for sustainable development and biomanufacturing. As researchers encounter these issues, innovation and sustainability harmonization in synthetic chemical biology will be critical in establishing the future of this emerging field (Biggs et al., 2020; Min-Seo et al., 2021).

SUMMARY

In chemical biology, the convergence of biochemistry of nature and synthetic systems holds a massive discovery space and innovation. Recent findings underscore the potential of synthetic approaches to uncover complex biological interactions and mechanisms, facilitating drug discovery and therapeutic design. An understanding of proteome-scale maps of protein-binding affinities, as in innovative chemical proteomics research, is integral to unraveling the human proteome. The finding of thousands of fragment-protein interactions bodes well for the development of probes capable of specifically targeting hitherto untargeted proteins, and thus plug some of the significant gaps in our therapeutic arsenal (Offensperger et al., 2024). The work serves to illustrate not just the potential of synthetic biology but also its applicability to improving our appreciation of biological systems at a mechanistic level. Further, the incorporation of machine learning approaches into chemical biology is a paradigm shift toward predictive models that will facilitate the streamlining of ligand discovery procedures. The integration of computational resources and experimental data enhances our ability to predict how small molecules act in biological systems. By the application of a large database of fragment-protein interactions, researchers can adapt their strategy towards the targeting of specific protein classes or functions, enabling crucial insight into their drug-like capacity. Systematic research and predictive modeling of fragment behavior enable more nuanced research into interactions, enabling a stepping stone between the synthetic and biological applications (Offensperger et al., 2024). Thus, this synergy is of extremely important implications not only for academic research, but for the pharmaceutical industry as it works to create more efficient and safer therapeutic medicines. Ultimately, one must reflect on the broader implications of these findings for sustainable operations and community health programs. By casting synthetic approaches against ancient wisdom, such as the wisdom contained in ethnobotanical studies, we are then able to develop means towards the synthesis of novel plant drugs. The documented relationships between native medicinal plants and medicinal applications are a tangible resource that remains underexploited in contemporary medicine. This connectivity among artificial systems and indigenous practices addresses the necessity to wed various disciplines of knowledge in order to drive innovations that are efficacious as well as eco-friendly. As the scientific world proceeds in this evolving climate, the demand for cross-cultural insight and interdisciplinary cooperation that will create a future in which healing modalities not only curve towards state-of-the-art technology but also towards the insight of nature is increasingly clear.

In comprehending the intricate interplay of chemical biology within synthetic systems, it is inevitable that microbiome research has an important role to play. The construction and use of sophisticated statistical packages, such as MaAsLin 2, provide robust methodologies for associating microbial community characteristics with diverse metadata, ranging from human health attributes to environmental conditions. The improved multivariable association model not only allows for increased detection of the meta-omics correlations but also maintains statistical power regardless of the complexity that comes with microbial information. MaAsLin 2's success in demystifying trends in large datasets, such as the Integrative Human Microbiome (HMP2) project, underscores its significance in exposing the complex interactions between microbial presence and chronic disease and underscores the significant meeting point between chemical biology and analytics in synthetic environments (Mallick et al., 2021). The finding of the microbiota-gut-brain axis has also transformed our knowledge of systemic interactions within biological frameworks. Recent studies highlight the microbial contribution to neurodevelopment and psychiatric illness, evoking gut microbiota-brain function interaction through various channels, including immunological mediators and metabolic derivatives. This bidirectional interaction highlights the complexity of biological control mechanisms and calls for interdisciplinary approaches in synthetic biology. The contribution of environmental factors, e.g., diet and stress, to microbial composition further emphasizes the multidomain character of these interactions. As scientists repeatedly elucidate the mechanisms of these interactions, there are increasing opportunities to develop microbial-based therapies for neuropsychiatric diseases, which also illustrate the profound roots of chemical biology in synthetic systems (Cryan et al., 2019). Finally, the influence of aging and microbial diversity on synthetic biology also deserves to be seriously considered. Since microbial diversity will generally decline with age, the contribution of these transitions to shaping health outcomes becomes more critical. The relationship between microbial community structure and conditions of aging presents both challenges and opportunities for intervention. Understanding these microbial transitions through work by synthetic systems and chemical biology researchers is important to develop a strategy for enhanced healthspan through designed intervention. Future studies have the ability to delve deeper into the mechanisms of the microbiota-gut-brain axis and move into new therapeutic approaches that capitalize on this information. In encouraging further investigation in these fields, the field of synthetic biology can pave the way for new solutions that enhance human health and well-being (Cryan et al., 2019).

REFERENCES

1. Amaan, A., Prekshi, G., & Prachi, S. (2024). Unlocking the transformative power of synthetic biology. *Archives of Biotechnology and Biomedicine.* https://www.semanticscholar.org/paper/b56b52e93625206c732537885b547b9c7ab305cd (accessed on 3rd October 2025).
2. Beal, J., & Rogers, M. (2020). Levels of autonomy in synthetic biology engineering. *Molecular Systems Biology, 16*. https://doi.org/10.15252/msb.202010019 (accessed on 3rd October 2025).
3. Berg, G., Rybakova, D., Fischer, D., Cernava, T., Champomier Vergès, M. C., Charles, T. C., Chen, X., et al. (2020). Microbiome definition re-visited: Old concepts and new challenges. *Microbiome, 8*. https://doi.org/10.1186/s40168-020-00875-0 (accessed on 3rd October 2025).
4. Biggs, B. W., Bedore, S. R., Arvay, E., Huang, S., Subramanian, H., McIntyre, E., Duscent-Maitland, C. V., et al. (2020). Development of a genetic toolset for the highly engineerable and metabolically versatile *Acinetobacter baylyi* ADP1. *Nucleic Acids Research, 48*, 5169–5182. https://doi.org/10.1093/nar/gkaa167 (accessed on 3rd October 2025).
5. Borchert, E., Hammerschmidt, K., Hentschel, U., & Deines, P. (2021). Enhancing microbial pollutant degradation by integrating eco-evolutionary principles with environmental biotechnology. *Trends in Microbiology, 29*, 908–918. https://doi.org/10.1016/j.tim.2021.03.002 (accessed on 3rd October 2025).
6. Bryant, J. M., Baumgarten, S., Glover, L., Hutchinson, S., & Rachidi, N. (2019). CRISPR in parasitology: Not exactly cut and dried! *Trends in Parasitology, 35*, 409–422. https://doi.org/10.1016/j.pt.2019.03.004 (accessed on 3rd October 2025).
7. Centore, R. C., Sandoval, G. J., Soares, L. M., Kadoch, C., & Chan, H. M. (2020). Mammalian SWI/SNF chromatin remodeling complexes: Emerging mechanisms and therapeutic strategies. *Trends in Genetics, 36*, 936–950. https://doi.org/10.1016/j.tig.2020.07.011 (accessed on 3rd October 2025).
8. Chahine, Z., Abel, S., Hollin, T., Barnes, G., Chung, J., Daub, M., Renard, I., et al. (2024). A kalihinol analog disrupts apicoplast function and vesicular trafficking in *P. falciparum* malaria. *Science, 385*. https://www.semanticscholar.org/paper/06f2a9e7cf8ea297ac37c2ad63200809be6f699a (accessed on 3rd October 2025).
9. Cryan, J. F., O'Riordan, K. J., Cowan, C. S. M., Sandhu, K. V., Bastiaanssen, T. F. S., Boehme, M., Codagnone, M. G., et al. (2019). The

microbiota-gut-brain axis. *Physiological Reviews, 99*, 1877–2013. https://doi.org/10.1152/physrev.00018.2018 (accessed on 3rd October 2025).

10. Firestone, J. W., Cohen, M. B., & Lutz, R. R. (2024). SynBioTrace: Integrating safety and security artifacts to build assurance cases for synthetic biology applications. *2024 IEEE 35th International Symposium on Software Reliability Engineering Workshops (ISSREW)*. https://www.semanticscholar.org/paper/7907a344fa119fbef0b3ef75daad7b9b84fcff55 (accessed on 3rd October 2025).

11. Garner, K. (2021). Principles of synthetic biology. *Essays in Biochemistry, 65*, 791–811. https://doi.org/10.1042/ebc20200059 (accessed on 3rd October 2025).

12. Hamany Djande, C. Y., Steenkamp, P., & Dubery, I. (2025). Chemical biology meets metabolomics: The response of barley seedlings to 3,5-dichloroanthranilic acid, a resistance inducer. *Molecules, 30*. https://www.semanticscholar.org/paper/05e82d5b23115d9c6f8472ea64f8db12dc2ffc41 (accessed on 3rd October 2025).

13. Hanczyc, M. (2020). Engineering life: A review of synthetic biology. *Artificial Life, 26*, 260–273. https://www.semanticscholar.org/paper/4dfd6e01e73a0b708ea91c3aa0b2e5d0fdca3ce6 (accessed on 3rd October 2025).

14. Hu, H., Nie, H. J., & Chen, X. (2025). Primary amine-based photoclick chemistry: From concept to diverse applications in chemical biology and medicinal chemistry. *Accounts of Chemical Research*. https://www.semanticscholar.org/paper/0d3d9db7d6614ad71e9859a36650955830f175d8 (accessed on 3rd October 2025).

15. Jia, W. (2025). Progress of biological whitening technology. *Journal of Clinical Medicine Research*. https://www.semanticscholar.org/paper/c66cfb048bf5289fe63c244fdd7b2d952e06cf56 (accessed on 3rd October 2025).

16. Joshi, S., Moody, A., Budthapa, P., Gurung, A., Gautam, R., Sanjel, P., Gupta, A., et al. (2024). Advances in natural-product-based fluorescent agents and synthetic analogues for analytical and biomedical applications. *Bioengineering, 11*. https://www.semanticscholar.org/paper/ddb9c261c4f4b824defcf35dfda8aea074304fb0 (accessed on 3rd October 2025).

17. Kallergi, A., Asin-Garcia, E., Martins dos Santos, V. A. P., & Landeweerd, L. (2020). Context matters. *EMBO Reports, 22*. https://doi.org/10.15252/embr.202051227 (accessed on 3rd October 2025).

18. Kuwahara, M., Li, Y., Rozners, E., & Murakami, H. (2013). Artificially

created nucleic acids and peptides/proteins in chemical biology. *Journal of Nucleic Acids, 2013.* https://www.semanticscholar.org/paper/7445771 6b449b0ffbc410dab16c67d9552a91907 (accessed on 3rd October 2025).

19. Li, H., Yang, Y., Hong, W., Huang, M., Wu, M., & Zhao, X. (2020). Applications of genome editing technology in the targeted therapy of human diseases: Mechanisms, advances and prospects. *Signal Transduction and Targeted Therapy, 5.* https://doi.org/10.1038/s41392-019-0089-y (accessed on 3rd October 2025).

20. Li, Q., Schenning, A., & Bunning, T. (2019). Light-responsive smart soft matter technologies. *Advanced Optical Materials, 7.* https://www.semanticscholar.org/paper/411c409d72c25865017ecc60239dad0360602e4b (accessed on 3rd October 2025).

21. Liu, F., He, L., Dong, S., Xuan, J., Cui, Q., & Feng, Y. (2023). Artificial small molecules as cofactors and bio macromolecular building blocks in synthetic biology: Design, synthesis, applications, and challenges. *Molecules, 28.* https://www.semanticscholar.org/paper/d09d45f39ab719d96a42853a435758eb50edbe77 (accessed on 3rd October 2025).

22. Ludwig, C., Gillet, L., Rosenberger, G., Amon, S., Collins, B. C., & Aebersold, R. (2018). Data-independent acquisition-based SWATH-MS for quantitative proteomics: A tutorial. *Molecular Systems Biology, 14.* https://doi.org/10.15252/msb.20178126 (accessed on 3rd October 2025).

23. Luo, Y., Abidian, M. R., Ahn, J. H., Akinwande, D., Andrews, A. M., Antonietti, M., Bao, Z., et al. (2023). Technology roadmap for flexible sensors. *ACS Nano, 17*, 5211–5295. https://doi.org/10.1021/acsnano.2c12606 (accessed on 3rd October 2025).

24. Mallick, H., Rahnavard, A., McIver, L. J., Ma, S., Zhang, Y., Nguyen, L. H., Tickle, T. L., et al. (2021). Multivariable association discovery in population-scale meta-omics studies. *PLoS Computational Biology, 17*, e1009442–e1009442. https://doi.org/10.1371/journal.pcbi.1009442 (accessed on 3rd October 2025).

25. Manjunath, M. S., Koshariya, A. K., Sharma, N., Rajput, A., Pandey, S. K., Singh, S., Kumar, R., et al. (2023). Exploring the use of aromatic compounds in crop growth and protection. *International Journal of Plant & Soil Science.* https://www.semanticscholar.org/paper/b2eb504368a46bc9d267140ebafddc0ada334233 (accessed on 3rd October 2025).

26. Martínez-Arribas, B., Annang, F., Díaz-González, R., Pérez-Moreno, G., Martín, J., Mackenzie, T., Castillo, F., et al. (2023). Establishment of a

screening platform based on human coronavirus OC43 for the identification of microbial natural products with antiviral activity. *Microbiology Spectrum, 12*. https://www.semanticscholar.org/paper/8891809c06dbeb9263a7b1bec7823764c34f0077 (accessed on 3rd October 2025).

27. Mishra, S. P., Bohidar, S., Padhi, S., Sethi, K. C., Manna, S., & Pattnayak, S. K. (2024). Geo-hydrobiology and geoinformatics mapping of ethnobotany in the Gandhamardan Hills, Odisha, India. *Journal of Scientific Research and Reports*. https://www.semanticscholar.org/paper/8d5369ccf80161daff71357e0224fb50521c1f88 (accessed on 3rd October 2025).
28. Muhamadali, H., Winder, C., Dunn, W., & Goodacre, R. (2023). Unlocking the secrets of the microbiome: Exploring the dynamic microbial interplay with humans through metabolomics and their manipulation for synthetic biology applications. *Biochemical Journal, 480*, 891–908. https://www.semanticscholar.org/paper/4ba5127a0d708619331b9a5da09a4e26f0eb8f69 (accessed on 3rd October 2025).
29. Muratov, E., Bajorath, J., Sheridan, R. P., Tetko, I. V., Filimonov, D., Poroikov, V., Oprea, T. I., et al. (2020). QSAR without borders. *Chemical Society Reviews, 49*, 3525–3564. https://doi.org/10.1039/d0cs00098a (accessed on 3rd October 2025).
30. Naaz, T., & Kim, B. S. (2025). Use of nicotinamide mononucleotide as non-natural cofactor. *Catalysts*. https://www.semanticscholar.org/paper/52f9865fb4d00261f1d480b16d3bb3cc62859325 (accessed on 3rd October 2025).
31. Nairn, A., & Piatti-Farnell, L. (2024). The artificial. *M/C Journal*. https://www.semanticscholar.org/paper/3b20c06de5410b831339f47adc4eeded1308689a (accessed on 3rd October 2025).
32. Natarajan, K. (2021). Several interdisciplinary of chemical biology. *Chemical Science, 12*, 1–1. https://www.semanticscholar.org/paper/9f56f76f8cf90b7244dff7d194bc7c2e139f2912 (accessed on 3rd October 2025).
33. Navid, E. L., & Einsiedel, E. (2012). Synthetic biology in the Science Café: What have we learned about public engagement? *Volume 11*. https://www.semanticscholar.org/paper/31ce0f6482ad24a489397acb46ad8da415d29b69 (accessed on 3rd October 2025).
34. Nielsen, J., Tillegreen, C. B., & Petranović, D. (2022). Innovation trends in industrial biotechnology. *Trends in Biotechnology, 40*, 1160–1172. https://doi.org/10.1016/j.tibtech.2022.03.007 (accessed on 3rd October 2025).
35. Offensperger, F., Tin, G., Duran-Frigola, M., Hahn, E., Dobner, S., am Ende, C. W., Strohbach, J. W., et al. (2024). Large-scale chemoproteomics

expedites ligand discovery and predicts ligand behavior in cells. *Science, 384*. https://www.semanticscholar.org/paper/4a25bf59d59d571aa782b2cf352572df62566d89 (accessed on 3rd October 2025).

36. Porcar, M., & Moya, A. (2010). Paving the way for synthetic biology-based bioremediation in Europe. *Microbial Biotechnology, 3*, 134–135. https://www.semanticscholar.org/paper/9bb150b575094e530c0e18589098da039cf58152 (accessed on 3rd October 2025).
37. Prakash, G., Parmar, B., & Bhatia, D. (2025). Structurally programmable, functionally tunable dendrimers in biomedical applications. *Biomaterials Science*. https://www.semanticscholar.org/paper/05dbf228a17182521be58c2446f334484ff921a0 (accessed on 3rd October 2025).
38. Ptacin, J. L., Caffaro, C. E., Ma, L., San Jose Gall, K. M., Aerni, H. R., Acuff, N. V., Herman, R. W., et al. (2021). An engineered IL-2 reprogrammed for anti-tumor therapy using a semi-synthetic organism. *Nature Communications, 12*. https://doi.org/10.1038/s41467-021-24987-9 (accessed on 3rd October 2025).
39. Qumsani, A. (2024). The contribution of microorganisms to sustainable development: Towards a green future through synthetic biology and systems biology. *Journal of Umm Al-Qura University for Applied Sciences*. https://www.semanticscholar.org/paper/e671bb338f7ea59821dec59cb4554c96342adc96 (accessed on 3rd October 2025).
40. Rebhan, L., Fürst, R., Schollmeyer, D., Serafim, R. A. M., & Gehringer, M. (2025). Go for gold: Development of a scalable synthesis of [1-(ethoxycarbonyl)cyclopropyl] triphenylphosphonium tetrafluoroborate, a key reagent to explore covalent monopolar spindle 1 inhibitors. *ChemistryOpen*. https://www.semanticscholar.org/paper/cdd700bf4fec3d0e2b6b0c4291da8454ebb4a53b (accessed on 3rd October 2025).
41. Riaz, A., Rasul, A., Sarfraz, I., Nawaz, J., Sadiqa, A., Zara, R., Khan, S. G., et al. (2020). *Chemical Biology Toolsets for Drug Discovery and Target Identification. IntechOpen eBooks*. https://doi.org/10.5772/intechopen.91732 (accessed on 3rd October 2025).
42. Rivers, M. A. J., & Lowell, A. N. (2024). Expanding the biosynthetic toolbox: The potential and challenges of *in vitro* type II polyketide synthase research. *SynBio*. https://www.semanticscholar.org/paper/10d5795f52cb1a6989e55df7f5ca2a9f441ae5a9 (accessed on 3rd October 2025).
43. Sarkar, D., & Maranas, C. (2019). Engineering microbial chemical factories using metabolic models. *BMC Chemical Engineering, 1*, 1–11.

https://www.semanticscholar.org/paper/23000ca87eddfc0c1d2e9fb94cba1e9a5a11df23 (accessed on 3rd October 2025).

44. Segneanu, A. E., Bejenaru, L., Bejenaru, C., Blendea, A., Mogoșanu, G., Biță, A., & Boia, E. (2025). Advancements in hydrogels: A comprehensive review of natural and synthetic innovations for biomedical applications. *Polymers*. https://www.semanticscholar.org/paper/193bc4b72d8bb7f4c7c6fe7c103e9e3b7672cd4f (accessed on 3rd October 2025).

45. Seo, M. J., & Schmidt-Dannert, C. (2021). Organizing multi-enzyme systems into programmable materials for biocatalysis. *Catalysts, 11*, 409–409. https://doi.org/10.3390/catal11040409 (accessed on 3rd October 2025).

46. Siddiqui, S. A., Ahmad, S., & Afaq, U. (2025). Innovations in biopesticides: The role of plant-based biopesticides in sustainable agriculture. *Journal of Biopesticides*. https://www.semanticscholar.org/paper/91fe520d425066508eccb95a25d2a1f9cfd53685 (accessed on 3rd October 2025).

47. Vindman, C., Trump, B. D., Cummings, C. L., Smith, M., Titus, A. J., Oye, K., Prado, V., et al. (2024). The convergence of AI and synthetic biology: The looming deluge. https://www.semanticscholar.org/paper/a5d4d9cacd4a16cd202f31dd8cf0b281490491ba (accessed on 3rd October 2025).

48. Xu, W., & Woo, C. M. (2025). Probing the E3 ligase adapter cereblon with chemical biology. *Accounts of Chemical Research*. https://www.semanticscholar.org/paper/349178953259af17970642a1c94ccaf75bc153c4 (accessed on 3rd October 2025).

49. Yang, H., Du, S., Ye, Z., Wang, X., Yan, Z., Lian, C., Bao, C., et al. (2022). A system for artificial light signal transduction via molecular translocation in a lipid membrane. *Chemical Science, 13*, 2487–2494. https://www.semanticscholar.org/paper/d11d5639f54e37e9471c4e98265eb95484966237 (accessed on 3rd October 2025).

50. Yi, W., Hu, S., Qian, X., Yan, W., & Li, Y. (2025). Synthetic biology-based engineering cells for drug delivery. *Exploration, 5*. https://www.semanticscholar.org/paper/3aaa8209c098e5e9327ba9e63f56f22e1a8ea274 (accessed on 3rd October 2025).

51. Zagnoni, M. (2012). Miniaturized technologies for the development of artificial lipid bilayer systems. *Lab on a Chip, 12*(6), 1026–1039. https://www.semanticscholar.org/paper/65c7960e521a29d53a200ee85a14d746bcb35401 (accessed on 3rd October 2025).

52. Zhang, H., Li, X., Liu, J., Lan, Y. Q., & Han, Y. (2024). Advancing

single-particle analysis in synthetic chemical systems: A forward-looking discussion. *Advanced Materials, 36*. https://www.semanticscholar.org/paper/571da9a61f741eb5bc0ceef8fbed62cfd24ae21a (accessed on 3rd October 2025).

53. Zhang, W., & Nielsen, D. (2014). Synthetic biology applications in industrial microbiology. *Frontiers in Microbiology, 5*. https://www.semanticscholar.org/paper/c9478e9d93a2464b71689e48606bf4e40af6d738 (accessed on 3rd October 2025).

54. Zhao, L. X., Fan, Y. G., Zhang, X., Li, C., Cheng, X. Y., Guo, F., & Wang, Z. Y. (2025). Graphdiyne biomaterials: From characterization to properties and applications. *Journal of Nanobiotechnology, 23*. https://www.semanticscholar.org/paper/8dcf6d7d5f3adf83778f3230737a5537aa34cdf1 (accessed on 3rd October 2025).

INDEX

D

W

X

Y

Z